AUGUST WEISMANN

THE GERM-PLASM

A THEORY OF HEREDITY

TRANSLATED BY

W. NEWTON PARKER AND HARRIET RÖNNFELDT

Elibron Classics
www.elibron.com

Elibron Classics series.

ISBN 1-4021-4395-8 (paperback)
ISBN 1-4212-8159-7 (hardcover)

This Elibron Classics Replica Edition is an unabridged facsimile of the edition published in 1898 by Charles Scribner's Sons, New York.

THE GERM-PLASM

THE GERM-PLASM

A THEORY OF HEREDITY

BY

AUGUST WEISMANN

Professor in the University of Freiburg-in-Baden

TRANSLATED BY

W. NEWTON PARKER PH.D.

Professor in the University College of South Wales and Monmouthshire

AND

HARRIET RÖNNFELDT B.Sc.

WITH TWENTY-FOUR ILLUSTRATIONS

NEW YORK

CHARLES SCRIBNER'S SONS

1898

TO THE MEMORY

OF

CHARLES DARWIN

NOTE BY TRANSLATOR

IN preparing the English edition of the present work, I have had the great advantage of being able to consult Professor Weismann personally with regard to many of the more difficult passages. Only those who have attempted to make a translation of an abstruse work from German manuscript, can appreciate the difficulties of rendering such a work into good English, and at the same time of keeping closely to the text. As the time in which the translation had to be prepared was a comparatively short one, I have been unable to revise the style as thoroughly as I could have wished, but trust that the author's meaning has been expressed with tolerable accuracy.

In the case of special technical terms which have no recognised English equivalents, I have in all cases added the German word in brackets the first time they are used. For the extremely useful and untranslateable word 'Anlage,' the somewhat awkward term 'primary constituent' has been used when it refers to the concrete vital units: in other cases, it has been rendered by 'rudiment;' or, when it has a more abstract meaning, by 'predisposition.' The words 'Eigenschaft,' 'Charakter,' 'Merkmal,' and 'Qualität,' are often used synonymously by the author, and have therefore been indiscriminately translated by 'characteristic,' 'character,' 'peculiarity,' and 'quality.'

I must express my thanks to Dr. G. H. Parker, of Harvard University, Cambridge, Mass., who kindly undertook a first revision of Chapters XIII. and XIV., and thereby rendered an earlier publication of the book possible; as well as to my friend and colleague Mr. Franck Arnold, for help in elucidating some of the more complicated sentences, and for many suggestions.

W. N. PARKER.

CARDIF, *Nov. 28th*, 1892.

" Naturgeheimniss werde nachgestammelt." — GOETHE.

PREFACE

ANY attempt at the present time to work out a theory of heredity in detail may appear to many premature, and almost presumptuous: I confess there have been times when it has seemed so even to myself. I could not, however, resist the temptation to endeavour to penetrate the mystery of this most marvellous and complex chapter of life as far as my own ability and the present state of our knowledge permitted.

Even though the present attempt may be very imperfect and incomplete, I cannot regard it as premature. Our knowledge has increased during the last twenty years to such an extent, that it does not seem to be altogether a hopeless task to inquire into the actual processes on which the phenomena of heredity depend. It is, moreover, very essential that we should possess a theory of heredity, worked out in such a manner as to suggest new problems, which in their turn will lead to new solutions.

Previous hypotheses have been insufficient in this respect, owing to the fact that they have not been worked out in detail. They are rather to be regarded as paving the way to future theories, by merely formulating explanatory principles without professing to apply them to all the different groups of phenomena which come under the head of heredity, by which means alone their true value can be tested. Even Darwin's theory of 'pangenesis' was inadequate in this respect: owing to the comparatively limited number of facts then at his disposal, it could not but be what we may call an *ideal* theory; that is to say, it is founded upon certain principles without inquiring how far they are based upon actual facts. In themselves, such theories can hardly be looked upon as suggestive, for *if once the assumed principle is accepted,* all the phenomena are thereby explained, and the matter is open to no further doubt.

Let us assume that the germ contains millions of the primary constituents ('Anlagen') of all the most minute portions of the body; moreover, that these constituents are always present at the right place and in the right combination during the process of development; and, further, that they are capable of giving rise in their turn to the parts or organs to which they severally correspond. Such a theory explains everything, or nothing — the premises alone can be attacked. No new problems can arise from it till it has been placed upon a sound basis; the premises must be shown to be correct, and it must be proved that the germ is actually composed of primary constituents, which by some means or other become combined into groups and are capable of giving rise to the various parts and organs in question. Then, and then only, would the theory serve as an incentive to further investigations into the phenomena of heredity of all kinds, and experiments might be made which would support or contradict it.

There is no doubt a natural tendency to base experiments upon certain preconceived ideas; but it is one thing to be guided solely by such phenomena as may at the moment appear of especial importance, and another to base operations upon the completed outline of a theory founded upon the principal data bearing upon the question. I have myself more than once abandoned a line of research undertaken in connection with the problem of heredity, because I felt that to proceed without the guidance of a theory more or less complete in itself, and developed on a basis of ascertained facts, would be little better than groping in the dark. The importance of such a theory lies primarily in its suggestiveness, by which alone it becomes a step towards the ideal at which we aim, viz., the formulation of *the true and complete theory.*

The growth of this book has been very gradual. What first struck me when I began seriously to consider the problem of heredity, some ten years ago, was the necessity for assuming the existence of a special organised and living *hereditary substance,* which in all multicellular organisms, unlike the substance composing the perishable body of the individual, is transmitted from generation to generation. This is the theory of *the continuity of the germ-plasm.* My conclusions led me to doubt the usually accepted view of the *transmission of variations acquired* by the body (soma) ; and further research, combined with experiments, tended more and more to strengthen my conviction that in point of fact no such transmission occurs. Meanwhile, the investigations of several distinguished biologists — in which I myself have had some share — on the process of fertilisation and conjugation, brought about a complete revolution in our previous ideas as to the meaning of this process, and further led me to see that the germ-plasm is composed of vital units, each of equal value, but differing in character, containing all the primary constituents of an individual. These '*ancestral germ-plasms*' ('Ahnenplasmen'), or '*ids,*' as I now prefer to call them, afforded additional matter wherewith to construct a theory of heredity, though much was wanting to render it complete.

In my last essay I certainly suggested the possibility of solving one of the most difficult problems in heredity — viz., the co-operation of the hereditary substance of the parents in sexual reproduction — by assuming the existence of these 'ids'; but I did not for a moment suppose that in doing so I had propounded a *complete and elaborated* theory of heredity, as some of my readers have thought to be the case; much still remained to be done first. I had as yet not touched upon such phenomena of heredity as have no direct bearing on the question of sexual reproduction, and had also abstained from any mention of the fundamental point of my theory of heredity — namely, the *constitution of the ids*. Although I pointed out that they must possess a complex structure which undergoes gradual and regular changes during the development of the individual from the egg-cell, I did not enter into any further details. This question remained in abeyance, for I was by no means sure whether the conception that I had formed on *à priori* grounds of the minute structure of the ids would prove tenable when viewed in the light of all the many phenomena of heredity. No conclusion could be arrived at respecting the structure of the ids till these phenomena had been individually considered.

All my investigations on the problem of heredity were so far only links, to be some day united into a chain which had as yet no existence. The question of the ultimate elements on which to base the theory was the very point on which I remained longest in doubt. The 'pangenesis' of Darwin, as already mentioned, seemed to me to be far too independent of facts, and even now I am of the opinion that the very hypothesis from which it derives its name is untenable. There is now scarcely any doubt that the entire conception of the production of the 'gemmules' by the body-cells, their separation from the latter, and their 'circulation,' is in reality wholly imaginary. In this regard I am still quite as much opposed to Darwin's views as formerly, for I believe that all parts of the body do not contribute to produce a germ from which the new individual arises, but that, on the contrary, the offspring owes its origin to a peculiar substance of extremely complicated structure, viz., the 'germ-plasm.' This substance can never be formed anew; it can only grow, multiply, and be transmitted from one generation to another. My theory might therefore well be denominated '*blasto-genesis*'—or origin from a germ-plasm, in contradistinction to Darwin's theory of '*pangenesis*'—or origin from all parts of the body.

My doubts as to the validity of Darwin's theory were for a long time not confined to this point alone: the assumption of the existence of *preformed* constituents of all parts of the body seemed to me far too easy a solution of the difficulty, besides entailing an impossibility in the shape of an absolutely inconceivable aggregation of primary constituents. I therefore endeavoured to see if it were not possible to imagine that the germ-plasm, though of complex structure, was not composed of such an immense number of particles, and that its further complication arose subsequently in the course of development. In other words, what I sought was a substance from which the whole organism might arise by *epigenesis*, and not by *evolution*.* After repeated attempts,

* The theory of 'evolution' or 'preformation' of the early physiologists supposed that all parts of the fully-formed animal or plant were present, in a minute form, in the germ. The rival theory of 'epigenesis'

in which I more than once imagined myself successful, but all of which broke down when further tested by facts, I finally became convinced that an epigenetic development is an *impossibility*. Moreover, I found an actual *proof of the reality of evolution*, which will be explained in the chapter on the structure of the germ-plasm. It is so simple and obvious that I can scarcely understand how it was possible that it should have escaped my notice so long.

It is gratifying to me to find myself at one with the great English naturalist Darwin, — as well as with de Vries and Wiesner, — at all events in the main point at issue; and this agreement seems to me to point to the possibility of solving in the end the problem of heredity, which might seem to be open only to the wildest speculations: we may now perhaps hope to succeed in recognising the *probable* explanations among the many *possible* ones, and in finally selecting from among these the *real* solution of the problem. This will assuredly be the work of time, and our approach to the truth will be a very gradual one. But our path is marked out; *reasoning supported by observation* will lead us to the goal. We are led by the observation of facts to form an opinion as to their bearing on each other. This gives rise to further problems and fresh investigations, which in their turn lead to a new interpretation. In this way light has before now been thrown on many a problem that seemed to baffle explanation. I need only mention the insight that we have now gained into the phenomenon of sexual reproduction.

---

taught that there is no preformation of parts in the germ, but that the fully-formed organism is produced by a gradual process of differentiation. It will be seen that the word 'evolution,' as here used, has no connection with the doctrine of descent with which it is usually connected. — W. N. P.

In the same way we shall succeed in obtaining a firmer and firmer grasp of the problem of heredity, which but a short while ago appeared so utterly unapproachable.

What in this particular question appears to afford additional promise of success is the fact that we can in a sense approach it from two sides;— namely, by observations, firstly, on the *phenomena* of heredity, and secondly, on the hereditary *substance* itself, with which we are now of course acquainted. We can now form an estimate as to whether an explanation of any particular phenomenon of heredity is of a merely hypothetical nature, or whether it may attain to the value of an established fact, inasmuch as we are in a position to judge, within certain limits at all events, whether it is consistent with the actual behaviour of the hereditary substance. Hitherto this has not been possible, and hence all previous theories, including both that of Darwin's gemmules and of Herbert Spencer's units, were up to a certain point purely speculative. We are now better off in this respect; and I have no doubt that further research will enable us to penetrate far more deeply still into the complicated processes connected with the idioplasm, if we are prepared to reason on the results of our observations, and to utilise every theoretical advance as an incentive to fresh questions regarding the processes in connection with the distribution of the mysterious nuclear substance.

We are still far from having attained a complete insight into the matter, but I trust nevertheless that the present attempt at a theory of heredity is no mere work of the imagination; and though it still be no more than an attempt, which will be followed by better ones, I venture to believe that time will prove it to contain more definite points, forming the centre of numerous possibilities, than many will for the present be prepared to admit. Nevertheless I am well aware that it is but the beginning of a theory, and for this reason I have presented it in the form of an inquiry rather than of an established system. My plan has been not so much to advance doctrines as to propound questions, and to answer them with a greater or lesser degree of certainty, or in some cases even to leave them to be decided by future researches. I do not regard my theory as a complete and perfect one, but trust that it is of such a nature as to be capable of improvement and further development.

It has been my endeavour to write as simply and intelligibly as possible; not as a specialist writing for specialists, but as one who desires to make his case clear to all interested in biological problems. For this reason a number of figures have been inserted, which, though perhaps superfluous for specialists, will, I trust, assist all who are less conversant with the subject, such as physiologists, medical men, and indeed all interested in natural science, to a clearer conception of the matters under discussion.

As a zoologist, I have naturally, in the first instance, considered the phenomena in their relation to animals, for every one must base his ideas on the facts most familiar to him. I have, however, done my best to lay due weight on the data afforded by the study of plants, and to take into account the views of botanists. It will be seen that the very facts which are furnished by certain hereditary phenomena in plants afford a strong support to certain fundamental points in my theory, and that even those which are at first sight in apparent contradiction, are in reality in perfect accordance with it.

It may perhaps be considered by medical men that I ought to have brought forward more evidence with regard to diseases. We certainly possess a rich material on which observations concerning the transmission of diseases might be based, and this I have made use of so far as seemed expedient. It must, however, not be forgotten that the transmission of so-called hereditary diseases is not always due to a true process of heredity, but in some cases, at any rate, results from an infection of the germ. Unfortunately, we are not always able to distinguish between these two causes; and as long as this is the case, the data furnished by diseases can only be used with great caution, as will be shown in Chapter XII.

The manuscript of this book was practically completed by the end of April last, but as the translation had then to be made, its publication was delayed for some months. This will account for the fact that no mention, or only a brief one, has been made of researches which have appeared in the interval. My sincerest thanks are due to the translator — Professor W. N. Parker, — whose task has been by no means an easy one: apart from the mere knowledge of the two languages, an intimate acquaintance with the facts treated of and with the whole science of biology is essential in order to render the meaning of this complicated subject clear, and at the same time to reproduce the original text with anything like accuracy. I am of course unable to judge how far Mr. Parker has succeeded in clothing my ideas in good English, but am glad to state that they have been given very correctly, so far as I can judge from those parts which we have discussed together.

In conclusion, I must express my warmest thanks to the Government under which I have the good fortune to live, for the efficient way in which they have seconded my endeavours, by releasing me from my academical duties during two winter sessions. My hearty thanks are also due to my friends and colleagues Professors Baumann, Lüroth, Wiedersheim, and Ziegler, as well as to Professor Goebel, of Munich, for information of various kinds; and I am no less indebted to Miss Else Diestel, who, in addition to much help of a technical nature, has also been at the great pains of preparing an alphabetical index.

I thus venture to bring into the light of day a work which is the fruit of many years labour and of many doubts; and even though but few of my results should remain unmodified, I hope nevertheless that my work has not been in vain; for even error, if it originate in correct deductions, must become a step towards truth.

AUGUST WEISMANN.

FREIBURG, I/BR.,
May 19th 1892.

CONTENTS

INTRODUCTION

PAGE

A.—HISTORICAL PART 1

B.—DESCRIPTIVE PART 20

PART I.—THE MATERIAL BASIS OF HEREDITY

CHAPTER I

THE GERM-PLASM—

1. The fundamental units 37
2. The control of the cell 45
3. The determinants 53
4. The id in ontogeny 60
5. Summary of sections 1–4, relating to the structure of the germ-plasm 75
6. The mechanism for the phyletic changes in the germ-plasm 77
7. The magnitude of the constituents of the germ-plasm . 85

PART II.—HEREDITY IN ITS RELATION TO MONOGONIC REPRODUCTION

CHAPTER II

REGENERATION—

1. Its cause and origin in the idoplasm 93
2. The phylogeny of regeneration 114
3. Facultative or polygenetic regeneration 126
4. Regeneration in plants 132
5. Regeneration in animal embryos, and the principles of ontogeny 134

CHAPTER III

PAGE

MULTIPLICATION BY FISSION—
1. Preliminary remarks 145
2. The process of fission in the *Naidæ* 146
3. The process of fission in the *Microstomidæ* 149
4. The phylogeny of the process of multiplication by fission 151

CHAPTER IV

MULTIPLICATION BY GEMMATION—
1. The process of gemmation in animals 154
   a. Cœlenterata 154
   b. Polyzoa 158
   c. Tunicata 160
2. The process of gemmation in plants 163
3. Comparison of the process of gemmation in animals and plants 166
4. The phylogeny of the process of multiplication by gemmation 168

CHAPTER V

ALTERNATION OF GENERATIONS IN ITS RELATION TO THE IDIOPLASM 173

CHAPTER VI

THE FORMATION OF GERM-CELLS—
1. The continuity of the germ-plasm 183
2. The germ-tracks 192
3. Historical account of the theory of the continuity of the germ-plasm 198
4. Objections to the theory of the germ-plasm . . . 202
5. Galls 218

CHAPTER VII

SUMMARY OF PART II. (CHAPTERS II.–VI.) 225

PART III.—THE PHENOMENA OF HEREDITY RESULTING FROM SEXUAL REPRODUCTION

INTRODUCTORY REMARKS ON THE NATURE OF SEXUAL REPRODUCTION 230

CHAPTER VIII

PAGE

MODIFICATIONS OF THE GERM-PLASM CAUSED BY AMPHIMIXIS—

1. The necessity of a halving of the germ-plasm 235
2. Proof that the essential part in the process of 'reducing division' consists in the extrusion of ids 240

CHAPTER IX

ONTOGENY RESULTING FROM THE UNION OF THE GERM-PLASM OF TWO PARENTS—

1. The nature of the offspring as determined by the process of fertilisation 253
2. The share taken by the ancestors in composition of the germ-plasm 256
3. The struggle of the ids in ontogeny 260
    - *a.* Plant-hybrids 260
    - *b.* Intercalary remarks on variation 271
    - *c.* The struggle of individual characters . . . 274
4. The force of heredity 290
5. Summary of Chapter IX. 293

CHAPTER X

THE PHENOMENA OF REVERSION IN THEIR RELATION TO AMPHIMIXIS—

1. Reversion to racial characters in plant-hybrids . . . 299
2. Reversion to individual characters in man 308
3. Reversion to the characters of ancestors far removed in animals and plants 316
4. Reversion to rudimentary characters 333
5. Preliminary summary of sections 1–5 335
6. Reversion in asexual reproduction 338
    - *a.* Reversion in the process of gemmation . . . 338
    - *b.* Reversion in parthenogenesis 344
7. Proof that the determinants become disintegrated into biophors 348

CHAPTER XI

DIMORPHISM AND POLYMORPHISM—

1. Normal dimorphism 352
2. Pathological dimorphism: hæmophilia 370
3. Polymorphism 374
4. Dich geny in plants 380

CHAPTER XII

PAGE

DOUBTFUL PHENOMENA OF HEREDITY—
1. 'Xenia' and telegony 383
2. The influence of temporary abnormal conditions of the parents on the offspring 386
3. The transmission of diseases 387

PART IV.—THE TRANSFORMATION OF SPECIES: ITS ORIGIN IN THE IDIOPLASM

CHAPTER XIII

THE SUPPOSED TRANSMISSION OF ACQUIRED CHARACTERS—
1. Difficulties in the way of a theoretical basis for this assumption 392
2. The hypothesis tested by facts 396
3. Climatic variation in butterflies 399

CHAPTER XIV

VARIATION—
1. Normal individual variation 410
2. Pathological variation 428
3. Summary of the above two sections and conclusions . 431
4. Variations on a larger scale 435
   a. The origin of these variations 435
   b. The transmission of these variations 444

SUMMARY AND CONCLUSION. 450

LIST OF FIGURES

Fig. 1. Diagram of nuclear division.
2. Two idants of *Ascaris megalocephala.*
3. Diagram of the cell-generations in the fore-limb of a newt.
4. Multiplication by fission in a marine worm (*Myrianida*).
5. Transverse section through the 'zone of gemmation' in a fresh-water worm (*Nais*).
6. An early stage in the formation of a bud in a Hydroid polype (*Eudendrium*).
7. Apex of a shoot of a fresh-water Alga (*Chara*).
8. Female water-flea (*Daphnia* , with summer eggs.
9. Nauplius larva of *Leptodora hyalina.*
10. Alternation of generations in *Bougainvillea.*
11. Strobilation of a Medusa.
12. Degeneration of a Medusa into a mere gonophore.
13. Early stages of development of the primitive germ-cells in *Sagitta.*
14. Early stages in the development of the primitive germ-cells in *Moina.*
15. Early stage of the embryo of *Rhabditis nigrovenosa.*
16. Diagram of the germ-track in *Rhabditis nigrovenosa.*
17. *Volvox* and *Pandornia.*
18. Process of fertilisation in *Ascaris megalocephala.*
19. Diagram to illustrate the composition of the idants out of individually different ids.
20. Formation of spermatozoa in *Ascaris megalocephala.*
21. Formation of ova in *Ascaris megalocephala.*
22. Diagram to illustrate the combination of idants in hybrids.
23. The two varieties of *Cypris reptans.*
24. *Bonellia viridis*, male and female.

INTRODUCTION

A.—Historical Part

HERBERT SPENCER was practically the first in the present generation to attempt a theoretical explanation of heredity when he propounded his theory of 'physiological units.' The regeneration of lost parts, *e.g.*, of a leg or the tail of a salamander, led him to the conception of these units, 'in all of which there dwells the intrinsic aptitude to aggregate into the form of that species; just as in the atoms of a salt there dwells the intrinsic aptitude to crystallise in a particular way.' He calls this aptitude the 'polarity of the organic units,' and defines the latter as being intermediate between the 'chemical units' or molecules and the 'morphological units' or cells. They must be 'immensely more complex than the chemical units,' and must therefore correspond to groups of molecules. It is very interesting at the present day, now that we have advanced somewhat further towards a theory of heredity, to summarise the various aptitudes and forces which Herbert Spencer thought it necessary to ascribe to his 'physiological units,' in order to arrive at an explanation of the phenomena. Although the sections on Heredity and Regeneration constitute only a small portion of his great work on the 'Principles of Biology,' and cannot therefore contain a detailed treatment of the phenomena of heredity, his opinions on this subject are evident.

Spencer considers, on the one hand, that the whole organism is composed of these units, which are all alike in kind, and on the other, that the germ-cells also contain small groups of them. The former supposition makes regeneration possible to each sufficiently large portion of the body, while the latter gives the germ-cell the power of reproducing the whole: inasmuch as the 'polarity' of the 'units' leads to their arrangement in such a way that the whole 'crystal'—the organism—is restored, or even formed anew. The mere *difference in the arrangement* of units alike in kind determines the diversity of the *parts of the body*, while the distinction between different species and that between different *individuals* is due to a diversity in the constitution of the units.

The units of an individual are therefore to a certain extent protean. They are capable of arranging themselves in an immense variety of ways, and so form the most diverse cells, tissues, organs, and parts of the body. But they only do this under the directing influence of the whole, in such a way that the whole forces the units of one part to arrange themselves in just such a way as is necessary for the perfection of that part,—a perfection required for the harmony of the whole. Spencer himself says very rightly, 'It seems difficult to conceive that this can be so, but we see that it *is* so.' As a matter of fact, groups of units removed from an organism possess the power of constructing the whole anew; and we are thus obliged to admit that the tendency to take a specific form is present in all parts of the organism. The 'units' are *physiologically variable* quantities, which in every case act in such a manner as the whole demands.

The assumption of these 'physiological units' does not suffice as an explanation of heredity: it proves insufficient even as interpreting the differentiation of organs in simple ontogeny, quite apart from the question of amphigonic heredity. But it has the merit of having utilised the smallest vital particles as constituent elements of the organism, and of having made them the basis of a theory of heredity.

Ernst Brücke was the first to admit the existence of small vital particles of this kind, and to give cogent reasons for so doing. Although he did not denote them by any special name in his extremely important paper entitled 'Elementar Organismen,'* he was the first to oppose the old theory of the cell, especially with regard to its fluid contents, and to show that its body must possess an *organisation*, quite distinct from the molecular structure of the organic compounds.

Darwin's theory of 'pangenesis' was stated in the final chapter of his great work on 'The Variation of Animals and Plants

* 'Wiener Sitzungsberichte,' Oct. 10, 1861, Bd. 44, ii., p. 381.

under Domestication,' which appeared only a few years after Spencer's 'Principles of Biology.' The enormous wealth of facts bearing on heredity which is accumulated in this book is in itself sufficient to show how the gifted author felt himself urged on all sides to consider this extremely difficult and complicated problem. For although Darwin modestly described his theory as a provisional hypothesis, his was, nevertheless, the first comprehensive attempt to explain all the known phenomena of heredity by a common principle. The theory has so often been discussed and is so well known that a brief account of its substance will suffice here.

A multicellular organism, whether animal or vegetable, is gradually built up by cell-division: but it is assumed that this method of multiplication is not the only one. Each cell possesses in addition, at each stage of its development, the power of giving off invisible granules or atoms, which, at a later period and under certain conditions, can develop again into cells similar to those from which they originated. Numbers of these 'gemmules' are being given off continually from all cells of the body and conveyed into the blood, and thus circulate through the body, finally settling down in some part, principally in those regions in which the development of offspring will take place later on, *i.e.*, in buds or germ-cells. As gemmules from all the cells of the body are aggregated in these cells, they invest the latter with the power of developing into a new and complete organism. This occurs as follows: — each gemmule reproduces the cell from which it is derived, and the gemmules of the different cells become active in the same order as that in which the corresponding cells followed each other in the ontogeny of the parent.

The germ is not by any means composed exclusively of gemmules which have been derived from the organism in which they were formed, but consists, at the same time, of a very large number of gemmules which are derived from parents and ancestors even of very remote generations; and hence a great many more gemmules take part in each case of ontogeny than there are cells formed. Each cell and each part is represented by a great variety of gemmules. A selection must therefore take place, as only *one* gemmule can form the required cell, and the rest must remain dormant. In this way a number of gemmules, which have been hitherto dormant, are transferred from one generation to the next: they may, under certain conditions, become active, and thus again bring into existence ancestral traits which had disappeared in the parents.

This is, in brief, the theory of pangenesis. It does not take into account the physical nature of the gemmules. They are capable of multiplying, and do so continually: but the question as to whether they have any definite arrangement, and if so, what the nature of that arrangement is, is not touched upon; nor is any mention made of the causes and mechanism by which it comes about that they are always in the right place and develop into cells at the right time.

I do not say this by any means as a reproach, but only to bring out clearly the speculative character of the whole hypothesis. Darwin did not go on to inquire whether all these assumptions were possible: he only asked what it was necessary to assume in order to explain this or that fact of heredity, without troubling himself to consider whether the assumption were borne out by facts or not. And he was right in doing so, for at the time when he propounded his hypothesis it was not possible to found any theory of heredity on the only sound basis, — that of a knowledge of the most minute cell-structure. I have already pointed out how extremely important and fruitful his theory of pangenesis has been: it drew attention for the first time to all the phenomena that needed explanation, and showed what assumptions must be made in order to explain them.

It will be shown later on that, in spite of the fact that a considerable number of these assumptions are untenable, a part of the theory still remains which must be accepted as fundamental and correct, — in principle at any rate, — not only now, but also for all time to come. I refer to the most general portion of these assumptions only, namely, that presupposing the existence of material particles in the germ which possess the properties of the living being, and each of which is to be regarded as the primary constituent ('Anlage') of one portion of the organism. I must honestly confess to having mentally resisted this fundamental point of the Darwinian doctrine for a long time. It appeared almost impossible to me that such an enormously large number of individual primary constituents as we must suppose to exist, according to Darwin's view, could be contained in the minimum of substance which, as will be shown hereafter, we have to regard as the actual bearer of heredity. I tried in several ways to arrive at a satisfactory epigenetic theory,* which, starting from a germ-substance of comparatively simple structure, should exhibit the various differentiations of the organism as due to regular changes brought about by the division of this primary structure. But the more I considered the problem as time went on, the more I was convinced that such a solution was impossible. And in this book I trust that I shall be able to give a satisfactory proof that only one theory of evolution in Darwin's sense, *i.e.*, the assumption of minute primary constituents in the germ, is in accordance with the facts; and the objection which for a long time prevented me from accepting this very simple assumption, disappears with the discovery that what is apparently impossible does really occur.

I certainly consider even now that Darwin's theory must be looked upon, and that he probably considered it, rather as an inquiry into the problem of heredity than as a solution of the problem. His assumptions do not, properly speaking, explain the phenomena. They are to a certain extent a mere paraphrase of the facts, an explanation of a purely *formal* nature, based on speculative assumptions, which were made not because they seemed possible, or even likely, but because they provided a formal explanation of all the phenomena on one principle. If we suppose that each cell arises from a special gemmule, and that these gemmules are present wherever they are wanted, it is easy to see how that structure, the origin of which we wish to explain, may appear in any given position. Further, when a large number of cells is to arise in regular succession from *one* egg-cell, the desired sequence of cells must of course result if we assume that the gemmules present become active in the required order. But this supposition does not really explain the phenomena. Even at the present day our explanations are imperfect enough, and are far from going to the bottom of the matter, but they differ from Darwin's provisional hypothesis in that they attempt to find out the actual facts concerned in the processes, and to arrive at a *real*, and not merely a formal, solution of the problem. The great naturalist's merit in having at once found the right foundation on which to base a real solution

* The indication of such a theory is given, *e.g.*, in the essay entitled 'Die Continuitat des Keimplasma's,' Jena, 1885, p. 38 *et seq.* (pp. 207 *et seq.* of the English translation).

is not diminished by the fact of his having been less startled by the consequences of his 'gemmule'-hypothesis when seeking for a purely formal explanation, than he would have been had he tried to adapt his hypothesis to the facts. The hypothesis, as stated by him, could not be regarded as a real solution of the problem of heredity, if only because it leaves unexplained the giving off of the gemmules into the blood, their circulation through the body, and intrusion into the germ- and other cells. All these are assumptions without a basis in fact. This is evidently the reason why modifications of the theory of pangenesis were repeatedly made very soon afterwards.

Before considering these modifications, I should like once more to state clearly the relation of Spencer's 'physiological units' to Darwin's 'gemmules.' Darwin himself considered the former to be closely related to his gemmules; and, in fact, he would have regarded Spencer's ideas as essentially coinciding with his own, had he not noticed certain passages in Spencer's book which seemed to point to something quite different.*

It will be apparent, I think, from what has already been said, that these two views are entirely different. What is common to both is that they assume the existence of minute living units, multiplying by fission: but the part taken by them in the constitution of the body is quite differently conceived. Spencer's units are the elements which exclusively compose the living body; while Darwin's gemmules only give rise to cells, *i.e.*, they are elements which are present for the special purpose of bringing about heredity, without anything being specified as to their share in the composition of the living body. As will be shown more clearly later on, Spencer's hypothesis is superior in this respect to Darwin's. On the other hand, Spencer's similar units are the bearers of all the characters of the species, owing to their complex molecular structure; while the Darwinian gemmules are primary constituents of individual cells, which are to be considered as differing in a manner corresponding to the difference of the individual cells. Spencer's theory is epigenetic, Darwin's evolutionary; in this respect the latter is, in my opinion, superior to the former.

---

* Charles Darwin, 'The Variation of Animals and Plants under Domestication,' 2nd ed., vol. ii., London, 1888; note on p. 371.

Galton * was the first to make an attempt to improve on the theory of pangenesis. In a short but suggestive essay he accepted the hypothesis of the gemmules, but rejected the doctrines of their circulation through the blood, and of the aggregation in the germ-cells of gemmules given off by the body-cells. Now as the gemmules which have been converted into body-cells are used up, it follows that the germ-cells can only contain those gemmules which are left — those, out of the enormous number contained in a germ-cell, which have not developed further. For each germ-cell, as both Galton and Darwin assume, contains each kind of gemmule in many modifications, originating from the different ancestors of the organism. The theory of the origin of the germ-cells from the remains of the germ mass not used up in ontogeny ('the residue of the stirp') has been compared to, and regarded as the precursor of, the conception of the continuity of the germ-plasm which I originated long afterwards. A certain resemblance does, it is true, exist between the two conceptions, but it will be shown in the section on the continuity of the germ-plasm that the similarity is only a superficial one.

Herbert Spencer defines heredity as the capacity of every plant and animal to produce other individuals of a like kind, and states expressly that in this fact, which is perfectly familiar to us, and for this reason seems to be a matter of course, lies the real essence and principle of heredity, 'the phenomena commonly referred to it being quite subordinate manifestations.' Thus the *blending of the individual 'characters' of the parents* in the children has, as a rule, been placed in the foreground in considering questions of heredity, and it has been overlooked that this is quite a secondary phenomenon, — important no doubt in many respects, and interesting in a high degree, but still only the result of a certain mode of multiplication, *i.e.*, sexual reproduction, and by no means an essential phenomenon of heredity. Darwin recognised this distinctly, and concerned himself primarily with the theoretical explanation of individual development (ontogeny). But the majority of writers on heredity, including Galton, have turned their whole attention to the blending of the qualities of the parents

---

* Francis Galton, 'A Theory of Heredity,' *Journal of the Anthropological Institute*, 1875.

in the children, — a problem which is doubtless well worthy of investigation, but which, at the same time, deals only with a side issue of the processes of reproduction. How little I underrate the significance of amphigonic heredity, even in its theoretical relations, will be evident in a later part of this book, in which I attempt to derive the existence of the germ-plasm from the phenomena of this form of heredity; but to me it seems dangerous to investigate heredity theoretically from the point of view of amphigonic descent exclusively, because one has here to deal with the most complex of all the phenomena, and the main point may easily be overlooked in a mass of confusing secondary considerations. Even Galton, in my opinion, allowed himself to be too much influenced by this aspect of the question. Excellent as are his later researches on the laws relating to the blending of characters of the parents in the children, I consider his theoretical deductions on the fundamental phenomena of heredity unsatisfactory. The few hints that he gives as to the cause of ontogeny seem to me by no means equal to Darwin's simple but truly penetrating and accurate deductions. It is quite conceivable that the phenomena of the blending of the characters of the parents in the children would be the most interesting to a statistician and anthropologist like Galton, but they have kept him within the limited range of these phenomena, and have prevented him from arriving at really general principles and at a comprehensive theory of heredity.

Galton has, however, the merit of having been the first to deny the circulation of the gemmules, and, in connection with this, to cast doubt upon the general validity of the doctrine of the transmission of acquired modifications. He certainly believes the latter to be 'faintly heritable,' and assumes, in order to explain this transmission, that no general 'circulation of the gemmules' takes place, but that each cell sets free some gemmules which get into the circulation and eventually penetrate into the sexual elements.

Galton's essay was published only a few years after the appearance of Darwin's theory of pangenesis; but it cannot be said that it exercised any influence on the subsequent development of the theory of heredity. Apparently it was not much noticed even in England, and on the Continent it remained unknown for a long time. This must be my excuse for being ignorant of the existence of this paper, and consequently for not referring to it in my essays which appeared nearly ten years later.* In one of these essays 'On Heredity' (1883), I contested at first in general terms not only the existence but also the theoretical possibility of the transmission of acquired characters, and tried to release the theory from the necessity of an explanation which deprived it of any further development. In this essay I further assumed the existence in the germ-cell of a reproductive substance, the *germ-plasm*, which cannot be formed spontaneously, but is always passed on from the germ-cell in which an organism originates in direct *continuity* to the germ-cells of the succeeding generations. The difference between the 'body' in the narrower sense (soma) and the reproductive cells was also emphasised, and it was maintained that the germ-cells alone transmit the reproductive substance or germ-plasm in uninterrupted succession from one generation to the next, while the body (soma) which bears and nourishes the germ-cells, is, in a certain sense, only an outgrowth from one of them.

A second attempt to improve upon the theory of pangenesis must be considered here. I have already referred elsewhere † to the interesting and ingenious book on 'The Laws of Heredity,' ‡ by W. K. Brooks. The author retains the fundamental points of this theory, viz., the formation of gemmules in all the cells of the body, their circulation through the latter, and their aggregation in the germ-cells or buds: he differs, however, from Darwin principally in ascribing to the male germ-cell a particularly strong power of attraction for the gemmules, so that it collects a special mass of them and stores them up. As this assumption is made chiefly for the purpose of explaining varia-

* These essays first appeared separately in the years 1881–91. The only complete edition of the collected essays which has hitherto appeared is the English translation, 'Essays upon Heredity and Kindred Biological Problems' (edited by Poulton, Schonland, and Shipley, Oxford, 1889), containing Essays I.—VIII. A second edition appeared in 1891 as Vol. I., and Essays IX.—XII. follow this year as Vol. II. A French translation of all these essays, with the exception of the last on 'Amphimixis,' &c., has also appeared with the title, 'Essais sur l'Héredité et la Selection Naturelle,' traduits par Henry de Varigny, Paris, 1892.

† 'The Significance of Sexual Reproduction in the Theory of Natural Selection.'—'Essays upon Heredity,' p. 326.

‡ W. K. Brooks, 'The Laws of Heredity a Study of the Cause of Variation and the Origin of Living Organisms' Baltimore, 1883.

tion, I shall postpone any further consideration of it to the section which treats of this subject.

In the following year, Nägeli's 'Mechanico-physiological Theory of Descent'* appeared. This book, which abounds in ingenious deductions and important suggestions, doubtless exercised a great influence on the views of that time. Its importance cannot be denied, even if, as I believe to be the case, only a small portion of its theoretical propositions can be retained. Many as are the fruitful ideas and anticipation of facts afterwards proved which we owe to Nägeli, his own theory of heredity has already become untenable. For this reason, and also because the theory is so well-known, I will not describe it fully here, but will only refer to the remarks which I made on the subject some years ago,† and to the recent detailed criticism by Wiesner.‡ Although I do not consider that Nägeli's hypothesis leads us towards a true theory of heredity, it nevertheless contains an important suggestion, that of the *idioplasm*, which gives us a further insight into the problem. I had already assumed the existence of a special reproductive substance — the germ-plasm — on the changes of which development depends, while heredity rests on its continuity: and now Nägeli independently postulated a special reproductive substance, an 'Anlagenplasma' or 'idioplasm,' which although much smaller in bulk than the rest of the living substance of the body — the trophoplasm ('Ernährungsplasma') — determines the detailed construction of the latter. The correctness of this conjecture has not as yet, so far as I know, been disputed, although it was very soon shown that Nägeli was wrong as regards the form in which he imagined the idioplasm to exist. He represented it as consisting of very fine parallel fibres which, by uniting into bundles and crossing each other so as to form a network, traverse the substance of the cell, and being continuous from cell to cell, pervade the whole body as a connected network.

At the time when Nägeli's book appeared, it was already suspected that the reproductive substance is not contained in

---

* C. v. Nageli, 'Mechanisch-physiologische Theorie der Abstammungslehre,' Munchen and Leipzig, 1884.

† *Vide* 'The Continuity of the Germ-plasm, 1885 (pp. 180 *et seq.*, 192, &c.).

‡ Julius Wiesner, 'Die Elementarstructur und das Wachsthum der lebenden Substanz,' Wien, 1892.

the *body* of the cell but in its *nucleus*, and several discoveries were made shortly afterwards which rendered it certain that the idioplasm is to be looked for in the 'chromosomes' of the nucleus, — those rod-like, coiled, or grain-like structures which are distinguished by their remarkable affinity for certain colouring matters. I shall return to the proof of this fact in the following section.

From this time onwards each subsequent theory of heredity was based on a firm foundation of fact. It was now not only known that the phenomena of heredity among the higher organisms are connected with a definite substance, but the seat of the latter had also been ascertained. I now therefore adopted this firm basis for my theory of the germ-plasm, if I may call the imperfect form in which it then existed by such a name: I localised the germ-plasm in the nuclear substance of the germ-cell, and supposed that ontogeny was due to a qualitative change in it, which hands the idioplasm on from one generation to the next by means of nuclear- and cell-division. But I soon went further. From the fact of sexual reproduction, which brings together equal amounts of paternal and maternal germ-plasm at each fertilisation, I inferred not only the composition of the germ-plasm out of a number of units, the '*ancestral germ-plasms*' ('Ahnenplasmen'), but also the necessity of *a reduction of the germ-plasm* each time to one-half of its bulk, as well as a reduction of the number of the ancestral germ-plasms contained in it.* The hypothesis of the 'reducing divisions of the germ-cells' has been thoroughly substantiated by subsequent observations: in fact it has even been proved that in many cases this reduction occurs exactly as I had foretold and had represented in a diagrammatic figure; † that is to say, by the non-occurrence of the longitudinal division of the chromosomes which occurs in the ordinary nuclear division, and by the distribution of these in the daughter-nuclei. This holds good for the ovum as well as for the sperm-cell in animals, and, as far as is known, in plants also. The germ-cell must in all cases by division get rid of half of its nuclear rods, — that is to say, of its germ-plasm, — in order to become capable of fertilisation. This fact supports the other assumption of the construction of the germ-plasm from ancestral germ-plasms, which are not minute vital particles — analogous to Spencer's

---

* 'On the Number of Polar Bodies,' &c., 1887. † *Ibid.*

physiological units — but rather bodies of a highly complex constitution, each containing all the primary constituents which are necessary to the formation of an organism. Each ancestral germ-plasm seemed to me to be of a 'special kind,' and just as many 'different kinds of idioplasm' are removed by the reducing division 'from the ovum as are afterwards introduced by the sperm-nucleus' on fertilisation. It will be seen that I retain this essential basis of the theory of the germ-plasm in its further development as presented here, and I trust that I may now succeed in refuting the objections which have been urged against the 'ancestral germ-plasms,' or, as I now call them, the '*ids*.' In any case it cannot be denied that they help to throw an important light on the subject.

De Vries has so far been my most powerful opponent as regards the ancestral germ-plasms, but his opposition is founded on the misunderstanding I have already referred to, for he looks upon them as the ultimate vital particles — an idea which was foreign to me from the beginning. Of this, however, I do not complain, as at that time I had left the question as to the construction of the ancestral germ-plasms unanswered.* This omission is supplied in the present book, and it will be shown that, although each ancestral germ-plasm is in my opinion a bearer of all the primary constituents required for the construction of an organism, my assumption does not exclude the possibility of its being composed of these constituents in the form of minute vital particles. The 'ancestral germ-plasm' is indeed a unit, but one of a higher order. For this reason alone it cannot be compared with Spencer's 'physiological units,' because the latter, as ultimate vital particles, compose the whole body; while the ancestral germ-plasms only form the nuclear matter, and merely serve the mechanical purpose of the processes of heredity.

De Vries has in a significant manner developed a theory of

* De Vries is also mistaken in ascribing to me the opinion that 'there is only one hereditary substance — only one material bearer of the hereditary tendencies in each individual.' The sentence quoted by him ('On the Number of Polar Bodies,' p. 355) does not deal with this question; it runs as follows: — From several reasons already stated 'at least one certain result follows, viz., that there is an *hereditary substance*, a material bearer of hereditary tendencies, and that this substance is contained in the nucleus of the germ-cell,' &c.

heredity in his essay on 'Intracellular Pangenesis.'* The opinions there expressed really contradict the title of the paper; for pangenesis, in Darwin's sense, means the development of gemmules throughout the body,—the composition of the hereditary substance from gemmules which are derived from all the cells of the body. This very point in Darwin's hypothesis is set aside completely by de Vries: the most characteristic part of it is removed, and what remains is of a more general nature, consisting of principles which, in one form or another, must form the basis of every theory of heredity, at the present day at any rate. Some ideas of his own, however, are then added, and it is these which give a characteristic stamp to his whole series of conceptions. If we regard his hypothesis, as de Vries himself does, as an alteration of the Darwinian theory of pangenesis, it is certainly a radical one, and is of such a kind as at one stroke to infuse new life into the latter, which had become untenable in its original form.

De Vries distinguishes two parts in Darwin's theory of pangenesis, one of which he rejects, while he retains the other. He calls the former portion the 'transport hypothesis,' meaning thereby the assumption of the origin of the gemmules in all the cells of the body, their separation from the cells, circulation in the blood, and ultimate aggregation in the germ-cells. And relying on my rejection of the heredity of 'somatogenic' characters, he shows that the assumption of the transportation of the gemmules from all the cells of the body to the germ-cells is superfluous. He thus does away with that portion of the hypothesis of pangenesis which makes it unacceptable to most people, and places the theory on a new and firmer foundation on which it is capable of further development.

De Vries nevertheless goes too far if he looks upon the 'transport hypothesis' as necessary only for explaining the transmission of somatogenic qualities. It must not be forgotten that the idea of the continuity of the germ-plasm did not exist in Darwin's time. How could the gemmules of all the cells of an organism enter its germ-cells unless they are formed in the body-cells, migrate therefrom, circulate through the body, and come together in the germ-cells? A direct connection between the fertilised egg-cell and the germ-cells of the organism

---

* 'Die Intracellulare Pangenesis,' Jena, 1889.

arising from it was not supposed to exist by any one at that time, nor does it do so except in isolated cases. The 'transport hypothesis' was therefore also necessary in order to explain the production of germ-cells of each kind, which must again contain the gemmules of the parents. Galton, who also rejected the 'transport hypothesis,' thus found himself in the peculiar position of being obliged to suppose that the germ-cells which the organism produces can only contain the unused remainder of the gemmules and their successors, *i.e.*, those gemmules which had been unable to take part in the construction of the organism, and which had remained dormant and were individually of a different nature from the other gemmules. He made use of this supposition to explain the difference between children of the same parents, but found himself obliged to resort to a very artificial assumption to account for the main problem of the resemblance between such children and their parents.

That part of Darwin's theory which de Vries retains is the existence of an hereditary substance composed of 'gemmules,' or minute vital particles which are capable of growth and multiplication by fission, and which become active consecutively in ontogeny, and so build up the organism. The theory is thus deprived of its merely speculative elements, and by transferring the gemmules, in accordance with the most recently ascertained facts, to the germ substance, which, as we know, is passed on by division from cell to cell, the theory of pangenesis is placed on a firm footing.

De Vries, however, was not content with simply modifying Darwin's theory of pangenesis in a negative manner by doing away with one — almost the greater — portion of it; he also reformed it positively by giving a new meaning to the 'gemmules.' There is an essential difference between Darwin's gemmules of *cells*, and de Vries's pangenes, which are gemmules of elements much smaller than cells — that is to say, of the smallest parts of which a single cell is composed. These pangenes are the bearers of the individual qualities or 'characters' of the cell.

The train of thought which led de Vries to imagine the construction of the hereditary substance from such 'bearers of the qualities' ('Eigenschaftsträger') of the cells is too interesting to be passed over. He bases this idea on the assumption of '*a mutual independence of the hereditary qualities*.' According to his view, all species consist of a sum of 'hereditary qualities'; very few, or none of these, are peculiar to any one species, the character of which is determined by the way in which they are combined. The *same* quality recurs in many species, but in different combinations. 'We constantly see how one and the same hereditary quality, or how a definite small group of such, may be combined with all kinds of other hereditary qualities; and how the different characters of individual species are due to the extreme variety of these combinations.' The different organs of a species stand in the same relation to one another in this respect, as do the different species themselves. They exhibit the same qualities, but in different combinations. The individual qualities which constitute a species 'can almost all vary independently of each other,' and can therefore be increased even by artificial selection according to the fancy of the breeder, without requiring a corresponding change in the remaining qualities of the species. But the qualities too are 'miscible in almost any proportion,' as experiments in hybridising are intended to show: 'in no other way can we so clearly demonstrate the secondary importance of a specific type ('Bild'), regarded as a whole as opposed to the independent factors which constitute it.' The qualities, or rather their material substratum, are therefore independent of one another, and miscible to almost any extent.

Those ultimate vital particles or pangenes, which de Vries substitutes for Darwin's gemmules, are therefore the bearers of constituent qualities of the species.

The fundamental idea of de Vries's whole deduction is doubtless perfectly correct. Some ten years ago, when I first began to devote my attention to the problem of heredity, I fully believed in the possibility of an epigenetic theory, but, as will be seen in the course of this book, have long since given up this idea as untenable. I too now believe that the hereditary substance is composed of primary constituents, and even trust that I can prove this assumption to be not only sound, but inevitable. But, at the same time, I do not imagine that it suffices as an explanation of the phenomena of heredity. According to de Vries, the germ-substance is formed of a number of different kinds of pangenes, of which as many are present as there are qualities in the species. He does not consider these pangenes as arranged in any definite grouping, but as freely miscible, in accordance with the assumed 'free miscibility of the qualities.' He contests as superfluous the assumption of *higher* units, such as might be formed by a certain number of pangenes in a definite order; and this view seems to me to be the weak point in his argument.

In the section on the control of the cell by the nuclear substance, I shall adopt what seems to me to be a remarkably happy idea on the part of de Vries, who supposes that material particles leave the nucleus, and take part in the construction of the body of the cell. These particles correspond to the 'pangenes,' they are the 'bearers of the qualities' of the cell, the specific character of which I believe to be stamped upon it by the nature, the different varieties, and the proportional numbers of these particles.

But does the character of a species depend only on these primary qualities of the cell? Are there not qualities of various degrees—primary, secondary, and so on? The pangenes are *primary* 'bearers of qualities'; their mere presence in the hereditary substance gives no indication, or at most, only a very slight one, as to the character of a species. If, for instance, 'chlorophyll-pangenes' are present in the egg-cell of a plant, the only conclusions we can draw as to the specific character of the latter are that it will have green cells of some sort: but we cannot thereby determine where they will be situated, or which portions of the plant will be green, and which variegated; or again, whether its flowers will be green, white, or of some other colour. Not until we were able to find groups of pangenes in the germ-substance, some of which were destined to give rise to leaves, and others to flowers, should we be able to say whether the latter will be green or otherwise.

In the course of his remarks, de Vries mentions the stripes of a zebra. How can these be hereditary if the different kinds of pangenes merely lie close together in the germ without being united into fixed groups, *hereditary as such?* There can be no 'zebra pangenes,' because the striping of a zebra is not a cell-character. There may perhaps be black and white pangenes whose presence causes the black or white colour of a cell: but the striping of a zebra does not depend on the development of these colours *within a cell*, but is due to the regular alternation of thousands of black and white cells arranged in stripes.

De Vries, in another place, refers to the long-stalked variety of the alpine *Primula acaulis*, which is due to reversion to a remote ancestral form of the species. In this case, again, the special peculiarity cannot depend on 'long-stalk pangenes,' because the possession of a long stalk is not an intracellular character. The specific form of the leaves and other parts of a plant is likewise not due to the character of the individual cells composing them: the serrated margin of a leaf, for instance, cannot depend on the presence of 'serration-pangenes,' but is due to the peculiar arrangement of the cells. The same argument would apply to almost all the obvious 'characters' of the species, genus, family, and so on. For instance, the size, structure, veining, and shape of leaves, the characteristic and often absolutely constant patches of colour on the petals of flowers, such as orchids, may be referred to similar causes: these qualities can only arise by the regular co-operation of many cells. The characteristics of the human race may be taken as another illustration. The peculiarities as regards the shape of the skull, nose, &c., cannot depend on the mere *presence* in the germ of pangenes, which are destined to form the hundreds and thousands of different cells constituting the respective qualities; but they must be due to a *fixed grouping of pangenes*, or some other primary elements of the germ, which is *transferable from generation to generation*.

The character of a species cannot depend only upon the number and relation of the pangenes in the germ. It is quite possible to conceive of two different species of totally different structure in which the pangenes of the germ were alike in nature and amount, the difference being solely due to the grouping of the pangenes in the germ. De Vries, it is true, traces 'systematic difference to the possession of *different kinds* of pangenes,' and considers that 'the number of similar pangenes in two species is the real measure of their affinity;'* but this statement seems to me to be somewhat at variance with his fundamental view, according to which 'a number of hereditary qualities constitute the character of each individual species, *though by far the greater majority of them recur in innumerable other species*.' Does he not, in so many words, emphasise the fact that the almost formidable number of different pangenes which are required for 'the construction of a single species' does not necessitate the existence of an inconceivably large multitude of

* *Loc. cit.*, p. 73.

different pangenes in the entire organic world, because 'the number of individual hereditary qualities required for the construction of the latter is relatively small when compared with the number of species'! Each species appears to us as an extremely complicated structure: the whole organic world, however, seems to be the result of innumerable different combinations and permutations of *relatively few factors*.

The idea which is here so clearly and decidedly expressed of the construction of innumerable species by various combinations of relatively few pangenes, shows that, even from de Vries's point of view, it is not the 'pangene *material*' as such, which is the main factor in determining the character of the species, but rather its *arrangement*, or as I shall afterwards express it, *the architecture of the germ-plasm*.

De Vries certainly speaks frequently of 'groups of pangenes,' but he only just touches upon this idea, and postpones entering into details until further discoveries are made with regard to the mechanism of nuclear division. Important as his fundamental view as regards the composition of the germ substance out of primary constituents undoubtedly is, it may easily seem to explain more than it really does; without assuming the formation of groups of such primary constituents for a number of orders each included in the other, even the simplest case of ontogeny cannot be explained, quite apart from reversion or any other complicated phenomenon of amphigonic heredity. Darwin's theory of pangenesis accomplishes more in this respect than does de Vries's modification of it, inasmuch as the former at least deals with the primary constituents of *cell*-structures. The mere presence of a certain collection of pangenes in the germ does not necessitate the formation in the offspring of similar cells to those which existed in the parent; for the character of the individual cell is determined by a definite selection of pangenes. If, indeed, it be assumed that the required pangenes always lie close together, and are always ready at hand whenever they are wanted, an explanation of any particular phenomenon of heredity is no longer difficult, but it seems to me that it would then be necessary to show how the nature of the germ can determine that the right primary constituents are always at the right spot.

As already stated, de Vries occasionally speaks of *groups* of pangenes, but, at the same time, he looks upon the view of there being any 'higher units' in the germ as a superfluous one. I can only explain this inconsistency by supposing that he regards the 'qualities' as independent and perfectly freely miscible, and, in fact, postulates a germ-mechanism which admits of their separation in any manner required. If this were really the case, and the primary constituents were not combined into fixed groups in the germ, how could composite characters composed of many different kinds of cells with a definite arrangement,—*e.g.*, the eye-like spot on a certain feather of a bird,—become a fixed specific character? I am of opinion that the view which entails an independence and uncontrolled miscibility of the qualities is a fallacy, originating in the conception of amphigonic reproduction as a necessary element in heredity. The chapters on amphigonic heredity, reversion, &c., will show how I imagine the idea of the uncontrolled miscibility of the separate qualities to have arisen.

It will frequently be apparent in the course of this book that my point of view is identical with that of the Dutch botanist in many of the most important particulars. I believe, however, that his 'pangenes' or similar minute elements do not suffice *in themselves* for the construction of a theory of heredity, but that something more must be added to make the phenomena comprehensible at any rate in principle.

The manuscript of the present book had already been written for some time when Wiesner's work on the elementary structure and growth of living substance * appeared. Although this monograph does not contain, and is not intended to offer, a theory of heredity, it is nevertheless of great importance in this respect, for it treats of the fundamental points of such a theory, viz., *the composition of living matter out of very small units.* Wiesner remarks that theories of heredity have hitherto always adopted units invented for the purpose, whereas the same units which make life possible at all, and which control assimilation and growth, must also be the agents in bringing about the phenomena of heredity. Spencer's 'physiological units,' Darwin's 'gemmules,' Haeckel's 'plastidules,' and my 'ancestral germ-plasms,' are all, in fact, elements of this kind, assumed for the explanation of the problem of heredity. De Vries stands

* J. Wiesner, 'Die Elementarstruktur und das Wachsthum der lebenden Substanz,' Wien, 1892.

alone in considering all living matter to be actually composed of his 'pangenes,' though I have already indicated that my 'ancestral germ-plasms' are also composed of similar primary units, which do not exist in them alone. The minute vital particles or 'plasomes,' adopted by Wiesner from Brücke, correspond in all essential points to the 'biophors' or bearers of life assumed by myself.

B.—Descriptive Part

By the term heredity is simply meant the well-known fact that living organisms are able to produce their like, and that the resemblance between a child and its parent, although never perfect, may nevertheless extend to the most minute details of construction and functions.

The fundamental phenomena of heredity are familiar in all existing organisms: the transmission of the character of the species from parent to offspring results whether the multiplication takes place by the halving of a unicellular organism or by the process which occurs in multicellular organisms, which consists in a complex succession of continually increasing groups of cells, *i.e.*, in development. These fundamental phenomena of heredity are, however, complicated in all the higher organisms by the connection of reproduction with that process which may be described as amphimixis.* This consists in the mingling of two individuals or of their germs, and owing to its constant connection with reproduction in multicellular organisms it is usually spoken of as 'sexual reproduction.' As will be shown in greater detail further on, the various phenomena, such as the blending of parental characters in the offspring, and reversion, depend exclusively on the hold which amphimixis has taken on the life of the species. Similar phenomena must occur amongst unicellular organisms, in which amphimixis is widely spread if not universal, in the form of conjugation, and is, therefore, not directly connected with reproduction. At present, however, we are ignorant of the details of heredity in these forms, and are therefore compelled to base our conclusions entirely upon what we know to occur in multicellular organisms.

---

* August Weismann, 'Amphimixis, oder die Vermischung der Individuen,' Jena, 1891. *See* 'Essays upon Heredity,' vol. ii., 1892.

These phenomena have only been observed in detail in the higher plants and animals, more particularly in man. In the case of the higher forms of life a large number of facts have now been accumulated which can be used for the purpose of theoretical analysis.

Although the study of heredity is greatly complicated by amphimixis, this mingling of the hereditary tendencies of *two* parents, and even the process of sexual reproduction which accompanies it, afford us a much deeper insight into the actual processes of heredity than we could ever have obtained in any other way. We may thus hope in time to penetrate further into its nature by carrying out more detailed investigations of the phenomena.

In order to do so, however, we must not forget that this form of reproduction is neither the only nor the original one, and that even in multicellular organisms reproduction is not necessarily connected with amphimixis; it must also be borne in mind that so-called asexual (*monogonic*) reproduction forms the basis of the amphigonic method. The fundamental phenomena of heredity had already shaped their course in the living world before the introduction of amphimixis, and have, therefoie, no connection with amphigonic descent and the complications arising from it. This fact has often been overlooked or left out of consideration, and thus the solution of the problem of heredity has been rendered much more difficult. A whole series of the phenomena of heredity can be investigated theoretically without considering the complications arising from amphimixis, though, in point of fact, it is always a factor, and thus the problem to be solved is very considerably simplified.

The natural course of such an investigation would be to pass from the simple to the complex, but it is not advisable at present to begin the study of heredity by a consideration of the simplest beings, and to ascend from the unicellular to the multicellular organisms. For besides the fact that we know nothing of the individual phenomena, — such as the transmission of the individual characters, — in the lower forms, the principal reason for not following the ordinary course in this case is the fact that amphigonic reproduction, or the processes of fertilisation and the complicated development of multicellular organisms, affords us, as already stated, a deep insight into the processes of heredity. The same is true in this case as in almost all physiological processes, — investigation cannot proceed from the simple to the more complex without taking into consideration the objects and processes for which it was first undertaken. It must, on the contrary, avoid the densely overgrown path and skirt the hedge which surrounds the enchanted castle of the secret of Nature, in order to see if there be not somewhere a gap through which it is possible to enter and obtain a firm foothold.

Such a gap in the hedge which encloses the secret of heredity may be found in the processes of fertilisation, if we connect them with the facts of heredity as observed in the organisms which have adopted sexual reproduction.

As long as we were under the erroneous impression that the fertilisation of the ovum by the spermatozoon depended on an *aura seminalis* which incited the egg to undergo development, we could only partially explain the fact that the father as well as the mother is able to transmit characters to the children by assuming the existence of a *spiritus rector*, contained in the *aura seminalis* which was transferred to the ovum and united with that of the latter, and thus with it directed the development. The discovery that development is effected by material particles of the substance of the sperm, the sperm-cells, entering the ovum, opened the way to a more correct interpretation of this process. We now know that fertilisation is nothing more than the partial or complete fusion of two cells, the sperm-cell and the egg-cell, and that normally only *one* of the former unites with *one* of the latter. Fertilisation thus depends on the union of two protoplasmic substances. Moreover, although the male germ-cell is always very much smaller relatively than the female germ-cell, we know that the father's capacity for transmission is as great as the mother's. The important conclusion is therefore arrived at that only a small portion of the substance of the ovum can be the actual hereditary substance. Pflüger and Nägeli were the first to follow out this idea to its logical conclusion, and the latter observer stated definitely that it is impossible to avoid the assumption that no more hereditary substance is contained in the egg-cell than in the male germ-cell, and that consequently the amount of the substance must be infinitesimal, for the sperm-cell is, in most cases, many hundred times smaller than the ovum.

The numerous and important results of the investigations of

many excellent observers on the process of fertilisation have now rendered it almost certain — in my opinion, absolutely so — that by far the larger part of the egg-cell does not consist of hereditary substance, and that the latter only constitutes a small portion even of the sperm-cell. From his observations on the egg of the star-fish, Oscar Hertwig had suspected that the essential part of the process of fertilisation consists in the union of the *nuclei* of the egg- and sperm-cells, and as it is now known that the hereditary substance is undoubtedly contained in the nucleus, this view has, in this respect at least, proved to be the right one. It is true that the nucleus of the male cell is always surrounded by a cell-body, and that Strasburger's opinion to the contrary is incorrect. We now know, through the researches of Guignard, that even in Phanerogams a small cell-body surrounds the nucleus, and that a special structure, the 'centrosome,' — which is absolutely essential for the commencement of development, — is contained within it. This structure will be treated of in further detail presently, but I must here lay stress upon my view, that *the 'centrosome' with its 'sphere of attraction' cannot in any case be the hereditary substance*, and that *it is merely an apparatus for the division of the cell and nucleus*.

Both in animals and plants, however, essentially the same substance is contained in the nucleus both of the sperm-cell and egg-cell: — this is the *hereditary substance of the species*. There can now be no longer any doubt that the view which has been held for years by Strasburger and myself is the correct one, according to which *the nuclei of the male and those of the female germ-cells are essentially similar*, *i.e.*, in any given species *they contain the same specific hereditary substance*.

The splendid and important investigations carried out by Auerbach, Bütschli, Flemming, and many others, on the detailed processes of nuclear division in general, and those dealing more particularly with the fertilisation of the egg in *Ascaris* by van Beneden, Boveri, and others, have given us the means of ascertaining more definitely what portion of the nucleus is the substance on which heredity depends. As already remarked, this substance corresponds to the 'chromosomes,' those rod-like, looped, or granular bodies which are contained in the nucleus, and which become deeply stained by colouring matters.

As soon as it had been undoubtedly proved that the nucleus,

and not the body of the cell, must contain the hereditary substance, the conclusion was drawn that neither the membrane of the nucleus, nor its fluid contents, nor the nucleoli — which latter had been the first to attract attention — could be regarded as such, and that the 'chromatic granules' alone were important in this respect. As a matter of fact several investigators, — Strasburger, Oscar Hertwig, Kollicker, and myself, — reasoning from the same data, arrived at this conclusion independently, within a short time of one another.

It will not be considered uninteresting or superfluous to recapitulate the weighty reasons which force us to this conclusion, for it is clear that it must be of fundamental importance in a theory of heredity to know for certain what the substance is from which the phenomena which are to be explained proceed.

The certainty with which we can claim the 'chromatin granules' of the nucleus as the hereditary substance depends firstly, on the process of amphimixis; and secondly, on that of nuclear division. We know that the process of fertilisation consists essentially in the association of an equal number of chromatin rods from the paternal and maternal germ-cells, and that these give rise to a new nucleus from which the formation of the offspring proceeds. We also know that in order to become capable of fertilisation each germ-cell must first get rid of half of its nuclear rods, a process which is accomplished by very peculiar divisions. Without entering into further particulars here, amphimixis may be described as a process by means of which one-half of the number of nuclear rods is removed from a cell and replaced by an equal number from another germ-cell.

The manner, however, in which the chromatin substance is divided in nuclear division strengthens the above view of its fundamental nature. This method of division leaves no doubt that it is a substance of the utmost importance. I need only briefly recapitulate the main points of the wonderfully complicated process of the so-called mitotic or karyokinetic cell-division, which follows a definite law even as regards the most minute details.

When the nucleus is going to divide, the chromatin granules, which till then were scattered, become arranged in a row, and form a long thread, which extends through the nucleus in an irregular spiral, and then divides into portions (*chromosomes*) of fairly equal length. The chromosomes have at first the form of long bands or loops, but afterwards become shortened, thus giving rise to short loops, or else to straight rods or rounded granules. With certain exceptions, to be mentioned later, the number of chromosomes which arise in this way is constant for each species of plant or animal, and also for successive series of cells. By the time the process has reached this stage a special mechanism appears, which has till now remained concealed in the cell substance. This serves to divide the chromatin elements into two equal parts, to separate the resulting halves from one another, and to arrange them in a regular manner. At the opposite poles of the longitudinal axis of the nucleus two clear bodies — the 'centrosomes,' each surrounded by a clear zone, the so-called 'sphere of attraction' — now become visible. The importance of these was first recognised by Fol, van Beneden, and Boveri. They possess a great power of attraction over the vital particles of the cell, so that these become arranged around them like a series of rays. At a certain stage in the preparation for division, the soft protoplasmic substance of the cell-body as well as of the nucleus gives rise to delicate fibres or threads: these fibres are motile, and, after the disappearance of the nuclear membrane, seize the chromosomes — whether these have the form of loops, rods, or globular bodies — with wonderful certainty and regularity, and in such a way that each element is held on either side by several threads from either pole. The chromatin elements thus immediately become arranged in a fixed and regular manner, so that they all come to lie in the equatorial plane of the nucleus, which we may consider as a spherical body. The chromatin elements then split longitudinally, and thus become doubled, as Flemming first pointed out. It must be mentioned that this splitting is not caused by a pull from the pole threads (spindle threads), which attach themselves to the chromatin rods on both sides; the division arises rather from forces acting in the rods themselves, as is proved by the fact that they are often ready to divide, or indeed have already done so, some time before their equatorial arrangement has taken place by means of these threads.

The splitting is completed by the two halves being gradually drawn further apart towards the opposite poles of the nuclear spindle, until they finally approach the centre of attraction or centrosome, which has now fulfilled its object for the present, and retires into the obscurity of the cell-substance, only to become active again at the next cell-division. Each separated half of the nucleus now constitutes a daughter-nucleus, in which it immediately breaks up, and becomes scattered in the form of minute granules in the delicate nuclear network, so that finally a nucleus is formed of exactly the same structure as that with which we started. Similar stages to those which occur in the aggregation of the chromatin substance in the mother-nucleus preparatory to division are passed through during the separation of the daughter-nuclei, but in the reverse order.

It is evident, as Wilhelm Roux was the first to point out, that the whole complex but wonderfully exact apparatus for the division of the nucleus exists for the purpose of dividing the chromatin substance in a fixed and regular manner, not merely quantitatively, but also in respect of the *different qualities* which must be contained in it. So complicated an apparatus would have been unnecessary for the quantitative division only: if, however, the chromatin substance is not uniform, but is made up of several or many different qualities, each of which has to be divided as nearly as possible into halves, or according to some definite rule, a better apparatus could not be devised for the purpose. On the strength of this argument, we may therefore represent *the hereditary substance as consisting of different 'qualities.'* The same conclusion is arrived at on purely theoretical grounds, as will be shown later on when we follow out the consequences of the process of amphimixis.

For the present it is sufficient to show that the complex mechanism for cell-division exists practically for the sole purpose of dividing the chromatin, and that thus the latter is without doubt the most important portion of the nucleus. Since, therefore, the hereditary substance is contained within the nucleus, *the chromatin must be the hereditary substance.*

De Vries's objection to this view is, in my opinion, only an apparent one; for it has not been asserted that 'the nucleus alone is the bearer of the hereditary characters,' as de Vries thinks, but that the nucleus alone contains the *hereditary substance*, or that substance which is capable of determining not only the character of a particular cell, but also that of its descendants. This is never contained in the cell-body, but always in the nucleus in multicellular organisms, and doubtless the same holds good for unicellular beings. It is quite possible that in certain lower Algæ a few of the structures in the cell—such

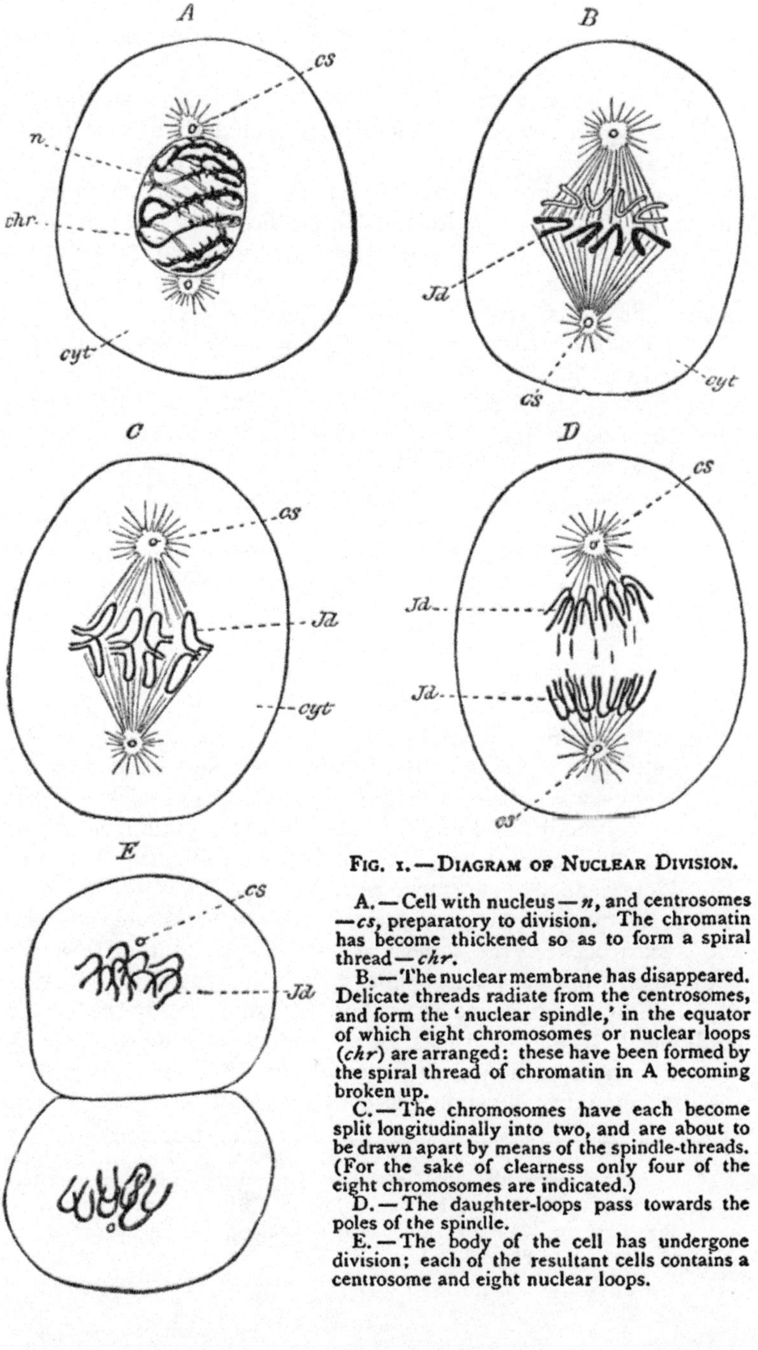

FIG. 1.—DIAGRAM OF NUCLEAR DIVISION.

A.—Cell with nucleus—*n*, and centrosomes—*cs*, preparatory to division. The chromatin has become thickened so as to form a spiral thread—*chr*.

B.—The nuclear membrane has disappeared. Delicate threads radiate from the centrosomes, and form the 'nuclear spindle,' in the equator of which eight chromosomes or nuclear loops (*chr*) are arranged: these have been formed by the spiral thread of chromatin in A becoming broken up.

C.—The chromosomes have each become split longitudinally into two, and are about to be drawn apart by means of the spindle-threads. (For the sake of clearness only four of the eight chromosomes are indicated.)

D.—The daughter-loops pass towards the poles of the spindle.

E.—The body of the cell has undergone division; each of the resultant cells contains a centrosome and eight nuclear loops.

as vacuoles and chlorophyll bodies—pass directly from the egg-cell into the daughter-cells, although this cannot by any means be considered as proved. In any case such a *direct* transmission plays only a very insignificant part in plants, and practically none at all in animals, for specific structures are not present in the egg-cells of animals: there may at most be deposits of nutrient material. These, however, are not living structures of the cell, but only passive chemical substances. So far from denying that the nucleus contains the hereditary substance, de Vries bases his whole theory on this incontestable fact. The last doubts on this point were dispelled by the experiments of Boveri,* who, after artificially removing the nucleus from the eggs of a species *A* of sea-urchin, and then pouring over them the sperm of another species *B*, found that these eggs developed into larvæ of the latter species. In this case therefore the substance of the maternal germ-cell acted as nutrient material only, whilst the paternal germ-cell impressed the character of the species on the larva. None of the maternal specific characteristics were transmitted, and in this case, at all events, the question of any 'heredity apart from the nucleus' is therefore excluded.

Several objections have been recently raised to my view that the nucleus is the seat of heredity. Verworn,† for instance, repeats the opinion, previously expressed by Whitman, that the cell-body, quite as much as the nucleus, must be looked upon as the hereditary substance, because the nucleus cannot exist without the cell-body; and also because, in his opinion, which is undoubtedly a correct one, the life of a cell consists in a continual interchange of substance between the cell and the nucleus. But is the question as to whether the closest physiological relations exist between the nucleus and the cell, so that neither can exist apart from the other, synonymous with that as to whether the hereditary substance is contained in the nucleus or in the cell-body? We must at least be allowed to make the hypothesis that the 'store of the primary constituents' ('Anlagenmagazin') of the hereditary substance is contained and preserved in the

* Boveri, 'Ein geschlechtlich erzeugter Organismus ohne mutterliche Eigenschaften.' Gesellsch. f. Morph. u. Physiol., Munchen, 16 Juli 1883.

† Max Verworn, 'Die physiologische Bedeutung des Zellkerns,' Bonn, 1891 (Archiv. f. ges. Physiol, Bd. 51).

nucleus; for, as has already been indicated, and will subsequently be shown more clearly, this substance can hardly be stored up in two different places, seeing that a very complicated apparatus is required for its distribution: a double apparatus would certainly not have been formed by nature if a single one suffices for the purpose. Only as long as the phenomena of heredity and the meaning of these phenomena are still far from being known, is it possible to hold such opinions as that which presupposes the distribution of the hereditary substance amongst both cell and nucleus. As soon as a further insight into these processes is obtained, it will no longer be possible to doubt that the structure of the hereditary substance must be so complex that we can only wonder how it could ever have been developed at all. We know that the nucleus contains a substance which, even with the imperfect means of observation at our disposal, is seen to be extremely complex, and that it becomes modified in a very remarkable manner after every cell division, only to be again transformed at the approach of the following division. We can, moreover, observe that the cell is provided with a special apparatus which evidently enables it to halve this substance very accurately. The statement that *this substance is the hereditary substance* can, therefore, hardly be considered as an hypothesis any longer.

It has also been supposed that fresh evidence against the view that the chromosomes are the hereditary substance, has been furnished by the recent observations of Fol * and Guignard,† which prove that the centrosome and its 'sphere of attraction,'— which belong to the cell-body, and constitute the apparatus for division,— pass into the ovum along with the sperm-nucleus in the process of fertilisation. Suppose I take two heaps of grain from different places, and load them on two carts, harness a horse to each cart, and drive them to the same place; does this prove that the horses consist of grain? They are merely the means by which the grain is transferred from one place to another, just as the centrosomes are the means whereby the sperm-nucleus is transferred to the ovum: whether they are any-

---

* 'Le Quadrille des Centres,' Archiv. Sc. Phys. et Nat., Genève, 15 Avril 1891.

† 'Sur l'Existence des Sphères Attractives dans les Cellules Vegétales,' Compt. Rend. Sc., 9 Mars 1891.

thing more than this, — *i.e.*, whether they contain hereditary substance, — still remains to be proved, and such a conjecture is hardly more probable than that the horses, besides being the means of transport of the corn, should actually consist of corn!

It might, however, be urged that the transference not only of the hereditary substance or *store of primary constituents*, but also of the centrosome or *means of transference* of this substance, implies the transference of the rate of cell-division, which is regulated by the centrosome and essentially decides the cell-sequence in the offspring, and which consequently also takes part in heredity. But I consider this also to be an incorrect deduction, because the periods of activity of the apparatus for division must obviously be dependent on the conditions of the cell itself; these conditions, however, apart from nutrition, depend on the ultimate specific structure of the cell. As, according to my view, this structure is impressed on the cell by the nuclear substance, the periodicity of cell-division must also be dependent on the nucleus. The law that *only a certain part of the nuclear matter is to be regarded as the hereditary substance appears to me to receive fresh support from all the more recent observations.**

Chromatin substance is not only contained in the nucleus of the germ-cells and of the fertilised ovum, but also in all the cells

---

* Many seem inclined to regard the process observed by Fol, which he described as 'Le quadrille des centres,' as a proof that the centrosomes nevertheless must — or at any rate might — contain a kind of hereditary substance. I believe, however, that this process is quite similar to that which occurs in every nuclear division, except that in fertilisation, — owing to the fact that the first segmentation nucleus receives a centrosome from each of the two conjugating cells, — it is a double, and not a single process. Each of these centrosomes divides and passes to the region of the two poles of the future spindle, just as would occur if only a single centrosome were present in the cells. I should be surprised if this were not the case, and if the centrosome of the egg-cell passed to one pole, and that of the spermatozoon to the other! Guignard is of the opinion that even if the nucleus is of great importance as regards the transference of transmissible qualities, we must nevertheless attribute to the 'sphères directives,' 'le rôle primordial dans l'accomplissement de la fécondation.' This is true if it only indicates that the beginning of embryonic development depends — as does every nuclear division — on the presence of the *apparatus* for division. But the view is not thereby refuted that fertilisation consists in the union of two nuclear substances.

of the entire organism in each phase of its development, at any rate as long as they are capable of multiplying, and are possessed of vitality. The chromatin in all the cells of the body is derived from that in the fertilised ovum, while the development of the body from the egg-cell is effected by a series of cell-divisions, each of which includes a division of the nucleus in the manner just described. In the process of ontogeny the chromatin of the first nucleus undergoes repeated subdivisions into two parts of equal volume, and it would very soon become so small as to be invisible even under the highest powers of the microscope, if it did not continue to grow, as does the cell-body. This occurs just as much in the case of numerous animal eggs to which no nutrient material is supplied during the development of the embryo, as in that of those which are nourished from the beginning, or of plants which as a rule begin to obtain their own nutriment at a very early stage. The chromatin, or hereditary substance of the fertilised ovum, enters upon a long and complex process of growth, which only ceases when no new cells are produced either for the formation of new parts, or to replace old ones, — that is to say, at the end of the life of the individual. This growing hereditary substance may be compared to a tree whose branches arise in strict dichotomy, except for the fact that the chromatin does not consist of *one* continuous mass, but of a number of separate particles not actually contiguous with one another; for at each cell-division the two halves of the chromatin rods separate never to unite again in *one* nucleus. Each is finally contained in a special nucleus, which is separated from the rest by being enclosed in a special cell-body. The question now arises as to whether all these fragments of the hereditary substance which compose the chromatin 'tree' of an organism are similar to, or different from, one another, and it can easily be shown that the latter must be the case.

In order to prove this, we take as our basis the well-grounded assumption that the chromatin in the nucleus of the fertilised egg is the substance on which heredity depends. Thus we know that the possibility of the offspring resembling its father, for example, in a thousand different physical and mental characters depends on the minute mass of a few chromatin granules in the nucleus of the sperm-cell, and that the characters of a fully formed organism depend as a whole, as well as in detail, on the arrangement, number, and nature of the cells which compose it. The influence therefore which the minute mass of paternal chromatin in the nucleus of the fertilised egg-cell exerts on the course of development, can only be such as to regulate the nature and the rate of multiplication of the cells in the body of the offspring in such a manner as to cause them to resemble the cells of the paternal body. *The chromatin is therefore in a condition to impress the specific character on the cell in the nucleus of which it is contained.*

As the thousands of cells which constitute an organism possess very different properties, *the chromatin* which controls them *cannot be uniform; it must be different in each kind of cell.*

The chromatin, moreover, cannot *become* different in the cells of the fully formed organism; the differences in the chromatin controlling the cells must begin with the development of the egg-cell, and must increase as development proceeds; for otherwise the different products of the division of the ovum could not give rise to entirely different hereditary tendencies. This is, however, the case. Even the two first daughter-cells which result from the division of the egg-cell give rise in many animals to totally different parts. One of them, by continued cell division, forms the *outer* germinal layer, and eventually all the organs which arise from it — *e.g.*, the epidermis, central nervous system, and sensory cells; the other gives rise to the *inner* germinal layer and the organs derived from it, — the alimentary system, certain glands, &c. The conclusion is inevitable that the chromatin determining these hereditary tendencies is different in the daughter-cells.

This holds good in all subsequent stages of ontogeny; the difference between the developmental tendencies of the cells resulting from the division of the ovum is in exact proportion to that between the chromatin substance of their nuclei. *Ontogeny, or the development of the individual, depends therefore on a series of gradual qualitative changes in the nuclear substance of the egg-cell.*

The fundamental principle of the view which has just been briefly sketched was put forward by me some years ago, and I then made use of the term *idioplasm* to represent the substance which is contained in the chromatin bodies of the nucleus, and which determines the nature of the whole cell. Oscar Hertwig also independently adopted this term, which had first been introduced by Nägeli with a somewhat different meaning. As stated in the first section of this book, Nägeli defined idioplasm as the guiding and controlling substance of the body, in contrast to the more passive and controllable trophoplasm. It is open to doubt whether the latter term should be retained, but the former is certainly happily chosen. It is true Nägeli did not mean to indicate any definite substance visible under the microscope when he used the word idioplasm, for the facts of nuclear division and fertilisation were then unknown. But these facts are so convincing that no doubt as to what is to be regarded as the idioplasm is any longer possible, and Nägeli's conception of an idioplasm forming a network, traversing and connecting the contents of all the cells in the organism, may be regarded as abandoned. We are therefore justified in transferring the term introduced by him to the nuclear substance which determines the nature of the cell.

We now therefore understand by the term idioplasm the nuclear substance controlling any particular cell. This is at the same time the hereditary substance, for it is never formed afresh, but is always derived from the idioplasm of another cell; moreover, it not only determines the *actual* characters of the particular cell, but also those of all of its descendants.

Hence we must assume a difference in the idioplasm not only in dealing with two cells differing in structure and functions, but also in all cases in which we know that different primary constituents are contained in two cells. This has often been overlooked in using the term 'embryonic cells' merely in the sense of equivalent elements 'which may give rise to any parts,' simply because they frequently resemble one another, assuming that they must therefore always be actually equivalent. It is quite true that the idioplasm of such cells appears similar, at least we can recognise no definable differences in the chromatin rods of two cells in the same animal. But this is no argument against the assumption of an internal difference. The perfect external resemblance between two eggs is not a sufficient reason why two identical chickens should be hatched from them. The eggs may have been produced by different mothers, or they may have been fertilised by two different males. We cannot perceive these slight differences in either case, and we could not even do so by attempting to analyse the idioplasm concealed in the nuclei of the two eggs by the aid of our most powerful objectives. Theoretical considerations will show later on that it must be so, and that the units of the idioplasm on which the nature of the latter depends are far too numerous, and therefore far too small, to be visible.

If therefore the two halves into which the chromatin rods are split in karyokinesis look exactly alike, and even if the divided portions of the granules (microsomes), of which the rods often visibly consist, resemble each other exactly, there is still no reason why they should not be different in their nature; in some cases one, and in others another occur.

We shall consequently in this connection have to assume two kinds of nuclear division which are externally indistinguishable from one another, in one of which the two daughter-nuclei contain similar idioplasm, while in the other they contain different kinds of idioplasm. These kinds of division we may speak of as *homœokinesis* and *heterokinesis*, that is, as a division into parts similar or dissimilar to each other with regard to the hereditary tendencies they contain ('erbgleich' and 'erbungleich'). The former must depend on a perfectly uniform distribution of the primary constituents in the two halves of the rods, and will consequently have been preceded by a duplication in the process of growth; in the latter this growth will be connected with a heterogeneous grouping of these constituents.

Although we cannot ascertain anything directly about the forces which cause this splitting of the chromatin rods, it may at any rate be asserted that they must be contained within the substance of the latter, and be connected with the actual development of the qualities of the idioplasm: for otherwise it could not be understood how the qualities, which are changed during the division of the nucleus, become separated sharply from one another and arranged in the two daughter-nuclei. And yet this must be the case if different cells with different kinds of idioplasm can all arise from *one* mother-cell, which is an undoubted fact.

It appears to me, therefore, that the regular ontogenetic changes of the idioplasm, as they begin with the division of the egg-cell and cease with the natural death of the organism, depend on purely internal causes, which lie in the physical nature of the idioplasm. In obedience to these, a division of the nucleus accompanies each qualitative change in the idioplasm, in which process the different qualities are distributed between the two resulting halves of the chromatin rods. I shall speak of the different kinds of idioplasm arising in this way as the *ontogenetic stages of the idioplasm*, or shortly, *the onto-idic stages*.

Hereditary substance, in the full meaning of the term—*i.e.*, that substance which contains all the primary constituents of the whole organism, is merely the idioplasm of the germ-cell, and it is advisable for practical purposes to denote this first onto-idic stage by the short term *germ-plasm*, which I suggested for it at a time when the idea of idioplasm had not been introduced. At that time I meant by the term 'germ-plasm' the hereditary substance of a germ-cell capable of development, without expressing any opinion as to its position or nature. We can now state that *the germ-plasm is the first ontogenetic stage of the idioplasm of an animal or a plant*, whether the cell, in the nucleus of which it is contained, is sexually differentiated or not.

We must next attempt to form an idea of *the constitution and nature of the germ-plasm*, and of the ontogenetic stages of the idioplasm, or onto-idic stages.

PART I

THE MATERIAL BASIS OF HEREDITY

CHAPTER I

THE GERM-PLASM

1. The Fundamental Units

Now that the conception of the germ-plasm as the hereditary substance contained in the germ-cells has been fully established, and since it has been shown in general terms that this form of the idioplasm must become changed during ontogeny and converted into the idioplasm of the cells which constitute the mature organism, we must attempt to form some idea of its nature; for it would otherwise be impossible to construct a theory of heredity. In attempting this, we shall for the present entirely neglect the complication due to sexual reproduction, and take as our starting-point a germ-plasm which does not contain the primary constituents of two parents, but those of one only, — that is to say, one which is constituted just as it would be in a species which had at all times multiplied asexually.

Before venturing to express an opinion concerning the constitution of the germ-plasm, and to derive therefrom the phenomena of heredity, I should like to premise that it is not my intention to attempt an explanation of life. It is necessary to distinguish between a theory of life and one of heredity. De Vries has pointed out very clearly that the former is impossible at present, but that it seems by no means impossible to arrive at a satisfactory explanation of the phenomena of heredity if one takes for granted the essential phenomena of life, — nutrition, assimilation, and growth.

These functions, together with the associated ones of sensation and movement, are connected in all organisms with which we are familiar, from the simplest unicellular forms to the highest plants and animals, with at least two different substances, viz., the idioplasm of the nucleus, — *i.e.*, the hereditary plasm in the more general sense, — and the protoplasm of the cell-body. These two differ as regards their functions, though they resemble each other in being composed of living substance: that is to say, the primary vital forces, nutrition and growth, are developed within them. As the term 'protoplasm' is used in a far too indefinite sense, I shall follow Nägeli's example, and call the vital substance of the cell the 'formative plasm' or *morphoplasm* (Nägeli's 'trophoplasm'), in contrast to the *idioplasm*. The latter is the active element in the process of formation, and the former the passive one. As we now know that the idioplasm is situated in the nuclei only, we cannot regard the cell-bodies which determine the form of all parts of the organism as mere 'nutrient plasm.'

Both forms of the living substance are included in the term 'protoplasm,' and we have now to decide how we are to imagine its constitution in detail. 'Protoplasm' has often been conceived as a 'modification of albumen'; till quite recently, in fact, this was the general idea. Brücke, however, pointed out a considerable time ago that albumen does not possess the power of assimilation, and has therefore no vitality; it has moreover been proved by the study of physiological chemistry that other substances besides albumen are also obtained from protoplasm, and that these cannot be assumed to be insignificant without further proof. Although compounds of sulphur and phosphorus, for instance, only exist in protoplasm in comparatively small quantities, we must not infer from this fact that they are of slight importance. In any case, we cannot say that protoplasm is a modification of albumen, because we can only examine it chemically when dead, and in this condition it has lost its most important properties, and has become changed in a manner which we need not here consider further. As de Vries expresses it, protoplasm is not a chemical, but a morphological conception. That is, it does not consist of a confused mass of certain chemical molecules, but of morphological units, which are themselves composed of molecules, or, as Brücke first expressed it, protoplasm is 'organised.' As I have shown in the historical introduction to this book, Herbert Spencer, and more recently de Vries and Wiesner, have assumed the existence of such organic units.

De Vries, moreover, points out that protoplasm possesses certain 'historical' properties besides its physical and chemical ones. It may certainly be doubted, as de Vries states, whether it will ever be possible to produce 'living protoplasm otherwise than in a phylogenetic manner,' that is to say, to make it artificially in the laboratory; but it cannot be admitted that this is so improbable, merely because the conception of protoplasm demands that it should be derived from pre-existing protoplasm. This would exclude for ever not only the possibility of its production in our laboratories, but also its logically inevitable and indispensable primary formation in the great laboratory of Nature. Most, in fact probably all, kinds of protoplasm with which we are acquainted possess historical qualities, not *in addition to*, but *within* their physico-chemical ones; that is, they contain special modifications of construction peculiar to themselves which arose in adaptation to the conditions of life, and have been transmitted for a long period of time. But protoplasm which does not yet possess 'historical' *i.e.*, inherited qualities, does not seem to me to be inconceivable. It would be the simplest form of living matter which, in virtue of its constitution, possessed the primary vital forces,—assimilation, metabolism, and so on. The *historical* qualities of the protoplasm, its special hereditary tendencies, are not connected with these primary vital forces. The latter must exist independently in all protoplasm.

All those writers * who have assumed the existence of units on which the vital forces of protoplasm depend, have pointed out that they are not chemical molecules, for the latter do not possess the power of assimilation and reproduction. Hence it follows that protoplasm is a complex substance which is not homogeneous, but which consists of different kinds of molecules. There is therefore no molecule of protoplasm, but we have to imagine that even in its simplest modifications, protoplasm invariably consists of *groups* of molecules, each of which is

* Brucke, Herbert Spencer, de Vries, and Wiesner.

composed of *different kinds* of chemical molecules. I shall call these units the '*bearers of vitality*' ('Lebensträger') or '*biophors*,' because they are the smallest units which exhibit the primary vital forces, viz., *assimilation and metabolism, growth and multiplication by fission.*

As living protoplasm cannot be subjected to chemical analysis, we cannot describe its chemical constitution more precisely; but what has so far been determined by the analysis of dead protoplasm certainly indicates that the albuminoids are not the only bearers of vitality, as has generally been assumed, but that other substances play a no less important part in living protoplasm,—a fact which has been insisted on by Hoppe-Seyler and Baumann. Besides albuminoids, compounds containing phosphorus, such as lecithin and nuclein, which are not related chemically to albumen, but enter into combination with it, are known to occur in dead protoplasm; and besides these, protoplasm also contains cholesterin, which is probably a product of destructive metabolism, and carbohydrates, such as glycogen, starch, inulin, and dextrine, as well as compounds of potassium.* Although we cannot at present guess from what chemical compounds in living protoplasm these bodies have been derived, there can be no doubt that 'a relation exists between them and the vital processes' (Hoppe-Seyler), and that albumen, or different kinds of albumen, do not alone bring about the vital processes, but that several other substances, such as salts, and compounds containing phosphorus, and more particularly water, are just as essential: in short, life depends simply on the interaction of molecules, differing chemically from one another, but *defined* within certain limits.

After long consideration, I have decided to designate such a group of molecules on which the phenomena of life depend by the special term '*biophor*.' This seemed to be advisable, because the various terms introduced previously by others were either left too vague for these minute vital particles to be identified with them, or if defined more exactly, were used with a different meaning. It would certainly be a mistake to make use of a name already introduced, in another sense from its original one. Herbert Spencer's † 'physiological units' are similar to the biophors

* Cf. Hoppe-Seyler, 'Allgemeine Biologie,' Berlin, 1877, p. 75 (Part I of the 'Lehrbuch der physiologischen Chemie').

† Herbert Spencer, 'Principles of Biology,' vol. i., p. 183.

and he looks upon them as being intermediate between the chemical units (molecules) and the morphological units (cells). But he supposes their function in heredity to be different from that which I ascribe to my biophors. Haeckel * understands by the term 'plastidule,' introduced by Elsberg,† the hypothetical ultimate particles of which 'protoplasm' is composed; he regards them as equivalent to the 'molecules' of inorganic matter, but supposes them to possess 'vital qualities' as well. Of course this definition is in itself insufficient proof, as de Vries very correctly remarks, that Haeckel's plastidules are not molecules in the physical sense; these very 'vital qualities' are the point in which they differ from them. I could not adopt Nägeli's term either, because a 'micella' differs essentially in its construction and properties from a biophor. It is defined as 'a minute crystal, microscopically invisible, consisting of a larger or smaller number of molecules, and is, when turgid, surrounded by a layer of water.' ‡ As regards the absolute size of the micella, Nägeli calculates that it may consist of one hundred molecules, or on the other hand, of only a single molecule of albumen. As in the case of Haeckel's plastidules, we have here therefore to deal with a unit the vital character of which does not depend on a peculiar grouping of several or even many different kinds of molecules. Indeed Nägeli draws attention in another part of his book (p. 63) to the unstable chemical composition of the proteids so far as can be made out by analyses, and very correctly considers it extremely probable 'that there are various molecules of albumen which differ from one another in containing unequal quantities of hydrogen, oxygen, &c.' This leads him to the further assumption 'that the micellæ of the proteids consist of a mixture of two or more different kinds of molecules of albumen. In each proteid the different molecules of albumen would be mixed in special proportions, and, further, each would contain special quantities of phosphates, salts of magnesia, lime, and so on.' This conception, however, hardly agrees with that of the 'crystalline'

---

* Ernst Haeckel, 'Die Perigenesis der Plastidule,' Berlin, 1876.

† Louis Elsberg, 'Regeneration; or, The Preservation of Organic Molecules, a Contribution to the Doctrine of Evolution.'—*Proceed. Am. Assoc. for the Advancement of Science*, Hartford Meeting, Aug. 1874.

‡ Carl Nageli, 'Mechanisch physiologische Theorie der Abstammungslehre' München u. Leipzig, 1884 p. 35.

nature of the micella, for crystals are not 'mixtures,' but chemically pure substances. And apart from this, we should be wrong in inferring from this passage that Nageli considers the vital properties of a micella dependent on the co-operation of *different* molecules united into a single group; for in the passage quoted above he also states that *one* molecule of albumen is sufficient for the constitution of a micella.

For this reason alone it will be seen that the conceptions of the biophor and of the micella do not coincide. They differ also as regards the mode of multiplication: the fundamental importance of this will become apparent later on. The biophors, as bearers of vitality, possess the power of growth and of multiplication by fission, *just as is the case in all orders of vital units on which direct observations have been made*, beginning with the microsomata, which constitute the chromatin of the nucleus, and passing through the chlorophyll granules, nuclei, and cells, up to the simpler plants and animals. Nägeli's micellæ also multiply, but the multiplication occurs 'by the free interposition of new micellæ, similar to, or identical with, those already present,' in the same manner as he supposed the addition of new particles to take place in a starch grain, or as crystals separate from the mother liquor. These new micellæ would certainly have to be formed by an influence, exerted by those already present, which cannot be further defined.

The 'pangenes' of de Vries correspond almost exactly to my biophors, for they are also accredited with the functions of growth and multiplication by division, and play a similar part in heredity. The biophors, as will be explained in the following pages, only differ from the pangenes in being constituents of higher units of the hereditary substance.

The minute vital particles or 'plasomes,' recently assumed by Wiesner, resemble both pangenes and biophors as regards their properties. The part they take in heredity is, however, only hinted at, and it is therefore better for me to use the special term biophor than to press the plasomes into the service of my theory of heredity.

The biophors play the same part with respect to heredity as that which de Vries ascribes to his pangenes, *i.e.*, they are the '*bearers of the qualities or 'characters*' of the cells;' or more accurately, *the bearers of the cell-qualities.* As all living matter consists of biophors, the differences in it can only depend on the differences in the biophors composing it; an animal cell containing, for example, transversely striped muscular substance, or delicate nervous or glandular structures, or again, a vegetable cell enclosing chlorophyll bodies, must contain several *different* kinds of biophors of which these various cell-structures are composed, and which constitute the germ-plasm of a species.

There must be a great number of different kinds of biophors, for otherwise they could not give rise to so great a variety of cells as exists in the organic world. Nor is it difficult to infer the possibility of an almost unlimited number of different kinds of biophors from their assumed composition.

As the biophors are not individual molecules, but *groups* of molecules, nothing prevents us from tracing a large number of variations in them to the widely varying *number of their molecules*. But even the *chemical constitution of the molecules* is not by any means necessarily the same in all cases, although the possible fluctuations are certainly confined within certain limits.

Numerous facts show that at any rate in the two main divisions of the organic world, the animal and vegetable kingdoms, several of the molecules composing the biophors differ chemically from one another, so that substitutions occur. Whereas glycogen is a constituent which is never absent from animal protoplasm, provided that the latter possesses amœboid movement, this carbohydrate has not yet been discovered in plants, in which, as Hoppe-Seyler suspects, it is probably replaced by amylum, dextrine, or gum. Similarly, the crystalline proteids in plants, which are known as aleurone grains, are chemically different from the yolk-granules in animals.

A difference in the biophors can, moreover, be conceived without a change in their atomic composition, by regarding as possible a *rearrangement of the atoms* in the individual molecules. The molecule of albumen in particular has, according to the conclusions of modern chemistry, a molecular weight of at least 1,000, so that innumerable isomeric molecules of albumen seem to be conceivable. It is, however, impossible to state how many of them actually exist.

In order to give as complete an explanation as possible of the phenomena of heredity with the aid of the biophors, the latter must be invested with the capacity for a further change, namely, *a rearrangement of the molecules*, analogous to the isomeric rearrangement of the atoms in a single molecule. This assumption is not unfounded, inasmuch as several instances of molecular compounds are known in chemistry, *e.g.*, the double salts and the water of crystallisation of salts, in which definite numbers of molecules are always present: this number is even retained in spite of substitution. Thus alum always contains twenty-four molecules of water of crystallisation, and this evidently indicates a degree of affinity between the molecules. We shall have to assume this property for the biophor also, for without it the latter would not be a real unit at all. We shall, moreover, be able to conclude that these degrees of affinity are of various kinds, and that the molecules can combine in many different ways and form groups, so that isomeric molecular compounds are formed. Such isomeric compounds, however, will possess other properties, just as in the isomeric arrangement of atoms in the individual molecule; and thus we conclude that the special properties of a biophor are to be considered dependent not only on the physico-chemical constitution of the molecule, but also very essentially on their position and relation to one another; so that one biophor can be changed into another by an alteration in the arrangement of its molecules.

According to this statement there are several kinds of biophors, the difference between which depends on either the absolute relative number of molecules, their chemical constitution (isomerism included), or their grouping; in fact we may say that the *number of possible kinds of biophors is unlimited*, just as is the number of conceivable organic molecules. We shall, at any rate, meet with no theoretical difficulties on this score, however large the number of different kinds of biophors may be which we require to explain the theory of heredity.

The biophors are not, I believe, by any means mere hypothetical units; they must exist, for the phenomena of life must be connected with a material unit of some sort. But since the primary vital forces — assimilation and growth — do not proceed spontaneously from either atoms or molecules, there must be a unit of a higher order from which these forces are developed, and this can only consist of a group consisting of a combination of dissimilar molecules. I emphasise this particularly, because a theory of heredity requires so many assumptions which cannot be substantiated that the few fixed points on which we can rely are doubly valuable.

These biophors constitute *all* protoplasm — the morphoplasm which is differentiated into the cell-substance, as well as the idioplasm contained in the nucleus. It will be shown subsequently in what manner these two kinds of protoplasm differ as regards their constitution, and I will only remark here that the idioplasm must have a far more complex structure than the morphoplasm. The latter, as the cell-substance of a muscle or gland-cell shows, can assimilate, grow, and also divide, but it is not able to change into anything *different from itself*. The idioplasm, on the other hand, is capable of regular change during growth; and ontogeny, or the development of the individual in multicellular organisms, depends upon this fact. The two first embryonic cells of an animal arise from the division of the ovum, and continually give rise to differently constituted cells during the course of embryogeny. The diversity of these cells must, as I have shown, depend on changes in the nuclear substance.

It now remains to be considered how we are to imagine this capacity on the part of the idioplasm for regular and spontaneous change. The fact in itself is beyond doubt, when once it is established that the morphoplasm of each cell is controlled, and its character decided, by the idioplasm of the nucleus. The regular changes occurring in the egg-cell and the products of its division in each embryogeny must then be referred to the corresponding changes of the idioplasm. *But what is the nature of these changes, and how are they brought about?*

2. The Control of the Cell

In order to answer the question which has just been asked, it will be necessary to consider the manner in which the idioplasm of the nucleus determines the characters of the cell. At present we only know that the idioplasm consists of a large number of different biophors of various kinds. To exert a determining influence on the minute structure of the cell-body and on the chemical composition of its different components, it must either be capable of exerting an emitted influence ('Fernwirkung') or else material particles must pass out of the nucleus into the cell-body.

Strasburger * has endeavoured to prove a dynamical effect of

E. Strasburger, 'Neue Untersuchungen über den Befruchtungsvorgang bei den Phanerogamen,' 1884, p. 111.

the nuclear matter. In his opinion 'molecular stimuli are transmitted from the nucleus to the surrounding cytoplasm, and, on the one hand, control the processes of metabolism in the cell, and on the other, give a definite specific character to the growth of the cytoplasm, this growth being caused by nutrition.' Although transmission of the molecular stimuli, proceeding from the nucleus to the rest of the cell, is certainly conceivable, de Vries has rightly shown that this is not a sufficient explanation of the phenomena, because it takes for granted the fundamental point of the matter requiring explanation. If the cell of any plant is to acquire the hereditary property of forming malic acid, those pangenes in the cell-body which can produce this acid could, it is true, come into play by molecular stimuli being transmitted to them from the nucleus; but this hypothesis takes their presence for granted, and the main question as to how these producers of malic acid get into the cell remains unanswered.

Haberlandt * has attempted to trace the control of the cell by the nucleus to the enzymatic action of the latter, *i.e.*, to the giving off from the nucleus of certain chemical compounds which cause the cell-substance to become changed in a given manner; but this explanation is regarded by de Vries as insufficient, because here again it is necessary to presuppose a definite differentiation of the cell-body.

De Vries himself gives a solution of the problem, and his hypothesis has, at any rate, the advantage of great simplicity and lucidity. He supposes that some of the pangenes which constitute the nuclear matter pass into the body of the cell through the nuclear membrane, and there form its parts and structures, of the qualities of which they are the special bearers.

Although I formerly inclined towards Strasburger's view, it always appeared to me rather as a formal than as a real explanation of the problem, and I regarded it more as a provisional formulation than as a solution of the difficulty. In my opinion de Vries's idea of the migration of minute, specific, vital particles from the nucleus into the cell-body affords an extremely happy solution of the apparently inexplicable manner in which

* G. Haberlandt, 'Uber die Beziehungen zwischen Funktionen und Lage des Zellkerns,' 1877.

the cell is controlled by the nucleus. It, moreover, fits in very well with my other views.

As long as I was engaged in seeking for an epigenetic theory of heredity, an explanation of this sort was naturally impossible, but as soon as I assumed that the germ-plasm consisted of biophors, the various kinds of which are required for the various characters of the respective cells, it was not only possible to suppose that the particles exerted an influence of this nature on the cell, but such an explanation of the phenomena became the most natural and satisfactory one. Much may of course be urged against this fundamental assumption, and it is not in itself a sufficient explanation; but it is not only fruitless to attempt a satisfactory explanation from the other point of view, but as will appear later on, de Vries's conception alone agrees with certain fundamental biological principles.

If the nuclear substance exerted an emitted influence on the cell-body so as to give rise to the structures characteristic of this particular kind of cell, they would be formed by a kind of '*generatio equivoca*'; they would have arisen by the operation of an external influence on the given substance in the cell, just as would be the case in primordial generation. Particularly favourable influences would have operated on certain combinations of inorganic substances in such a way as to give rise to a vital particle.

We know nothing of such a primordial generation as far as our experience extends, and even if it must be considered to be logically necessary, we have every reason to suppose that it has no share in the origin of those forms of life with which we are acquainted, but that these always arise by division from others similar to themselves. Moreover, what is true of the independent organisms familiar to us must also hold good for *all the different orders of vital units* which have united to form higher organisms, for each of the earliest and lowest organisms must have been neither more nor less than *the equivalent of one biophor*. If, then, in order to explain the presence of life on the earth, we must assume that such individual biophors arose at one time by primordial generation, they must have been capable of reproduction by division immediately after their origin, for such multiplication is caused directly by the primary forces of life, — assimilation and growth. We can only imagine the very simplest biophors as having been produced by primordial generation: *all subsequent and more complex kinds of biophors can only have arisen on the principle of adaptation to new conditions of life;* they must have been developed gradually by the long-continued co-operation of heredity and selection. All these biophors of a higher order, which are adapted to the special conditions of existence and which in endless varieties form organisms as we see them around us, possess '*historical*' qualities; they can, therefore, only arise from others like themselves, and cannot be formed spontaneously. This fact is confirmed by experience. Not only does a cell always arise from a cell, and a nucleus from a nucleus, as de Vries, and more recently Wiesner, have shown, but all the other constituents which occur in the cell-body and determine its structure never arise, so far as we know, by '*generatio equivoca*,' or, as de Vries expresses it, 'neogenetically.' They are always produced by the division of similar structures already present. This is apparently true of the green chromatophores and the 'vacuoles' of plant-cells, as well as of the 'sphere of attraction,' or centrosome, which controls the division of the nucleus: the same must also hold good for those invisible vital units, the various kinds of biophors, which have arisen during the course of the earth's history by gradual adaptation to continually new conditions of life.

If then, each vital unit in all organisms, from the lowest to the highest grade, can only arise by division from another like itself, an answer is given to the question with which we started; and we see that the structures of a cell-body, which constitute the specific character of the cell, cannot be produced by the emitted influence of the nuclear substance, nor by its enzymatic action, but can only arise owing to the migration of material particles of the nucleus into the cell-body. *Hence the nuclear matter must be in a sense a storehouse for the various kinds of biophors which enter into the cell-body and are destined to transform it.* Thus the development of the 'undifferentiated' embryonic cell into a nerve-, gland-, or muscle-cell, as the case may be, is determined in each case by the presence of the corresponding biophors in the respective nuclei, and in due time these biophors will pass out of the nuclei into the cell-bodies, and transform them.

To me this reasoning is so convincing that any difficulties we meet with in the process of determining the nature of the cell hardly come into account. We are still far from being able to describe in detail the entire histological process of the differentiation of a cell. The passage of invisible 'biophors' through the pores of the nuclear membrane is probably just as admissible an assumption as that of the independent power of motion thereby necessitated in these bearers of vitality; but the histological structure of a cell is not completed by the mere emission into the cell-body of a few kinds of biophors with great powers of multiplication. Numerous questions suggest themselves in this connection, all pointing to the fact that forces are at work of which we are at present ignorant. The immigrating biophors are the mere material which forms the histological structure of a cell. only when subjected to the guiding forces — presumably those of attraction and repulsion — which must be located in the biophor.

We can as yet form no more exact conception of this process than we can of the manner in which the biophors already contained in the cell-body behave in respect to those which have migrated into it from the nucleus. Presumably a struggle of the parts occurs, in which the weaker are suppressed and serve as nutritive material for the stronger ones. But although much remains to be decided by future investigation, the main point at issue, at any rate, viz., that the nature of the cell is really decided by the elements of the nucleus, is definitely established. By the nature of the cell must be understood not only the histological structure of the cell as a whole and its mode of reacting to external influences, but more particularly its *mode of division* in respect of time and place. It is true that the cell-*body* itself and its apparatus for division (the centrosome) primarily determine whether a cell is to divide sooner or later, and into equal or unequal parts; but these processes always depend finally on the nucleus, which controls the cell-body and impresses on the latter its definite nature.

The most plausible objection which can be urged against the migration of the particles of the idioplasm into the cell-body is that the substance of the latter is chemically quite different from that of the nucleus. Their behaviour as regards taking up colouring matters is certainly different, as the terms chromosome and chromatin indicate; but even if a difference in their chemical composition could be inferred from this fact, it would still fail to constitute a decisive proof against the hypothesis of migration:

for it is well known that the affinity of the chromosomes for colouring matter varies markedly at different periods, and this indicates that slight changes, which are beyond our control, take place in the constitution of this substance, and are sufficient to cause its most striking reaction with regard to colouring matters to disappear for a time. Chemical analysis of the substance contained in the nucleus has certainly established the presence of 'nuclein'; but although it is probable from Miescher's * excellent observations on the sperm of the salmon that nuclein is derived from the nuclei of the sperm-cells, it is not by any means certain from what part of the nucleus it originates: if one supposes that over 48 per cent. of the dried sperm consists of nuclein, it is doubtful whether this is contained in the small mass of chromatin which we see in the form of chromosomes.

Another recent observation may be mentioned here, which proves at any rate that matter is actually transferred from the chromosomes of the nucleus into the cell-body just at the time when the characteristic structure of the cell-body is being formed. I refer to Ruckert's observations on the remarkable *alteration in the size of the chromosomes* of the nucleus during the growth of the ovum of the dog-fish.† One of the youngest ova observed in the ovary—which measured 2 mm. in diameter—contained from 30 to 36 chromosomes, each of which was 12 microns ‡ long, and 2 cubic microns in bulk: later on, in nearly ripe eggs, the length of a chromosome reaches 100 μ, and its cubic contents 7,850 cubic μ, or more accurately, since it has meanwhile become doubled by division, 15,700 cubic μ. Still later, just before the formation of the first polar body, when the ovum is ripe and has attained its full size, the length of the individual chromosome diminishes to 2 μ, and the cubic contents of a double rod to 3 cubic μ. Rückert infers from these facts that the chromosomes give off a great amount of substance to the ovum during the gradual ripening of the latter, and we can only agree with him on this point. But the question arises as to how this transference of substance takes place,—

* Miescher-Rüsch, 'Statist. u. biolog. Beitrage zur Kentniss vom Leben des Rheinsalm,' 1880; Schweiz. Literatursamml. z. internationalen Fischereianstell. in Berlin.

† J. Ruckert, 'Anat. Anzeiger,' 10th March, 1892.

‡ A micron (μ) is the $\frac{1}{1000}$ of a millimetre.

whether it occurs in the ordinary way, fluid nutrient material being given off, and then assimilated by the cell-body, or in some other manner. There seems to me to be no reason why we should not assume that *minute, specific, vital particles*, and not merely nutritive substances, are produced by the chromosomes during the growth of the egg, and are then emitted through the nuclear membrane into the cell-body. Further facts must be ascertained before we can attempt to explain the details of the curious morphological transformations which the chromosomes undergo during this period. We are already, however, in a position to state that the extremely interesting processes described by Rückert must have a wide significance, and must occur in all cells which become histologically differentiated as well as in all animal ova. But they cannot appear so distinctly in these other cells, for no animal cell grows to such an enormous size as does the egg-cell. I shall again refer to the process in a later section, in order to emphasise one of the consequences which results from it still more strongly.

Let us now suppose with de Vries that the nature of a cell depends on the extrusion of minute vital particles of different kinds from the nucleus into the cell-body, and that these subsequently multiply and become regularly distributed and arranged in groups according to the forces of attraction and repulsion situated within them. On this supposition, heredity could be simply and easily accounted for in unicellular organisms, for in them multiplication depends on a division of the whole body and of the nucleus into two parts, and thus each product of the division receives a similar supply of latent biophors which form its nucleus, and from which it can then provide the necessary material to the cell-body.

As the influence of amphimixis is not taken into account in the present connection, I may here leave out of consideration the fact that the nucleus may be differentiated into two different kinds of nuclei. This arrangement is practically universal amongst the highest unicellular forms — the Infusoria — and is merely an adaptation for conjugation. In the unicellular forms heredity will therefore depend, firstly, on the fact that all the different kinds of biophors which are required for the construction of the body are present in the nucleus in a latent condition, and in definite proportions — very probably they have also a definite style of architecture; and secondly, on the periodical or occasional migration of these biophors into the cell-body, where they multiply, and become arranged in obedience to the forces acting within them. The difficulty of ascertaining the actual mode of arrangement is nowhere greater than in the case of the higher unicellular forms. How is it possible that the nucleus should always allow only those kinds of biophors to migrate which are required to replace those structures lost by division? And why do these biophors always move either in the direction of the missing oral region, or towards the posterior end of the body, according to which parts are wanting in the two daughter-animals? For the present these questions are unanswerable; and in the meantime we must be content with having shown how the materials for the construction of the cell-substance are transmitted from mother to daughter, and in what way they are placed at the disposal of the forces acting in the cell-body.

The experiments made by Nussbaum * and Gruber † on the artificial division of Infusoria prove that the nucleus really controls the cell-body. These observers found that only those portions which contained a part of the nucleus were capable of giving rise to a complete animal; the other pieces lived for a time, and then perished. One of Gruber's observations also tends to show that when regeneration of missing parts occurs, the nucleus sends out invisible material particles into the cell-body. He cut a large *Stentor* which was preparing for division transversely into two parts, so that the posterior portion contained no trace of the nucleus, and then observed that regeneration of the missing parts nevertheless took place, especially in the oral region. If the control of the cell depended on the emitted influence of the nucleus, this regeneration would be totally inexplicable; if, however, biophors proceed from the nucleus into the cell-body when regeneration is to take place, this might have already occurred in an animal preparing for division, as this one was before it was artificially divided.

The descendants of unicellular animals are similar to their ancestors: two daughter-cells are produced by the division of

* Nussbaum, 'Ueber die Theilbarkeit der lebenden Materie,' Archiv. f. mikr. Anat., 1886.

† Gruber, 'Ueber kunstliche Teilung bei Infusorien,' 'Biol. Centralblatt,' Bd. iv.; and 'Beitrage zur Kentniss der Physiologie und Biologie der Protozoen,' Ber. d. naturf. Gesellsch. zu Freiburg i Br., 1886.

the mother-cell, and thus the nuclear substance is always composed of different kinds of biophors. But how does this apply to multicellular forms in which so large a number of different kinds of cells, each presupposing a different structure of the nuclear matter, arises from the germ-plasm of the ovum? Thus we find ourselves brought back to the question asked at the end of the last section: — on what do the regular series of changes in the germ-plasm during ontogeny depend?

3. The Determinants

As has just been shown, the nuclear matter of an Infusorian must be composed of a great number of different kinds of biophors, each of which corresponds to the primary constituent of a definite portion of the unicellular organism. If the cells of a multicellular animal were represented in the germ-plasm by all the kinds of biophors occurring in them, such an enormous aggregation of biophors would result that, even if they were extremely small, the minute quantity of matter in the germ-plasm would not be able to contain them. It was this consideration more than any other which for many years made me persevere in my attempt to discover an epigenetic theory of heredity. I thought that it must be possible to imagine a germ-plasm which, although highly complex, nevertheless did not consist of such an inconceivably large number of separate particles, but which was of such a structure as to become changed in a regular manner during its growth in the course of ontogeny, and, finally, to yield a large number of different kinds of idioplasm for the control of the cells of the body in a specific manner.

Hatschek,* too, has recently put forward the view that 'the egg-cell may be supposed to contain a relatively small number of qualities,' and that this number is not larger than that which is to be assumed in the case of any other histologically differentiated cell of the body. The diversity in structure seen in multicellular organisms is due, in his opinion, to the fact that in spite of the limited diversity as regards the qualities contained within a single cell (including the ovum), a far greater complication of the body as a whole is attained by the variation of these few qualities ('des einen Grundthemas').

* B. Hatschek, 'Lehrbuch der Zoologie,' 2te Lieferung, Jena, 1889, p. 232.

If in considering a theory of heredity we had only to deal with an explanation of the transmission of an *unalterable* structure from the parent to the offspring from generation to generation, there would be theoretically no objection to the assumption of such a structure of the germ-plasm. We have, however, to deal with the transmission of parts which are *variable*, and this necessitates the assumption that just as many independent and variable parts exist in the germ-plasm as are present in the fully formed organism. It is impossible that a portion of the body should exhibit an independent variation capable of transmission unless it were represented in the germ-plasm by a special particle, a variation in which is followed by one in the part under consideration. If this were represented, together with other parts of the body, by one particle of the germ-plasm, a change in the latter would be followed by a variation in all the parts of the body determined by it. *The independently and hereditarily variable parts of the body therefore serve as an exact measure for determining the number of ultimate particles of which the germ-plasm is composed: the latter must contain at least as great a number as would be arrived at by such a computation.*

An example may make it clear that the independently variable parts are not identical with those which are merely hereditary.

It is well known that butterflies pass through a metamorphosis in the course of development, the stages of which are independently variable from the germ onwards: that is to say, a variation in the caterpillar is not necessarily followed by one in the butterfly, and *vice versâ*. The caterpillars of a species may be dimorphic, some being green, and others brown, but both of these forms nevertheless give rise to butterflies with a similar coloration. If, therefore, the phyletic modifications depend on changes in the minute structure of the germ-plasm, there must be at least *two* independently variable units in the germ-plasm of such a butterfly; for if there were only one, the butterfly as well as the caterpillar would be affected by a variation in it. But a comparison of nearly related species shows us that the individual parts of the caterpillar or butterfly must also be variable from the germ onwards: the limbs, for instance, of two species may be very similar, while their wings are different, and even the separate parts of the wings may vary independently of one another. We must therefore assume that the germ-plasm contains a large number of units, on the variation of which the independent changes of certain parts of the body depend.

In all the higher animals the number of these units must be very large, because the parts which are independently variable from the germ onwards is large also.

A consideration of the individual and hereditary characters in the human species will show most clearly how great this number may be. I know of a family in which a depression of the size of a pin's head in the skin in front of the left ear has been transmitted through three generations. This slight abnormality must therefore have been contained potentially in the germ-plasm of the respective individuals, and their germ-plasm must differ from that of other people in the slightly abnormal form of the element which determines this peculiarity. We are logically compelled to assume a particular element of the germ-plasm for each peculiarity of this sort, not because heredity may be manifested in details so minute, but because *the transmission of such details may be independent*. If all people possessed such a depression in front of one ear, we could not thereby conclude that it must be represented by a special element in the germ-plasm merely because it is hereditary. It might conceivably be represented, together with the skin of half the face, by *one* element or biophor, which in the course of ontogeny became divided into a number of secondary ones of divers sorts, one of which proved to be abnormal and came to be situated at that particular spot in the skin. What compels us to accept the above assumption is the fact that all people do not possess this depression, and that two persons might conceivably resemble one another in all other respects except in the possession of this abnormality. The germ-plasm of both these persons would be almost identical, but not *perfectly* so, for it would contain a certain element which differed in the two cases. This simply means that *this particular character which is independently variable from the germ onwards is also represented by a special element in the germ-plasm*. It would not have been possible to infer this from its transmissibility alone. A hundred different characters might conceivably be determined by a single element in the germ-plasm; the whole hundred would then be transmitted as soon as the determining element was present in the latter, but not one of them would be independently variable from the germ onwards; but if the determining element varied, all the hundred characters would vary at the same time. The capacity for transmission and that of independent variation from the germ onwards are distinct from one another.

The germ-plasm must consequently be composed of as many units as there are transmissible parts in the body which are independently variable from the germ onwards. Each of these units cannot be smaller than a biophor, and they can therefore not be simple molecules within a biophor; for variation is a biological conception, and a biological element does not presuppose a one that is merely physical.

What parts of the body of a multicellular organism are represented in the germ by special particles of the minimum value of one biophor? Is each cell, or even each part of a cell? Darwin adopted the former, and de Vries the latter of these two alternatives. Darwin's gemmules are germs of *cells*, so that every cell of the body would be represented in the ovum by these units; while de Vries's pangenes are in a sense germs of the characters or structures ('Zellorganen') of the cell. There is no doubt that the hereditary variations in plants and animals manifest themselves in alterations of the individual parts or structures of the cell, and not only in the *number*, relative arrangement, and the changes in the form, size, and nature of the cells as a whole. The variegated varieties of our ornamental plants possess similar cells to those of their ancestral forms, but the green colour of the leaf is absent in certain of the cells: the red tint of the leaves of the copper beech, and other varieties of plants, depends on the red colour of the sap in a certain layer of cells, and this colour is transmissible. The coloured pattern on a butterfly's wing or a bird's plumage depends on cellular elements which were probably all alike in remote ancestors, but which afterwards became gradually changed by hereditary variations in the individual components or in the structure of the cell. Although the entire phyletic transformation of a species does not by any means alone depend on its *intra*-cellular variation, the latter has, nevertheless, constantly accompanied the other variations, and has shared to a greater or less extent in the transformation of the species. Hence it cannot be doubted that even in multicellular forms not only the cells as a whole, but also their parts, are determined from the germ onwards.

It seems therefore impossible to avoid the stupendous assumption that each of the millions of cells in a multicellular organism is represented in the germ-plasm by several or many different kinds of biophors. There is, however, a simple and natural way out of this dilemma, as soon as we inquire whether *every* cell of a plant or an animal is independently variable at all, and whether consequently it must be represented by special elements in the germ-plasm.

I shall designate the cells or groups of cells which are independently variable from the germ onwards as the '*hereditary parts*' or '*determinates*,' and the particles of the germ-plasm corresponding to and determining them, as the '*determining parts*' or '*determinants*.' It is evident that many of the cells in the higher animals are not represented *individually* in the germ-plasm by a determinant. The millions of blood-corpuscles which are formed during the life of a Vertebrate might possibly be controlled in the germ-plasm by a *single* determinant. At any rate no disadvantage to the species would result from this, because the capacity for being independently determined on the part of the individual blood-corpuscles, or even individual thousands of them, would be of no value to the animal. They are not localised: one of them has the same value as another, and their variability therefore might well be controlled from a single point. In conformity with the law of economy, Nature would not have incorporated more determinants than was necessary into the germ-plasm.

Thus there are probably many groups of cells in the higher animals, the constituents of which are not represented individually in the germ-plasm. All the nerve-cells of the brain do, it is true, possess their special determinants, as otherwise the transmission of such fine shades of mental qualities in man would be inexplicable; but it can matter little whether each fibre of a muscle, or each cell of the epidermis or of the epithelial lining of the alimentary canal, has its special determinant: in the last-mentioned cases larger or smaller *groups* of cells are presumably controlled by a single determinant. The manner in which the epithelium of the alimentary canal is renewed amongst insects may perhaps be taken as pointing to this assumption. In flies and butterflies, for instance, as I have proved long ago, the alimentary canal of the larva undergoes disintegration, and that of the imago, which has a very different structure, is developed out of its remains. Kowalewski and van Rees have since shown that the process takes place as follows: — the formation of portions of the new alimentary canal begins in certain cells which are separated by fairly regular intervals; these then spread until they come into contact with one another. The idioplasm of the new intestinal cells is consequently only contained in these formative cells, and it is natural to suppose that each of them contains only *one* kind of determinant.

The same appears to be the case with the hair of mammals. Every hair does not possess a special determinant in the germ, but more or less extensive regions of the hairy covering are represented each by one determinant. These regions are not large, as is shown by the stripes and spots on the coat of such animals as the tiger and leopard. The recurrence in the son, on exactly the same part of the head as in the parent, of an abnormal tuft of white hair, has been observed in the human subject.

Similar hereditary parts or determinates may be observed in butterflies, in which the colours on the wings often form very complicated lines and spots of slight extent but of great constancy. Such regions are often limited to quite a few scales (cells): *Lycæna argus*, for instance, possesses a black spot on a particular part of the anterior wing consisting of only ten scales, while the surrounding parts are blue. In this case we may therefore conclude that the black cells are represented in the germ-plasm by at least one determinant. The determination may possibly be carried out in still further detail in this instance, and each cell in the black spot may be determined from the germ onwards; and possibly it is only the constant intermingling of two hereditary tendencies in sexual reproduction, and the consequent variability in the number of scales, which prevents us from recognising the fact. We can at any rate, however, find instances of the determination of single cells in other species of animals. For example, in many Crustaceans a number of sensory organs are situated on the anterior antennæ: each of these corresponds to one cell. The number, position, and form of these 'olfactory' setæ is determined exactly for each species. The Ostracod *Cypris* possesses only *one* olfactory seta on each antennule, while in the common fresh-water species of *Gammarus*, there are about twenty of these structures, each of which is separately attached to one of the consecutive joints of the feeler. In many blind Crustacean , which live in the dark, the number of these setæ is greater than in the case of related forms which possess the sense of sight. And though in all these instances individual deviations occur, we may nevertheless suppose them to be hereditary, for otherwise the increase in the number of olfactory setæ incident on a life in darkness, could not have been established as a specific character.

In smaller and simpler organisms each individual cell may well have been determined from the germ onwards, and not merely with the result that the number of cells is a definite one, and the position of each definitely localised: the determination may also have caused individual peculiarities of each cell, in so far as they depend on changes in the germ-plasm at all — *i.e*, are 'blastogenic,' — to reappear in the corresponding cell in the next generation, just as in the case of a birthmark in the human subject which recurs in precisely the same place on the same side of the body. This may also be true of animals as simple as the *Dicyemidæ* or the *Tardigrada*, although it is not possible to prove it positively.

In all the more highly differentiated animals there can be little doubt that the number of determinants is always very much less than that of the cells which are the factors in the process of ontogeny. If we compare this statement with Darwin's assumption of the presence of a gemmule — or rather of several gemmules — for each cell, it is evident that the germ-plasm is thus to some extent relieved of a burden.

We must not forget, however, that a cell may vary as regards transmission not only as a whole but also in its parts, so that not *one* but several biophors must be assumed for each determinant of a cell or group of cells; we must, in fact, suppose just as many to be present as there are structures in the cell which are variable from the germ onwards. We ought, properly speaking, to speak of these bearers of qualities, which correspond to de Vries's pangenes, as determinants also, for they determine the parts of a cell. As the name of biophor has been given to them, however, it is better to retain this term, and to define a *determinant as a primary constituent of a cell or group of cells.* Thus a determinant is always a group of biophors, and never a single one.

It may now, I believe, be proved without difficulty that the biophors determining a cell not only lie close together in the germ-plasm so as to form a group, but that *they also combine to form a higher unit.* The determinant is not a disconnected mass of different biophors, but *a vital unit of a higher order than the biophor, possessed of special qualities.*

The fact that the determinants must possess the power of multiplication is in itself a sufficient proof of this. We know how greatly the nuclear matter contained in the fertilised egg-cell increases in volume during development, and this can only be due to the multiplication of its vital particles, the biophors. Such a multiplication could never occur with as much precision and regularity as is necessary for the preservation of the character of a certain cell, if the biophors which determine it were scattered at random instead of being definitely separated from those of other cells. Hence the multiplication of the biophors must occur within the fixed limits of the determinant, and must be preliminary to the division of the determinant itself. And consequently the latter is also a vital unit.

In accordance with our assumption, which can scarcely be refuted, a single determinant of the germ-plasm frequently controls entire groups of cells: this is a further proof that the determinants as such must multiply. This is only possible if they do so in the process of ontogeny. It is very probable, moreover, that the nucleoplasm of any cell in the body never contains *one* specimen only of the determinant controlling it, but several; otherwise, how could such a cell be visible at all under our microscopes? Biophors, at any rate, are far beyond the limit of vision, and even determinants can hardly come within it.

Thus the assumption made by the gifted propounder of the theory of pangenesis is so far justified. 'Gemmules' of cells really exist, and multiply by fission; but they are not the ultimate vital units, nor are special gemmules of all the cells of the body already present in the germ-plasm.

We have next to deal with the question as to how these two elements of the germ-plasm, which have now been formulated, are instrumental in the process of ontogeny.

4. The Id in Ontogeny

We can now make an attempt to solve the problem stated at the close of the last section concerning the way in which the germ-plasm is capable of giving rise to the various kinds of idioplasm required in the construction of the organism.

As we have seen, the germ-plasm contains the primary constituents of all the cells in the body in its determinants, and it only remains to inquire how each kind of determinant reaches the right part in the right number. Although we do not know what forces are called into play for this purpose, the elements of the germ-plasm now formulated, and the processes and course of ontogeny, nevertheless enable us to draw certain conclusions as to the structure of the germ-plasm and the nature of the changes it undergoes; and I trust that these conclusions will not lead us too far from the truth.

We can, in the first place, state with certainty that the germ-plasm possesses a *fixed architecture, which has been transmitted historically*. In working out the idea of determinates, it was stated that probably not nearly all the cells of the higher organisms are represented in the germ-plasm by special determinants: possibly all the blood-corpuscles, or the thousands of fibres in a particular muscle, for instance, are represented each by *one* determinant. But it does not therefore follow that all the cells of a similar kind which exist in the body can be represented by *one* common determinant: this would be equivalent to abandoning the conception of determinants altogether. If, for instance, all the transversely striped muscles of a Vertebrate were represented in the germ-plasm by a *single* determinant, each variation in the latter would also produce a corresponding change in *all* the muscles, and the independent variation of which each individual muscle is actually capable would then be impossible.

Several, or even many, similar determinants must therefore exist in the germ-plasm of an animal. Muscle-cells and nerve-cells are repeatedly formed even in the fully developed organism, and, in so far as they can vary individually at all from the germ onwards, will be represented by identical or by very similar determinants in the germ-plasm.

If such *identical* determinants represent a single fixed cell or group of cells, they cannot be situated anywhere in the germ-plasm, nor can they change their position according to varying influences: the determinants must be definitely localised, for otherwise, they would not be certain to reach the right cell and the right position in the course of ontogeny. I have already mentioned the olfactory setæ of *Gammarus*, which are situated individually on particular segments of the feeler. Each of these can vary hereditarily, and thus it is necessary to assume special determinants for them in the germ-plasm; these, however, will all be similar to one another. This is also true of the black spots on the wings of certain butterflies, already referred to. In *Lycæna Argus*, for instance, there is a spot on that part of the wing which is known to entomologists as 'cell 1 b,' and this spot is independently variable: it may be larger or smaller, and the variations in it can be transmitted quite independently of the numerous other black marks on the wing. The particular spot referred to may have disappeared entirely in another species of *Lycæna*, while a precisely similar spot in 'cell 4' has become much larger. We have also decided indications that homologous parts in the two halves of the body in bilaterally symmetrical animals can vary independently of one another. The human birthmark mentioned above was always inherited on the left side, and never on the right.

If each determinant occupies a fixed position in the germ-plasm, *it cannot have an indefinite or variable size and f rm, but must form a complete unit by itself*, from which nothing can be removed, and to which nothing can be added. In other words, we are led to the assumption *of groups of determinants*, each of which represents a separate vital unit of the third degree, since it is composed of determinants, which in their turn are made up of biophors. These are the units which I formulated on different lines long ago, and to which the name of *ancestral germ-plasms* was then given. I shall now speak of them as '*ids*,'* a term which recalls the 'idioplasm' of Nägeli.

I assume that just as the individual biophor has other qualities than those of the determinant, which is composed of biophors, so also does the id possess qualities differing from those of its component determinants. The fundamental vital properties — growth and multiplication by division — must however be attributed to the id as to all vital units. Several reasons,

---

* I have already used this term in my essay on 'Amphimixis' ('Amphimixis, oder die Vermischung der Individuen,' Jena, 1891, p. 39). In my earlier essays the ids were spoken of as 'ancestral germ-plasms,' the meaning and derivation of which term will be explained in the chapter on amphigonic heredity.

more especially those furnished by the phenomena of heredity in sexual reproduction, lead us to assume that the germ-plasm does not consist of a *single* id, but of several, or even many of them, and this assumption must be made even in the case of asexual reproduction.

I shall therefore assume that *each idioplasm is composed of several or many ids, which are capable of growth and multiplication by division.* If animals existed, in the whole series of ancestors of which sexual reproduction had never occurred, these ids would be exactly similar to one another. But in all cases every id of the germ-plasm contains the whole of the elements which are necessary for the development of all subsequent idic stages. Theoretically, therefore, *one* id would suffice for ontogeny.

We assume that *the changes in the id of germ-plasm during ontogeny* consist merely in a regular disintegration of the determinants into smaller and smaller groups, until finally only *one* kind of determinant is contained in the cell, viz., that which has to determine it. It is highly improbable that all the determinants in the id of germ-plasm are carried along through all the idic stages of the ontogeny. In discussing regeneration and gemmation later on, I shall have to show that, under certain circumstances, groups of determinants are supplied to certain series of cells, and that these are not actually required for determining the cells; this arrangement, however, depends, I believe, on special adaptations, and is not primitive, at any rate not in the higher animals and plants. Why should Nature, who always manages with economy, indulge in the luxury of providing all the cells of the body with the whole of the determinants of the germ-plasm if a single kind of them is sufficient? Such an arrangement will presumably only have occurred in cases in which it serves definite purposes. The enormous number of determinants contained in the germ-plasm also stands in the way of such an assumption, for in the higher animals they can be reckoned by hundreds of thousands at the very least; and although we may assume that they all remain in a latent condition in every cell, and so need not interfere with the activity of the determinants which control the cell, they nevertheless deprive the active determinants — which we must also suppose to exist in large numbers — of a considerable space.

If we wished to assume that the whole of the determinants of the germ-plasm are supplied to all the cells of the ontogeny, we should have to suppose that differentiation of the body is due to all the determinants except *one* particular one remaining dormant in a regular order, and that, apart from special adaptations, only *one* determinant reaches the cell, viz., that which has to control it. This latter supposition is undoubtedly less likely than the former.

If however we do make this assumption, the question then arises as to what factors can cause the gradual disintegration of the id of germ-plasm into smaller and smaller groups of determinants,—that is to say, into ids which contain fewer and fewer *kinds* of determinants.

This disintegration I believe to be due to the co-operation of three factors: these are—*the inherited architecture of the germ-plasm*, in which each determinant has its definite position; *the unequally vigorous multiplication of the various determinants*; and possibly also, *the forces of attraction* which are situated within each determinant, and result from its specific nature as a special and independent vital unit. The architecture of the germ-plasm has already been discussed in general terms: for the present, at any rate, we can hardly conjecture the actual details of its structure. In order to do so, it would be necessary to suppose that hundreds of thousands, or millions, of determinants, which are all definitely localised, take part in the formation of the higher organisms. The fact that the right and left halves of the body can vary independently in bilaterally symmetrical animals, points to the conclusion that all the determinants are present in pairs in the germ-plasm. As, moreover, in many of these animals, *e.g.* the frog, the division of the ovum into the two first embryonic cells indicates a separation of the body into right and left halves, it follows that the id of germ-plasm itself possesses a bilateral structure, and that it also divides so as to give rise to the determinants of the right and left halves of the body. This illustration may be taken as a further proof of our view of the constant architecture of the germ-plasm. An id is evidently not constituted like the sediment of a complicated and well-shaken mixture, in which the heavier particles come to lie at the bottom and the lighter ones at the top; nor is it constituted in such a manner that the respective positions of the particles are only determined independently by the forces acting on them and between them momentarily. Its structure may be compared to that of a complicated ancient building, the stones of which we may suppose to be alive, so that they can grow and increase, and thus cause displacements and fissures in the walls, in which process the forces of attraction present within these living stones take part. *The historical transmission of the architecture of the germ-plasm forms the basis of the entire ontogenetic development of the idioplasm.*

If however the id has a right and left half in bilateral animals, we must not thereby infer that it is merely a miniature of the fully-formed animal, and that therefore we are once more dealing with the old theory of preformation. Quite apart from all conjectures as to the detailed architecture of the id of germ-plasm, it is at any rate certain that the arrangement of the determinants in it is quite different from that of the corresponding parts in the fully-formed organism. This is proved by a study of development, and need scarcely be treated of in detail here. Any one with a knowledge of animal embryology knows how great a difference there is between the mode of development of the parts from one another in the embryo and their respective relation in the mature organism. The early stages of segmentation of the ovum show that groups of determinants have been formed in the id of germ-plasm, and that these, moreover, correspond to the parts of the body which arise from one another consecutively, though they can have no resemblance to them either in form or in their degree of perfection.

In some worms the two first blastomeres do not give rise respectively to the right and left sides of the body, but to the entire ectoderm and endoderm. In these cases the id of germ-plasm must break up into two groups, one of which contains all the determinants of the ectodermal organs, and the other all those of the endoderm: it is evident that this arrangement has no analogy to that which obtains as regards the organs of the fully-formed animal. If in any species we knew the 'value in primary constituents' ('Anlagenwerth') — if I may use such a term — of each cell in the ontogeny, we could give an approximate representation of the architecture of the germ-plasm; for, beginning with the last formed cells, we could infer the nature of the determinants which must have been contained in each previous mother-cell, passing gradually backwards to the ovum; thus we should reach the two first blastomeres, and finally the egg-cell itself. The groups of determinants which are present at each stage would thus be known, and we might in imagination then arrange them in such a way that it would be possible to picture their disintegration into the respective series of smaller and smaller groups.

Such a representation of the architecture of the id of germ-plasm would, however, never be an accurate one, because its parts must be subjected to incessant slow displacement during the growth of the idioplasm and in the course of development.

This brings us to the *second factor* which takes part in the ontogeny of the idioplasm, viz., the uneven *rate of multiplication of the determinants*. An id of germ-plasm composed entirely of *similar* determinants, would have to retain its original architecture even during vigorous growth and continued division; just as would be the case in one of the lowest forms of life — a Moner — consisting of a number of identical biophors, which must remain the same throughout all the divisions which it undergoes. In a germ-plasm consisting of a number of different determinants, a perfectly even rate of multiplication cannot be assumed in the case of all of them. For the difference between two determinants depends presumably on the differences in the nature, number, or arrangement of their constituent biophors, and the latter differ again in their molecular structure, *i.e.* in their essential physico-chemical properties. Hence the determinants will behave differently as regards their reaction to external influences, — more especially in respect of their rate of growth and increase, — according to their constitution. The same conditions of nutrition will therefore stimulate one to a faster, and another to a slower, growth and corresponding multiplication, and thus an alteration in the proportional numbers in which the individual kinds of determinants are present in the germ-plasm must occur continually in the course of embryogeny; for the latter is connected with a constant growth of the idioplasm, and therefore also with a continual increase of the determinants. This must cause a disarrangement in the architecture of the germ-plasm, in which process the third factor concerned in these changes, viz., *the forces of attraction in the determinants*, may take part.

The assumption of such forces can scarcely be avoided. For it is very probable, *a priori*, that vital units do act upon one another in different degrees, and this view is supported by a consideration of the processes of nuclear division, together with the distribution of the primary constituents in ontogeny.

So far I have not touched upon the question as to what observable parts of the idioplasm are to be regarded as ids. This point cannot be decided with certainty at present, but I have elsewhere expressed the opinion that those rod-like, loop-like, or granular masses of chromatin in the nucleus, — the chromosomes, — are to be considered equivalent, not to single ids, but to series or aggregations of ids. I have therefore proposed to call the chromosomes *idants*,[1] in order to keep up a certain uniformity in the nomenclature. It is probable that the ids correspond to the small granules hitherto called 'microsomata,' which are known to form the individual idants in many animals: we may mention as an example, *Ascaris megalocephala*, as in it the nuclear structure is best known. These microsomata, although lying very close together in *one* row, are nevertheless separated by a thin layer of intermediate substance; the whole idant cannot therefore be equivalent to one id, for the latter is a clearly defined vital unit possessing a fixed architecture, and cannot consist of completely separated parts.

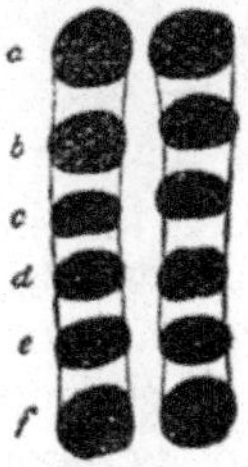

FIG. 2.
Two Idants with their contained Ids of *Ascaris megalocephala*. (After Boveri.)

The great variety as regards size, number, and form of the chromosomes in different species of animals, indicates that they possibly have not always a similar morphological value. As however there is no reason for assuming that the number of ids must always be the same in all species, and as, on the contrary, it is much more probable that their number varies greatly, it is impossible to make use of the above fact as a decisive argument. We can only state that the individual chromosome or idant in all probability represents a different number of ids in different species.

Division of the nucleus depends on the longitudinal splitting of the idants, in which process each of the spherical ids — assuming these to correspond to the microsomata — becomes halved. Each half then becomes rounded off, and passes, together with the idant to which it belongs, into one of the two daughter-nuclei.

In the ordinary process of cell-division in tissues, which results in the formation of daughter-cells similar to those from

[1] 'Amphimixis,' pp. 39, 40.

which they arose, the ids produced by the division naturally contain precisely similar determinants; in embryogeny, on the other hand, divisions occur which ensure that the two daughter-nuclei contain combinations of determinants which are usually entirely different from one another. We have an example of such a nuclear division in the segmentation of the ovum in the case, for instance, of certain worms already referred to, in which two cells are formed by the first division of the egg-cell, one of which contains all the determinants of the internal, and the other all those of the external germinal layer. A division of this latter kind we may speak of as *differential* or *dissimilar as regards heredity* ('erbungleich'), in contrast to the former, which is *integral* or *similar as regards heredity* ('erbgleich'). As in the case of the entire idants, the ids are split by an internal force, and are not pulled apart mechanically by the threads of the 'nuclear spindle' which are attached to them. Flemming has shown that this splitting often takes place long before the spindle-threads become active. The forces of attraction in the determinants must therefore take part in this process, just as they must be assumed to act between the biophors which constitute the body of a dividing cell.

It appears to me, therefore, that the inherited architecture of the id of germ-plasm undergoes a gradual change, owing to the uneven rate of multiplication of the determinants, and that it is further regulated by the forces of attraction which we must suppose to act between them. We might represent the architecture of the id by a very complicated geometrical figure, which gradually becomes changed during the growth of the id; this change does not occur in the first division, the preparation for which has been accurately made in the original figure, but in the subsequent stages of ontogeny. As the greater number of these divisions is connected with a diminution in the number of kinds of determinants, the geometrical figure representing the id gradually becomes simpler and simpler, until finally it assumes the simplest conceivable form, and then each cell will contain the single kind of determinant which controls it. The disintegration of the germ-plasm is a wonderfully complicated process; it is a true 'development,' in which the idic stages necessarily follow one another in a regular order, and thus the thousands and hundreds of thousands of hereditary parts are gradually formed, each in its right place, and each provided with the proper determinants.

The construction of the whole body, as well as its differentiation into parts, its segmentation, and the formation of its organs, and even the size of these organs,—determined by the number of cells composing them,—depends on this complicated disintegration of the determinants in the id of germ-plasm. *The transmission of characters of the most general kind—that is to say, those which determine the structure of an animal as well as those characterising the class, order, family, and genus to which it belongs—are due exclusively to this process.* The slight differences only, namely those which distinguish species from species, and individual from individual, depend partly on the characters of the *individual cells.* De Vries has overlooked this in his attempt to explain all the facts of heredity by the theory of '*intra*-cellular' pangenesis. As was mentioned in the 'Historical Introduction' to this book, it must be borne in mind that most of the 'characters' of any of the higher forms of life result not from the characters of the individual cells, but from the way in which they are combined. On the other hand, the construction of a living organism is not conceivable unless we presuppose the determination of the characters of each cell.

We have therefore to give an explanation of this concluding part of the process of ontogeny; this has already been done to a great extent above, in the section treating of the control of the cell by the idioplasm. I there assumed, as de Vries has also done, that this determination depends on the migration of minute vital particles from the nucleus into the cell-body. We have now seen by what means the biophors characteristic of any particular cell reach that cell in the requisite proportion. This results from the fact that the biophors are held together in a determinant which previously existed as such in the germ-plasm, and which was passed on mechanically, owing to its ontogenetic disintegration, to the right part of the body. In order that the determinant may really control the cell, it is necessary that it should *break up into its constituent biophors.* This is an inevitable consequence of the assumed mode of determination of the cell. We must suppose that the determinants gradually break up into biophors when they have reached their destination. This assumption allows, at the same time, an explanation of the otherwise enigmatical circumstance, that the rest of the determinants, which are contained in every id except in the last stages of development, exert no influence on the cell. As each determinant consists of many biophors, it must be considerably larger than a biophor, and is probably therefore unable to pass out through the pores of the nuclear membrane, which we must suppose to be very small and only adapted for the passage of the biophors. Although it is impossible to make any definite statement with regard to the internal structure of the determinants, it must be owing to this structure that each determinant only breaks up into biophors when it reaches the cell to be determined by it. We may suppose that, just as one fruit on a tree ripens more quickly than another, even when the same external influences act on both, so also one sort of determinant may mature sooner than another, although similar nourishment is supplied to both.

It must not, however, be overlooked, that a difference in the time of maturation of the determinants in the embryogeny of animals is chiefly to be assumed only in the case of the actual embryonic cells; for the histological differentiation of the cells of the body, and the differentiation of the parts of the latter, occur at about the same time; that is, not until the organs already exist as definite groups of cells. This is equivalent to saying that the disintegration into biophors occurs when the id only contains the *single* determinant which controls that particular kind of cell. It is well known how suddenly the histological differentiation of the cells occurs in the embryogeny of an animal. For a long time the various parts and tissues are very similar to one another, though not perfectly so, and then histological differentiation suddenly sets in. This is very markedly the case as regards the transversely striped muscles of Arthropods and Vertebrates, in which the contractile substance is first seen as a mere narrow ring around the cell, and then gradually becomes thicker, so as to replace the greater portion of the cell-body, — just as one would expect if it were caused by muscle-biophors which had migrated into the cell-substance and there multiplied.

The assumption of a 'ripening' of the determinants, which though not simultaneous, is yet exactly regulated, nevertheless remains indispensable; or, to express it differently, we must assume that the determinants pass through a strictly regulated period of inactivity, at the close of which the disintegration into biophors sets in. The determinants certainly continue to grow and multiply without interruption during this period, as may be deduced from the fact that the amount of the nuclear substance in the individual cell does not decrease during embryogeny, although such an enormous increase in the number of cells takes place. No accurate and methodical observations have at present been made with regard to the comparative size of the chromosomes in the various stages of development and in the different organs of the body, but it may nevertheless be taken as certain that the entire mass of the nuclear substance grows considerably during embryonic development. It appears to me, however, to follow from the observations of Rückert I have already referred to concerning the chromosomes of the ovum of the dog-fish,* that *the most marked growth of the determinants takes place immediately before, and during, their activity.* During the period of growth and histological differentiation of the egg in this fish the idants grow enormously, and towards the completion of these processes they gradually decrease in size, until finally, when the ovum is ripe, they have become almost as small as they were originally.

This may be expressed, in the terms we have adopted, as follows: *the determinants which control the histological structure of the egg† multiply enormously during the growth of the ovum, into the body of which they transmit their numerous biophors.* After this has occurred, only those determinants of the germ-plasm are left which have in the meantime been inactive, and which have only increased to a slight extent; these are thus contained in those idants which are not much larger than they were in the young egg-cell. From the beginning of ontogeny and onwards, one determinant after another becomes active, and during their activity they also multiply. It has for a long time appeared to me probable that the determination of a cell does not take place, as one might suppose, by the agency of a single determinant, but by that of many determinants of

---

* Anat. Anzeiger, 10th March 1892.

† These determinants of the ovum correspond to the 'oogenetic nucleoplasm' of my earlier essays, and constitute the substance which determines the growth and histological differentiation of the egg. For a long time I believed that this substance was extruded from the ovum at the close of the period of maturation by means of the polar bodies. We now see that such an extrusion is not required, as this substance is used up in the differentiation of the egg.

a similar kind; and I imagine that that kind of determinant which has to control a particular cell,.multiplies considerably by division befoie—and perhaps even during—the process or determination. This view is completely borne out by Rückert's interesting observations.

Every cell during the whole period of ontogeny is, however, controlled—not only as regards its structure, but also in respect of its mode of division—by a *single* determinant only. The inactive determinants remain without exerting any influence on the cell-body; they however determine the architecture of the id, and therefore the further formation of the embryo also. For, indeed, the mode of disintegration of the id into smaller groups of determinants is necessitated by its architecture.

I have above attributed to the determinants forces of attraction which take part in the configuration of the structure of the ids. Such forces must be present, for otherwise the id could not possess a definite architecture; but I do not wish it to be understood that these forces are the principal factors in the arrangement of the determinants. They are concerned in connecting together the parts of which the determinants are composed, and not in their continual rearrangement during the course of ontogeny. It is primarily always the inherited definite architecture of the id of germ-plasm which results mechanically in the idic figure of the subsequent stages; disarrangements in this architecture are due to the unequally vigorous increase of the various kinds of determinants, all of which naturally are definitely determined beforehand. The arbitrary or accidental action of the forces of attraction takes no part at all in this process.

I must emphasise this view particularly, in contrast to that of Galton, who speaks of 'repulsions and affinities' of the gemmules which compose the 'stirp.' He compares the masses of these gemmules, which undergo active and incessant changes of their mutual positions owing to attraction and repulsion, to a swarm of flying insects, in which 'the personal likings and dislikings of an individual may be supposed to determine the position that he occupies in it.' With this view I can by no means agree, for it rests on the assumption that the germ-substance is composed of many *homologous* gemmules ('competing germs') which struggle for the supremacy, only those which are successful determining the character of the future organism. From the very first Galton takes into consideration the complications or the germ-substance caused by sexual reproduction, which, as will be shown subsequently, are due essentially to the fact that the germ-plasm contains *many*, and not a *single* specimen of each primary constituent, and that these are present in various modifications. It is this struggle between the *homologous* primary constituents which Galton refers to in the passage just quoted, which indicates that first one, and then another, reaches the desired spot, without any definite order being observed. This conception appears still more plainly in another passage, in which he compares the germ-plasm (the 'stirp') to a nation, and those gemmules 'that achieve development,'—*i.e.*, become transformed into the corresponding parts of the body—'to the foremost men of that nation, who succeed in becoming its representatives.'

Excellent as these similes are in themselves, I cannot help thinking that they lead to error if intended as an explanation of ontogeny. If we take up the position which Galton occupies with regard to the essential part of the theory of pangenesis, we must suppose that a large number of gemmules—many more than are necessary for the construction of the body—are contained in the stirp; that is, in the germ-substance of the fertilised egg. For only *one* gemmule is required for each cell of the body, but nevertheless a large number are present; and these, so to speak, struggle for the precedence, the successful gemmule alone becoming converted into the cell which is to be formed. In this conception the fact is entirely overlooked that ontogeny itself cannot possibly depend on this struggle, but would take place just the same if only *one* gemmule were present in the 'stirp' for each cell, and that the cause for the progress of development must therefore be sought elsewhere than in the rivalry between homologous gemmules: it must b due to the right *succession* of the gemmules. Galton considers that the 'purely step-by-step-development' assumed by Darwin in his theory of pangenesis is insufficient, but I think, nevertheless, that Darwin's opinion is the more correct one.

Neither does Galton's simile of the swarm of insects seem to me to be appropriate as an explanation of the struggle between homologous gemmules derived from different ancestors. Even if the gemmules in the 'stirp' were in perpetual motion, and if on this depended the decision as to which of them obtained the privilege of taking part in the formation of the organism, how could one explain the existence of identical twins, about which we have received such valuable information from Galton himself? How would it be possible for the *exactly* corresponding gemmules in two individuals in the flying and ever-changing swarm always to reach the most favourable position, even if the 'stirp' contained precisely similar gemmules?

In a subsequent section I shall attempt to show that this struggle between homologous but individually different primary constituents can be proved in quite another manner in connection with the idioplasm. It was necessary to mention Galton's view here, in order to show that the forces of attraction and repulsion, assumed by him, are introduced for an entirely different purpose from that which I have stated with regard to the similar forces in connection with the biophors of the idio-plasm.

Two physiological conditions of the elements of the idioplasm exist, — an active and an inactive. In the former, these elements become disintegrated into their constituent parts; while in the latter, they remain entire, although they are capable of multiplication. When determinants are active, they become disintegrated into biophors, and are then capable of controlling the cell in the nucleus of which they are situated. The activity of entire ids depends on a disintegration into determinants, which, though certainly successive, is often very slow; it must be contrasted with the inactive state, which in both elements of the idioplasm depends on the fact that their constituent parts do not become separated from one another, but remain in their primarily entire condition. In the immature ovum, for instance, only *one* kind of determinant — the 'oogenetic' determinant — is active, and this controls the growth and histological differentiation of the egg; all the other kinds remain inactive, as do also the ids which are formed from them. Only when fertilisation has occurred do they become active, — that is to say, one kind of determinant after another begins to separate itself from the entire id. We shall see later on, however, that ids of the germ-plasm also exist which remain inactive even after fertilisation has occurred, and are passed on from cell to cell in what we may call an *unalterable* ('*gebundenem*,') condition, so as to form subsequently the germ-cells of the embryo. We know as little about the cause of this condition as we do about that of the state of the brain during sleep, or of the latent period of certain fertilised animal eggs, which, after beginning to undergo development, remain inactive at a certain stage for months. The facts with which we are acquainted, however, render the assumption of an active and an inactive state of the ids and determinants unavoidable, as will become more evident in the course of this book. A similar assumption has been made by all those who have formulated vital units: thus Darwin has assumed these conditions in connection with his 'gemmules,' and de Vries with regard to his 'pangenes.'

Two forms of heredity, which we call homotopic and homochronic, may be deduced from the theory given above. As the individual determinants — from the germ-plasm onwards, throughout all the stages of ontogeny — take up a definite position in the id, they must reach the right place in the body, and there cause the development of a structure corresponding to that of the parent. As, moreover, the period of maturation of each determinant is decided by the nature of the latter, the determinant will become active in the individual and will cause the formation of any particular part of the body at the same stage of development as in the parent. Exceptions to this rule occur in the case of abnormalities, and also in that of phylogenetic displacements.

5. Summary of Sections 1–4, relating to the Structure of the Germ-plasm

According to my view, the germ-plasm of multicellular organisms is composed of ancestral germ-plasms or *ids*, — the vital units of the third order, — each nuclear rod or *idant* being formed of a number of these. Each id in the germ-plasm is built up of thousands or hundreds of thousands of *determinants*, — the vital units of the second order, — which, in their turn, are composed of the actual bearers of vitality ('Lebensträger'), or *biophors*, — the ultimate vital units. The biophors are of various kinds, and each kind corresponds to a different part of a cell: they are, therefore, the bearers of the characters or qualities' ('Eigenschaftsträger') of cells. Various but perfectly definite numbers and combinations of these form the determinants, each of which is the primary constituent ('Anlage') of a particular cell, or of a small or even large group of cells (*e.g.*, blood-corpuscles).

These determinants control the cell by breaking up into biophors, which migrate into the cell-body through the pores of the nuclear membrane, multiply there, arrange themselves according to the forces within them, and determine the histological structure of the cell. But they only do so after a certain definitely prescribed period of development, during which they reach the cell which they have to control.

The cause of each determinant reaching its proper place in the body depends on the fact that it takes up a definite position in the id of germ-plasm, and that the latter, therefore, exhibits an inherited and perfectly definite architecture. Ontogeny depends on a gradual process of disintegration of the id of germ-plasm, which splits into smaller and smaller groups of determinants in the development of each individual, so that in place of a million different determinants, of which we may suppose the id of germ-plasm to be composed, each daughter-cell in the next ontogenetic stage would only possess half a million, and each cell in the next following stage only a quarter of a million, and so on. Finally, if we neglect possible complications, only *one* kind of determinant remains in each cell, viz., that which has to control that particular cell or group of cells. This gradual disintegration of the id of germ-plasm into smaller and smaller groups of determinants in the subsequent idic stages does not consist in a mere division of the id into portions, but — as occurs in all disintegrations of vital units — is accompanied by displacements in the groups of these units, brought about by the unequally vigorous multiplication of the various individual determinants, and regulated by the forces of attraction acting within them. In spite of all the alterations in the arrangement of the determinants which must occur, owing to the differential nuclear divisions together with unequal growth of the various kinds of these units of the second order, the original position of each determinant in the extremely complex structure of the id of germ-plasm renders it necessary that it should take up a definite and fixed position in each idic stage; and also that it should traverse the precisely regulated course from the id of germ-plasm, through perfectly definite series of cells, to the cell in which it reaches maturity in the final stage of development. In this cell it breaks up into its constituent biophors, and gives the cell its inherited specific character. *Each id, in every stage, has its definitely inherited architecture; its structure is a complex but perfectly definite one, which, originating in the id of germ-plasm, is transferred by regular changes to the subsequent idic stages.* The structure exhibited in all these stages exists potentially in the architecture of the id of germ-plasm: to this architecture is due, not only the regular distribution of the determinants, — that is to say, the entire construction of the body from its primary form to that in which its parts attain their final arrangement and relation, — but also the fact that the determinant, of a small spot on a butterfly's wing, for example, reaches exactly the right place; and that, to take another instance, the determinant of the fifth segment in the feeler of a *Gammarus* reaches this particular segment. The determination of the character of the individual cell depends on the biophors which the corresponding determinant contains, and which it transmits to the cell.

6. The Mechanism for the Phyletic Variations in the Germ-plasm

The *causes* of phyletic development will be treated of in the chapter on Variation: the present section merely gives an account of the mechanism existing in the idioplasm in connection with this process. I shall here attempt to show how the phyletic changes in the idioplasm follow mechanically from its assumed ultimate structure.

Since all parts of the organism are determined from the germ onwards, permanent variations in these parts can only originate from variations in the germ. Each phyletic variation must therefore be due to a variation in the structure of the id of germ-plasm. If we suppose, with Darwin, that the transformation of species is a gradual one, originating in individual variations which become increased and directed by selection, it follows that the corresponding process in the idioplasm cannot be due to a sudden and complete variation in the entire id, but must begin with changes in the individual biophors or in individual determinants and groups of determinants also, and must then extend gradually to more numerous groups, until finally the nature of the id becomes entirely, or to a great extent, changed.

The basis of the process must be sought in the variability of the biophors, which is followed in turn by that of the units of a higher order, — the determinants and ids. These variations are not by any means confined to the *structure* of the individual cell, but concern primarily the *number* of cells of which an organism consists. A leaf of a plant, or a bird's feather, may increase considerably in size during the course of phylogeny, without a change necessarily occurring in the cells which form these parts. The variation will depend primarily on a multiplication of the respective determinants. If the primitive eye of a lower animal consisted of a single cell, constituting a visual rod, and the power of multiplication of its determinants gradually increased in the course of phylogeny, the number of identical determinants which would arise during development by the multiplication of the *single* determinant in the germ-plasm would gradually increase so as to suffice for two cells instead of one. The eye would then possess two visual rods, and if the power of multiplication increased still more, a whole group of visual rods would be controlled by *one* determinant. We are unable to conjecture on what internal variations in the determinant such an increase in the power of multiplication depends; but the fact that every individual cell in the body does not possess a special determinant, while large groups of cells are controlled by a single one, proves that such variations must be possible.

Such a very simple phyletic variation, resulting in the local increase of the number of cells, will be followed by a further variation as soon as the multiplication of the determinant of, *e.g.*, an undifferentiated sensory cell, is not confined to the later stages of ontogeny, but occurs also in the germ-plasm itself; that is, when the doubling of the determinant has already taken place in the id of germ-plasm. For in this case the group of sensory cells, which have become developed phyletically from the originally single cell, will now be controlled by two determinants, each of which can vary independently of the other, and can transform the group of cells under its control. Thus one of these groups might give rise to auditory cells, and the other to gustatory or olfactory cells.

Thus the increase in the differentiation of the body depends primarily upon the multiplication of the determinants in the id of germ-plasm, but this differentiation is only rendered complete by variations in the determinants of similar origin taking place in different directions. The mere addition of a new ontogenetic stage can very easily be conceived without an increase occurring in the determinants of the id; but as soon as the double number of cells which are present in the new idic stage have to become differentiated in various ways, the differentiation must be preceded by a doubling of the determinants in the id of germ-plasm. A higher degree of differentiation will therefore be primarily connected with an increase in the number of cells of which the organism is constituted. It is known that the extreme prolongation of development, due to the constant addition of new generations of cells at the end of ontogeny, can be neutralised by the abbreviation and reduction of the ontogenetic stages: this process may be also to some extent understood if we trace it to its origin in the structure of the idioplasm. The reduction in the number of generations of cells from two or more to one, depends on the fact that the process of multiplication and rearrangement of the determinants takes place more rapidly during these particular stages, than does that of cell-division; so that several idic stages, each of which formerly characterised a *particular* stage of the cell, pass into one another during the *same* stage of the cell. The respective idic stages have not here disappeared completely: they only follow one another more rapidly, and therefore disappear as recognisable stages in development.

In lowly organised beings the differentiation of the body may become increased by a simple *reduction of the hereditary parts or determinates*, without an increase taking place in the cell-generations. If a determinant which controls a region consisting of a hundred cells divides into two, each of which only controls fifty cells, the two resulting groups of cells can vary independently of each other from this point onwards, and may give rise to very different structures. In this way a continued division of the determinants, and consequently also a constantly increasing differentiation of the species, may occur, without necessitating an increase in the total number of cells present in ontogeny.

Each additional differentiation denotes an increase in the degree of organisation. But the phyletic development of the organism is by no means invariably connected with an increase, or, in fact, with any other change in the degree of organisation.

The species of a genus, and often the genera of a family, cannot be distinguished from one another by the number of cells composing them, nor by an increase in the variety of these cells, but only by qualitative differences in the structure of the various parts. Hence the phyletic development of living beings cannot simply be due to the augmentation of the number of determinants in the id of germ-plasm, but must also be attributed to a change in the *nature* of the determinants and in that of their component biophors.

The structure of the idioplasm which we have here assumed, also offers an explanation of the phenomena of *parallelism between ontogeny and phylogeny*, which depend on the law of biogenesis as well as on *the relegation of the final characters* to earlier and earlier ontogenetic stages in the course of phylogeny. Let us first consider the former of these phenomena. We have assumed that each ontogenetic stage is characterised by a definite 'determinant figure,' *i.e.*, a sort of geometrical structure composed of the determinants. The nature of each individual cell is certainly controlled by those determinants in the nucleus which have reached maturity,—that is to say, have arrived at the stage in which they break up into biophors and migrate into the cell-body. But the manner in which the embryonic development of an animal occurs does not by any means depend only on the histological structure of the individual cells in each stage,—it rests to a much greater degree on the manner in which these cells divide and on the rate of their division, and also primarily on the way in which the 'unripe' determinants of the nuclear substance, which are still latent, are grouped together and distributed by means of the cell-divisions. *This distribution of the primary constituents* among the different cells is of the first importance in determining the character of the ontogeny; and one could easily imagine a case of animal embryogeny in which ten or twenty generations of similarly constituted 'embryonic cells' followed each other, and in which a perfectly definite distribution of the primary constituents (determinants) had nevertheless occurred, although only now apparent for the first time. It is well known how close a resemblance exists between the cells of the embryo in various stages in the case of the higher animals.

The regular distribution of the determinants which are still latent or 'unripe' must therefore decide the course of ontogeny;

and the manner of this distribution finds expression in the architecture of each idic stage, or, as I have expressed it, in each 'determinant-figure.'

It is obvious that the same geometrical figure may be constructed out of different elements, just as the same form of crystal may be produced from molecules of a different nature. Thus the resemblance between the ontogenetic stages of nearly allied species is to be explained by the degree of similarity between their respective 'determinant figures,' which persists although the individual determinants constituting the figure differ more or less from one another. As the study of development shows, an explanation is thus offered of the fact that the earlier ontogenetic stages are so very much alike in allied species, and that the differences only appear later on; for in the early idic stages, the differences as regards the nature or power of multiplication of single determinants, or groups of determinants, can exert no marked influence, because the entire number of determinants is still very large, and thus the architecture of the id will be practically the same in corresponding stages. But the further ontogeny advances, and the smaller the groups become into which the determinants separate, the greater also will be the diversity in the architecture of the id, and in the further distribution of 'unripe' determinants resulting from this architecture. Thus a certain part will be longer or shorter, a spot of colour larger or smaller, and the final stages of ontogeny — in which the cells possess only *one* determinant — will differ according to the degree of difference which obtains in the respective determinants. This explains the fact that the segmentation cells in allied species are frequently exactly alike, and also that the resemblance between many mammalian embryos in their earlier stages, though not complete, is nevertheless a very close one.

The law of biogenesis, as far as it applies at all, depends on the fact that phyletic development is partly due to new ontogenetic stages being added at the end of ontogeny. In order that these new stages may be reached, the stages which were previously the final ones must be passed through in each ontogeny. This may be expressed in terms of the idioplasm as follows: — the determinants of the id of germ-plasm become endowed with a greater power of multiplication, so that each one of them causes the addition of one or more cell-generations to the end of the ontogeny. At the same time, the determinants in the germ-plasm increase in number, and each of them becomes differentiated in a fresh manner. As, however, every two new determinants always follow the same course from the id of germ-plasm to the final stage in ontogeny as was taken by the single original determinant, they will pass through the same determinant figures as before, and only lead to the formation of new structures in the final stages, when they become separated from one another.

The ontogenetic stages of the ancestors will be repeated less accurately the nearer development approaches its termination.

The disappearance of a character or of a part which has become useless, may also be traced to the mechanism of the idioplasm. The group of determinants which gives rise to a particular character, will have to be removed entirely from the germ-plasm if the corresponding part is to disappear completely. But this is a very complicated process, and one of long duration as regards more complex organs, such as, for instance, the limbs of Vertebrates. For the determinants which take part in the formation of an extremity are very numerous, and of many different kinds; and moreover, they cause the rudiment of the limb to appear very early in ontogeny. Hence the determinants will have to suffer successively many retrogressive and simplifying changes before a noticeable reduction of the organ occurs. The functionless and rudimentary wings of the Australian Kiwi (*Apteryx*), which are concealed by the plumage, possess all the bones of the perfect wing, though these are very much reduced in size. This is to be explained by supposing that the entire group of determinants for the wing still remain in the id of germ-plasm, but that it has decreased in strength, — that is to say, its elements no longer increase so rapidly, — and they therefore can only control smaller groups of cells. If the process of degeneration continued, the organ would not only grow smaller and smaller, but its component parts would also disappear at different rates, and, losing their characteristic form, would appear as indistinguishable rudiments. Such a degeneration has occurred in certain species of whales, in which the rudiments of the posterior extremity lie concealed beneath the skin; while in other species, the form of the separate bones has been to some extent preserved, and those of the thigh and shank can still be plainly distinguished. In these cases, many of the determinants which were formerly present must have disappeared entirely from the id of germ-plasm, and the remainder must have lost the power of multiplication to a greater extent than has occurred in the case of the wing of *Apteryx*.

We know, however, that even in such animals as snakes, in which the extremities have in most cases disappeared completely in previous geological periods, the rudiments of the limbs arise in the form of 'muscle-buds' in the earlier stages of development, and then disappear very shortly afterwards.* This fact may be expressed in terms of the idioplasm as follows:—the power of multiplication in the small remnant of the group of determinants of the extremity which still exists in the id of germ-plasm, has decreased so considerably that it only suffices for these early embryonic stages. The youngest determinants, and consequently the most recent hereditary structures, are the first to disappear, the loss of the older ones taking place gradually, until even the oldest of all are no longer present. This must be due to the manner in which the determinants increase, although the actual connection between the two phenomena is not apparent. It may perhaps be traced to the fact that those determinants which are the youngest phyletically are destined for the latest ontogenetic stages, in which only therefore they become 'ripe,' and undergo disintegration into biophors. If then, their power of multiplication decreases considerably during the process of degeneration, the number of determinants required for the control of any particular group of cells will not be reached, nor will the determinants even become ripe. Although still present, they are unable to exert any influence; whereas the determinants of the older phyletic stages which are still passed through, ripen in the earlier stages of ontogeny.

The process of degeneration of an organ may be represented as depending on the fact that the determinants first become changed in such a manner as to cause a decrease in their power of multiplication, and this then leads to a very gradual reduction affecting an increasing number of determinants belonging to the group in question. At the same time, the power of mul-

---

* *Cf.* J. van Bemmelen, 'Over den oorsprong von de vorste ledematen en de tongspieren bij Reptilen.' Kon. Akademie de Wetenschappen te Amsterdam, 30th June 1888.

tiplication in the remaining determinants also diminishes, so that the groups which they constitute gradually extend a less distance into the ontogeny, until finally they drop out of it altogether.

It must not be understood that I have given a mechanico-physiological explanation of the process of degeneration because I have connected it with the theory of determinants. As long as we know practically nothing about the forces which act within and among the biophors, it will be impossible to offer an explanation of this kind. I have only attempted to show that this doctrine does not contradict the facts, but that, on the contrary, it agrees with them up to a certain point. The phenomena of degeneration have not hitherto been considered from this point of view. When a deeper insight into the actual phenomena has been obtained, we may perhaps be able to make further theoretical deductions, and it would then be possible to develop the theory of determinants more fully.

A few words may now be said as regards *correlated variations*. Darwin has shown what an important part these variations play in the transformation of species, and how changes which we must consider to be primary are followed by a number of others in various parts of the organism. Thus an increase in size in a stag's antler necessitates a thickening of the skull, and a strengthening of other parts, viz., the muscles of the neck, the spines of the cervical vertebræ, the ligamentum nuchæ, and even the thoracic skeleton and fore-limbs. Referring all these variations to the processes which take place in the idioplasm, they will be seen to depend on changes in the corresponding groups of determinants in the id of germ-plasm, which cannot be due *directly* to the change and increase in the group of determinants of the antler: they must have arisen secondarily, owing to the occurrence of variations in the determinants upon which selection could act. There is also an entirely different kind of correlation, in which the variation in one part is accompanied by that in another, the latter having no anatomical or functional connection with the former. Thus Darwin states, for instance, that cats with blue eyes are generally deaf, and that pigeons with feathered legs have a web between the outer toes.

I do not think such correlations can be traced to a connection of the parts by means of the nervous system: it is perhaps more likely that they are due to *the contiguity of the determinants in the id of germ-plasm* of those parts which vary correlatively. It will be shown later on that local differences in nutrition occur in the id, and that these may cause changes in the determinants affected by them. If, now, the determinants controlling regions of the body which are far apart, are situated close together in the id, they might easily be affected simultaneously by influences producing variation. But the perfectly definite architecture of the id of germ-plasm, on which we base our argument, does not only permit of a vicinity of the determinants of parts of the body far removed from one another, but actually requires it. For, according to our assumption, the id of germ-plasm is not a representation of the body in miniature, but a structure of a special kind, in which the individual component parts are arranged in the order in which they are passed on subsequently in the process of ontogeny to their final destination, viz., to the determinates or hereditary parts. This however requires that the determinants of the ectoderm should be closely adjacent to those of the endoderm in the id, if they are to be distributed to a primary ectoderm and a primary endoderm cell in the first division of the ovum. A cell-division which leads to the separation of widely differing groups of determinants, admits of a close aggregation of these different groups in the id of the mother-cell. This may give some slight insight into the above-mentioned phenomena of correlation.

7. The Magnitude of the Constituents of the Germ-plasm

The assumption that the germ-plasm is composed of biophors, determinants, and ids, implies the existence within a narrow space of a large number of ultimate vital units (biophors) in all the higher organisms. The question arises whether a sufficient number of these units can be contained within an id. Although I believe it is at present quite impossible to obtain anything like a reliable answer to this question by calculating the relative sizes of the elements of the germ-plasm, it may perhaps not be uninteresting to attempt to make such a calculation.

In order to solve the problem with any approach to accuracy, it would at least be necessary to know the sizes of a biophor and of an id, and also the number of determinants in a given species. Unfortunately, however, we are completely ignorant on these points, nor do we even know how many molecules take part in the construction of a biophor: even the computed size of the molecule is somewhat uncertain.

The diameter of a molecule has been estimated at between the $\frac{1}{1,000,000}$th and the $\frac{1}{1,000,000}$th of a millimetre by four different lines of reasoning, 'founded respectively on the undulatory theory of light, on the phenomena of contact electricity on capillary attraction, and on the kinetic theory of gases.'* O. E. Meyer has calculated the size of a molecule 'from the properties and behaviour of vapours. From the constant of friction and the comparison between the space occupied in the liquid and gaseous conditions, together with the deviations from Boyle and Mariott's law, we can approximately calculate, firstly, the volume of all the particles contained within a given space; secondly, that of a single particle; thirdly, the number of particles; and finally, the weight of a single particle.' The result of such a calculation agrees with that given above.

If we take the average diameter of a molecule to be $\frac{1}{2,000,000}$th mm., and reckon that each biophor, which we will suppose to be a cubical structure, is composed of 1,000 molecules, the biophor would measure 10 molecules in length. A row of 200 biophors would therefore measure 1 μ, and 8,000,000 biophors would occupy the space of 1 cubic μ. A human blood-corpuscle measures 7.7 μ. in diameter; if we imagine it to be enlarged so as to form a cube of 7.7 μ. in diagonal length, this space would contain 703,000,000 biophors. Let us further assume that those portions of the cell which, according to the facts at our disposal, must contain the idioplasm, viz., the chromosomes, are mostly a great deal smaller than the nucleus in which they are situated, and that the germ-plasm is composed not of *one* but of several ids, each of which contains all the biophors required for the construction of the entire body, it will then be evident that only a limited number of biophors can be contained in one id.

The chromosomes in the germ-plasm of *Ascaris megalocephala* are the largest which are at present known to us. Each

* Sir William Thomson, 'Popular Lectures and Addresses,' Vol. I., 1889, p. 148.

nucleus in this animal contains two or four rod-like chromosomes (see fig. 2), each of which is composed of 'six thickened granular or disc-shaped portions, which become deeply stained with colouring matter, and which are separated by portions staining less deeply' (Boveri). If we connect this fact with the hypothetical composition of the germ-plasm out of ids, it follows that an id cannot in any case be larger, and is probably smaller, than one of these granules or microsomata. It cannot be larger, because the id is a unit which is capable of division into two daughter-ids, but which cannot remain permanently separated into several parts by a different kind of intermediate substance. If we suppose the id to be as large as it can possibly be, — that is to say, to correspond in size to a microsome, — it will measure, according to Boveri's drawing and scale of enlargement, .0,008 mm., or not quite 1 μ in diameter. Only the terminal granules of the rods, however, are as large as this; the greatest diameter of those in the middle measures .0,006 mm., while their shorter diameter is about .0,003 — .0,004 mm. The terminal granules, looked upon as spherical bodies, would be capable of containing about two million biophors of the size given above.

This number is certainly a very considerable one, and it would apparently be sufficient to make up the number of determinants in such a lowly organised animal as *Ascaris*. But even in Arthropods the number of determinates, and therefore that of the determinants also, is considerably greater. Each of the olfactory setæ on the feelers of Crustaceans, which were mentioned above, must be capable of being determined from the germ onwards; and this is also true of the spots and lines on a butterfly's wing, each of which represents at least *one* determinant, and in case of all the large markings several, or even many, of these units. If we consider that the pattern on the wing is often very complicated, and frequently differs on the under and upper surfaces, it is evident that hundreds of determinants must exist for this pattern alone. But there are, again, several peculiarities in the structure of the wing-scales, and thus it is probable that almost every scale can vary independently from the germ onwards. In some males of the family *Lycaenidæ*, *e.g.*, *Lycaena adonis*, small guitar-shaped odoriferous scales (the 'androconia' of Scudder) are distributed regularly amongst the colour-scales, while these are entirely absent in other nearly allied species, such as *Lycaena agestis:* hence we must conclude that these androconia have arisen by the transformation of ordinary scales. This, however, presupposes the independent variability of the scales which are to become changed phyletically, and consequently also their capability of being determined from the germ onwards. Were this not the case, a single scale could never have varied from the others *hereditarily*. In *Lycaena adonis* there are 30,830 scales on the upper surface of the wing.* If each of these is to be looked upon as corresponding to a determinate, the enormous number of about 240,000 determinants of the germ-plasm would result merely from the scales covering the wings, provided that the upper and under surface of the four wings possess each about the same number of scales.

I have endeavoured by direct experiment to ascertain the lowest limit to the size of a determinate,—that is to say, the size of the smallest determinates for a particular character of a certain species. For this purpose I selected one of the Ostracoda, *Cypris reptans*, which multiplies parthenogenetically, and in which it is easy to compare the different green spots on the shell in the mother and daughter. It appears that the larger spots are strictly transmitted, though this is not the case as regards the very small ones, which consist of only one or two pigment-cells. The form of these larger spots, which consist of fifty or a hundred pigment-cells, also varies to some extent, so that the number is here also somewhat inconstant. If parthenogenetic reproduction could be looked upon as being *purely uniserial*, it might be inferred that the determinates are not in this case single cells, but groups of cells. Unfortunately, however, this experiment cannot be considered conclusive, for—as will appear later on—the germ-plasm is here composed, just as in the case of sexual reproduction, of dissimilar, and not of *similar* ids, and consequently variations in heredity may thus arise.

We must conclude, even from the external coloration, that a very considerable number of determinates exists in the case of the higher Vertebrates. Thus most, if not all, of the contour-feathers of a bird must be controlled by special determinants in the germ-plasm, for they are independently variable hereditarily.

* My assistant, Dr. V. Häcker, was good enough to make this calculation for me.

The number of wing- and tail-quills is nevertheless definitely fixed for every species of bird, and each of these feathers possesses a definite form, size, and coloration. We must assume that *more than one* determinant is necessary for an entire feather, for a feather is formed from thousands of epidermic cells, which are not by any means all similar to one another, either as regards form, mode of combination, or colour. Many feathers are striped, while others have a brilliant ornamental spot at the tip; as in the case, for instance, of the peacock, many humming-birds, and certain birds of paradise. The cells to which these stripes and spots owe their origin, must contain determinants which differ from those of the rest of the cells which take part in the construction of the feather. We must therefore conclude that at least *one*, and often several, determinants of the germ exist for each of these two kinds of cells; for, as is well known, ornamental spots of this kind are often formed of several colours, and are very complex.

It would be erroneous to suppose that the contour-feathers are not determined individually in such birds as the raven, in which the plumage is all of one colour; but in such cases the qualitative differences refer less to colour than to form and size. The fact that each part of the feather is determined hereditarily, even as regards its colour, is proved by the variation which occurs, and which in individual species has resulted in certain feathers being partially or entirely white, or being brilliantly coloured, as in the case of the bird of paradise, which is allied to the raven. One need only look through a collection of humming-birds, and compare the females, which are so often plainly coloured, with the wonderfully variegated males, in order to become convinced that almost every contour-feather can vary in almost any direction as regards coloration, form, size, and minute structure.

As has already been remarked, the internal organs are apparently by no means so specially determined from the germ onwards as are the external parts: their determinants must therefore control larger regions of cells, as in the case, *e.g.*, of blood-corpuscles and epithelial cells. The number of determinants in the germ-plasm of the higher animals is nevertheless an enormous one, and it might certainly be doubted whether such a large number of biophors as must be required for the construction of an id of the germ-plasm could be contained within a single id.

It is impossible, as we have already seen, to obtain a satisfactory answer by means of a calculation. But let us assume for the moment that we possess reliable data as to the number of determinants and the size of an id in a particular species. We will further assume that each determinant is composed of, let us say, fifty biophors, and each biophor of a thousand molecules, and that the average diameter of a molecule is $\frac{1}{2,000,000}$th mm. Supposing we found that all these units could not be contained within an id of the size we have assumed, we should be forced to conclude that one or more of these quantities had been overestimated. This result would not weaken the theory of determinants, for minute particles *must* exist in the germ-plasm for each hereditary and independently variable part of the body. I therefore consider it fruitless to attempt a more accurate estimation of the number of determinants in individual species, and to endeavour to find a support for this fundamental theory by means of such calculations. The theory is correct in any case, although our conception of the structure of the germ-plasm may be very incomplete.

The object of making the above calculation was simply to arrive at this result. The germ-plasm is an extremely delicately-formed organic structure, — a microcosm in the true sense of the word, — in which each independently variable part present throughout ontogeny is represented by a vital particle, each of which again has its definite inherited position, structure, and rate of increase. *A theory of evolution appears to me to be only possible in this sense.* The constituents of the germ-plasm are not miniatures of the fully-formed parts, or even particles existing solely for the formation of the corresponding parts in the body. But each of these particles (the biophors and determinants) has a definite and important share in the preceding stages of development, for it takes part in determining the architecture of each idic stage, and consequently also assists in the further ontogenetic disintegration and distribution of the determinants amongst the subsequent cell-stages. All the more essential differences in the structure of organisms depend on this fact. The determinants are particles on whose nature that of the corresponding parts in the fully-formed body depends, whether the latter consists of a *single* cell or of several or many cells. The assumption of such particles is inevitable in a theory of heredity, and it alone necessitates an almost inconceivable complexity in the architecture of the germ-plasm. But if we suppose that the number of ultimate particles of which the germ-plasm is composed is less than the number of parts in the body which are independently variable hereditarily, it would then follow that several minute parts of the body must become changed simultaneously with the variation of *one* of these particles, — that is to say, the number of determinants then become too small theoretically.

PART II

HEREDITY IN ITS RELATION TO MONOGONIC REPRODUCTION

INTRODUCTORY REMARKS

IN the following part those phenomena of heredity will be considered which do not result directly from the composition of the germ-plasm as already described, but which would also occur if there were no such thing as sexual reproduction. In considering the phenomena of the *regeneration* of lost parts, of multiplication by *fission* and *gemmation*, of the *production of unicellular germs*, and of the *continuity of the germ-plasm*, it will materially facilitate the attainment of clear results if the investigation is conducted throughout as though these phenomena occurred in organisms in which the process of multiplication is, and always has been, entirely an asexual one. The complications resulting from sexual reproduction can be considered afterwards, and it will then be easy to connect them with all these phenomena of heredity.

CHAPTER II

REGENERATION

1. Its Cause and Origin in the Idioplasm

It does not follow directly from what has already been said with regard to the structure of the germ-plasm, that lost parts can be more or less completely replaced. The only deduction that can be made so far is, that all the parts of which the entire organism is composed are formed *once* during the development of the organism from the egg: no explanation is given of the fact that individual parts can be produced a *second time*, when they have been lost by the action of external influences. During ontogeny, the determinants of the part in question pass from the ovum into the segmentation-cells, from these into embryonic cells of a later stage, and finally into those cells which constitute the fully formed part. If this part is forcibly removed from the organism to which it belongs, its determinants are removed along with it: this follows from what has already been assumed with regard to the ontogenetic stages of the idioplasm. We must now therefore attempt to explain the fact that a part of the body can nevertheless be reconstructed.

If the capacity for regeneration were possible at all, it is obvious that it would have to be introduced by Nature, for its physiological importance is apparent. The power of replacing larger or smaller parts of the body must in all cases be useful to the organism, and is often indeed indispensable to its further existence. Arnold Lang * is certainly right in considering the faculty of regeneration in animals to be one of the arrangements for protection which prevent the species from perishing. The capability of completely restoring those parts of the body which have become injured by the bite of an enemy, forms a more

* 'Ueber den Einfluss der festsitzenden Lebensweise auf die Thiere,' &c. Jena, 1888, p 108.

efficient protection in many of the lower animals — more especially in polypes and worms — than would the possession of shells, stings, poison-organs, and all other kinds of weapons, or even protective coloration. For although all these arrangements certainly serve as a protection from many enemies, and from various dangers, they are not *always* effective, and therefore the capability of restoring losses of substance would certainly be extremely valuable in any case. This fact must not be forgotten in any inquiry with regard to the question of regeneration. If we consider how highly important regeneration is from a physiological point of view, its wide and even general distribution in the animal kingdom need not surprise us, and we shall be able to understand why it has been introduced even into the course of normal life: for the functions of certain organs depend on the fact that their parts continually undergo destruction, and are then correspondingly renewed. In this case it is the process of life itself, and not an external enemy, that destroys the life of a cell. I refer, of course, to the process of *physiological regeneration*.

Our knowledge of histology is not yet sufficient for us to be able to determine what tissue-cells in the higher animals become worn out by use during life, and have therefore to be continually replaced; but it has been proved in many cases that the wearing away of the cells goes on incessantly, and that life could not last if these cells were not constantly replaced. Such a constant loss and renewal of the cells occurs in the cases of the epidermis of the higher Vertebrates, the human finger-nails, blood-corpuscles, hairs and feathers, claws and hoofs, the epithelial lining of the respiratory and other passages, and even in the antlers of stags. In all these cases a continual or periodic wasting away or shedding of groups of cells occurs normally, and a corresponding replacement of these cells is one of the normal functions of the body, and is therefore provided for.

It is not difficult to explain the simplest of these cases of physiological regeneration theoretically. If a tissue such as the human epidermis, for instance, consists of *one* kind of cell only, it is only necessary, in order that regeneration may take place, that all these cells should not be thrown off simultaneously, and that the tissue should be composed of cells of various ages, the youngest of which, under certain influences of nutrition and pressure, always retain the power of reproduction, and so form a stock in which the necessary substitutes for the older cells can constantly be produced. The whole supply of the corresponding determinants is not therefore removed from the body simultaneously by the loss of the worn-out cells, for the young cells which remain contain determinants of the same kind. In the human epidermis, this stock of young cells constitutes the so-called 'rete Malpighii' or 'mucous layer,' in which new cells are constantly being formed by division; these, in proportion as they become older, are gradually pushed upwards mechanically from the deeper into the superficial layers, while the deepest layer of all consists entirely of young cells which are capable of division.

No special theoretical assumption need be made to explain this process. We must only suppose that the first formed epidermic cells are endowed in advance with a capacity for reproduction during many generations. It must be assumed that the reproductive power of a cell is regulated by the idioplasm, because the power and rate of multiplication are essential qualities of a cell, and, as we have seen, are controlled by the nuclear substance. But we cannot at present even form a conjecture as to which qnalities of the idioplasm the degree and rate of the capacity for reproduction are due. We must be satisfied with attributing to the cells which form the epidermis of the embryo an idioplasm possessing a definite reproductive power, which gradually decreases. We can further only suppose that the idioplasm retains its constitution during life, or, in other words, that the determinant of a particular part of the epidermis is always retained in the permanent stock of young cells. Regeneration depends simply on a regular increase of those cells which contain epidermic idioplasm.

The nature of the epidermis is not the same in different parts of the human skin: thus it differs on the volar and on the dorsal surfaces of the fingers; and, again, on the two basal and on the ungual segments. But this fact does not stand in the way of the theoretical explanation of regeneration, for the determinants of different parts must differ somewhat from one another Even in places where two or more dissimilar parts are situated close together, the retention of the limits between them, during their continual regeneration, may be explained simply by the fact that the different regions of the tissue are regenerated by formative cells possessing different determinants.

Many tissues, even in the highest animals, when they have suffered an *abnormal loss of substance*, are renewed in precisely the same way as in the cases of physiological regeneration already mentioned. Thus in mammals, for instance, portions of muscular tissue, of epithelium covering an organ or lining the duct of a gland, and of bone, can be replaced by cellular elements of a similar kind; and recent researches in pathological anatomy render it almost certain that all these regenerative processes originate in the cells of the tissue which is to be replaced. Hence these tissue-cells retain the power of multiplying by division, but they only begin to exercise this power in response to certain external stimuli, more particularly to that which is produced by a loss of substance in their immediate vicinity. Thus epithelial cells multiply around a defect in the epithelium; and in an injured muscle, the nuclei multiply and cause the surrounding protoplasm to be transformed into cells, which become spindle-shaped, and give rise to muscle-fibres. In both these cases we must merely attribute to the idioplasm the capacity for multiplication: the cells in question only begin to divide when influenced by a stimulus due to the loss of substance, or, as it would be expressed in the language of modern pathology,* 'by the removal of the resistances to growth.' Thus in these very simple cases of the abnormal loss of parts, the rest of the tissue gives rise to a stock of determinants from which replacement of the part can occur.

In more complicated tissues, the process of regeneration is less simple. Thus Fraisse has shown that in the Amphibia *the entire epidermis, together with the slime-glands and the integumentary sense-organs*, is regenerated by the epidermic cells in the vicinity of the defect. In this case also, the new material is furnished by the deeper uncornified layers of the epidermis. But the newly-formed cells do not all develop into the same kind of tissue. The main mass of them gives rise to the stratified epidermis, while others 'unite to form pearl-shaped masses in the deeper part of the epidermis, the cells becoming grouped around an imaginary centre.' 'Connective tissue-cells then migrate from the cutis, and these masses, each consisting of from ten to twenty cells, thus become marked off from the epidermis.' 'At the same time pigment-cells wander into the

---

* Cf. E. Ziegler, Lehrbuch der pathologischen Anat mie,' Jena, 1890.

epidermis, and finally the development of smooth muscle fibres takes place.'* New integumentary sense-organs arise in a similar way. A number of young cells become arranged so as to form a rounded mass in the deeper portion of the newly-formed epidermis: these then become elongated in a direction vertical to the surface of the epidermis, the central element undergoing differentiation into sensory cells, while the peripheral ones form an investment around these.

It is evident that the process is rendered more complicated in this case by the fact that the young epidermic cells, formed by the proliferation of those already present, give rise to cells of various kinds, viz., to ordinary epidermic cells, to gland cells, and to sensory and 'investing' cells; and the complication is further increased by all these cells being arranged and localised in a perfectly definite and more or less prescribed manner. We certainly must not assume that the formative cells which undergo these various differentiations are really identical, although they may appear so. It cannot possibly depend on external influences alone whether one of these subsequently becomes transformed into a horny, glandular, or sensory cell; for we cannot assume the existence of such a regular and localised difference in the external influences. The various differentiations of the formative cells must therefore depend on their own nature — that is to say, on the determinants contained within them, which have hitherto been latent but which have now become ripe, and have impressed a specific character upon each cell. *These formative cells must have contained different sorts of determinants from the first.*

Fraisse compares the processes which can be observed in the regeneration of the skin in Amphibia with those which occur in the embryogeny of this class, and shows that they are essentially similar. We shall therefore be justified in imagining these processes — which are invisible to us even under the highest power of the microscope — to be homologous with those which take place during the development of the embryo.

We can thus further assume that the stratified cells in the 'mucous layer' of the epidermis, although apparently all alike,

* Cf. Fraisse, 'Die Regeneration von Geweben u. Organen bei den Wirbelthieren, besonders bei Amphibien u. Reptilien.' Cassel and Berlin, 1885.

— as are those cells which form the first rudiment of the embryonic integument, — must nevertheless possess several kinds of determinants. We can hardly venture to say whether the three kinds of determinants with which we are here concerned are all present together in the formative cells, and only become distributed amongst special cells when regeneration sets in, or whether they are distributed amongst special cells from the first. Either arrangement is possible. Hence we may assume that some of the young formative cells contain determinants for the glands, and others those for horny or sensory cells, and that the proportional numbers and topographical arrangement of these are definitely fixed from the first. A precisely similar assumption is also necessary in the case of embryogeny.

If, for instance, the sensory organs of the lateral line in a fish or amphibian occur only along the lateral lines and their branches, we must suppose that the subdivision of the idioplasm of the ectodermic cells occurs during the development of the epidermis in such a way that the cells containing the determinants of these sensory organs come to be situated only along the lateral lines, and only in definite places on these lines. If now, all the formative cells of the sensory organs do not undergo further development at once, but some of them, on the contrary, remain undeveloped in the immediate neighbourhood — *i.e.*, in the deep layer of young cells — until a necessity for regeneration arises, we can understand in principle why a similar topographical arrangement and numerical relation of the sensory organs to the remaining epidermic elements occurs in the case of regeneration, as well as in that of the primary formation of the epidermis in the embryo.

The idioplasm of the cells does not alone decide what will happen in regenerative processes of this kind. This is shown by the fact that the occurrence of regenerative cell-multiplication depends on a loss of substance, and that the cells cease to proliferate as soon as the defect is made good. The stimulus to the further proliferation of the cells ceases at the same time. These facts, however, only give us a very vague insight into the causes of the limitation of the regenerative process; and we shall presently see that the above explanation is insufficient in more complicated cases of regeneration, and that we must, indeed, assume in addition the existence of other regulating factors, which are situated within the active cells, and not outside them.

It is well known that the limbs of a salamander grow again after they have been cut off, and we owe our accurate knowledge of the regenerative processes concerned mainly to the researches of Götte * and Fraisse.† The investigations of these observers show that the regeneration of the limbs and their formation in embryogeny take place in a similar manner: the individual parts and segments of the extremity become developed in the same order, and are formed of similar cell-material in each instance. Both here and in the case of the epidermis described above, the regeneration is *palingenetic*.

If we take as our basis the law, which holds good at any rate as regards Vertebrates, that in regeneration each specific tissue can only reproduce its own specific cells, we can test the theory of regeneration by taking as an example a single tissue of an extremity. It is certain as regards the bones, for instance, that regeneration always proceeds from the injured bone, or rather from its periosteum. If the extremity is disarticulated from the shoulder-girdle, for example, and the bones are uninjured, these latter do not become re-formed. Although it cannot be denied that the various tissues which are required for the regeneration of the entire limb have an influence upon one another, especially when pressure is exerted by one part on another situated near to it, it is clear that the formation of new bones depends entirely on the bones present in the stump of the amputated limb, which not only determine the quality of the tissue, but also regulate the size and shape of the bone which is to be formed anew. These last-mentioned facts are the most important of all in explaining the phenomenon of regeneration of a limb. From what has already been said, it is evident that the bony tissue, including the periosteum, *can* be formed from the cells of the corresponding pre-existing parts. All that is necessary in order that the process may take place is a supply of cells, capable of proliferation, which contain 'bone-idioplasm,' and which are incited to multiply by the stimulus due to the injury in the tissue surrounding them. The regeneration of the epidermis may be explained in a similar manner. But as regards these bones, it is not merely the production of bony tissue of a definite

---

* Gotte, 'Uber Entwickelung u. Regeneration des Gliedmassen-Skeletts der Molche,' Leipzig, 1879.

† Fraisse, *loc. cit.* on p. 97.

structure which has to be considered, but *the formation of a definite number of bones of a definite shape and size, arranged in a definite series*, must also be taken into account. What assumptions must we make in order to explain such an accurately prescribed and complex mode of construction of these parts? If the fore-limb of a newt (*Triton*) is cut off between the shoulder and elbow, not only does the lost portion of the humerus become formed afresh, but the radius and ulna, and all the bones of the wrist and hand, are regenerated accurately, even as regards the number of segments. It seems hardly possible that so complex a structure could be produced merely by the co-operation of proliferating cells, and it might be supposed that an invisible power — a *spiritus rector* or a *vis formativa* — must be present to direct their mode of increase and arrangement. We are nevertheless probably right in assuming that no such external direction takes place, and that the complex structures in living beings are produced merely by the agency of the forces which are present in the individual cells.

We can understand these processes to some extent in the case of embryogeny if we base our reasoning on the principle of the gradual transformation of the idioplasm, which has already been treated of in connection with ontogeny. This principle may be roughly illustrated with respect to the skeleton of the anterior extremity in the following manner.

When the fore-limb of a Triton begins to arise as a small blunt elevation of the skin, it consists of cells of the external and middle embryonic layers. The whole of the former, and that portion of the latter which forms the cutis, may be left out of consideration; they together give rise to the integument. The rest of the mesoderm now forms a mass of cells which have not yet begun to undergo differentiation, and which individually do not apparently differ essentially from one another. They must, nevertheless, be very different as regards the primary constituents which they contain, for some of them will subsequently give rise, for instance, to muscles, others to connective tissue or to blood-vessels, and others, again, to bones. These cells, which are so differently predisposed, must therefore contain various determinants, which, when they obtain control over them in the course of further cell-divisions, impress on the subsequent generations the character of muscle- or bone-cells.

Each of these kinds of cells must be present from the first in a perfectly definite number, and must occupy a perfectly definite position.

Let us follow out this line of reasoning with regard to one system of organs, namely, the bones, and assume for the sake of simplicity that only a single bone-forming cell is present in the first rudiment of the limb. This cell would virtually contain the entire skeleton of the limb; and we should have to attribute to its idioplasm the power not only of giving the succeeding cells of a certain number of generations the character of bone-forming cells, but also of determining the entire sequence of these cells as regards quantity, quality, and mutual arrangement, as well as the rhythm in which the divisions will follow one another. For the particular point at which an interruption occurs in the continuity of the bone, and consequently also the boundary line between two segments of the bony chain, might essentially depend, indeed, on this rhythm.

We must therefore suppose that the composition of the idioplasm of the first primordial bone-cell of the limb causes all these sequences to take place: in other words, *the idioplasm must contain the determinants of all the succeeding bone-cells*. This may be illustrated by the following diagram (fig. 3), in which the actual processes, which concern hundreds of thousands of cells, are represented as greatly abbreviated, and the different generations of cells are indicated arbitrarily by a genealogical tree, which, however, does not by any means always represent their actual connection.

Each primary cell of the individual bones is represented in the figure by a circle, and is supposed to be so simple that it can be controlled by *one* determinant. Thus the primary cell of the entire series of bones is controlled by determinant 1, but also contains the determinants 2-35 in its ids. In the first cell-division this cell divides into two, — the primary cells of the upper arm (humerus), and of the fore-arm and hand. The former contains determinant 2, and its further division is indicated by the cells containing determinants 2*a*-2*x*. The latter contains the remaining determinants 3-35, which become separated into smaller and smaller groups in each cell-division, until finally each cell only contains a *single* determinant. The diagram only represents the main bones of the extremity, — the individual carpals are omitted.

Let us now return to the question of regeneration. If each cell in the fully-formed bone only contains that kind of idioplasm which controls it, and which is therefore the molecular expression of its own particular nature, it would be impossible to understand how the regeneration of the bone could be effected — when, for instance, it had been cut through longitudinally.

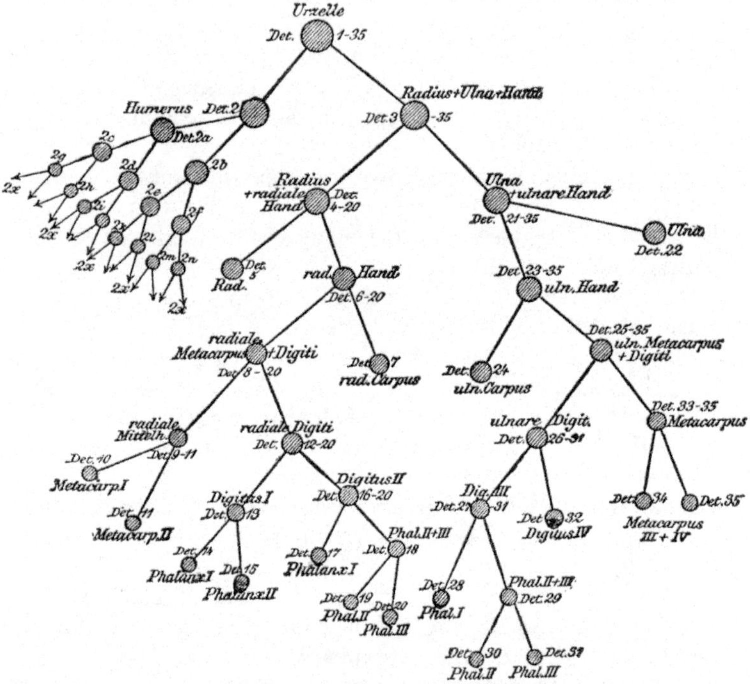

FIG. 3. — DIAGRAM OF THE CELL-GENERATIONS IN THE FORE-LIMB OF A TRITON.

Supposing that a stimulus, produced by the injury, caused the cells of the injured part to undergo multiplication: bony tissue would then, indeed, be developed, but a bone of a definite shape and size would not necessarily be formed. The formation of a definite bone can only take place if the proliferating cells possess, in addition to their active determinants, a supply of determinants which control the missing parts which have to be renewed. If therefore we wish to suppose that Blumenbach's '*nisus formativus*' is situated in the idioplasm of the cell, it appears necessary to assume that each cell capable of regeneration contains an *accessory idioplasm*, consisting of the determinants of the parts which can be regenerated by it, in addition to its primary idioplasm. Thus, for instance, the cells in the bone of the upper-arm must contain, in addition to their controlling determinant 2, the determinants 3-35 as accessory idioplasm, which can cause all the bony parts of the fore-arm to be formed anew; the cells of the radius, again, must contain the determinants 4-20 as accessory idioplasm for the reconstruction of the radial portion of the wrist and hand.

This theoretical illustration may be looked upon, indeed, as representing the phenomena as they occur in reality. It is very possible that the required accessory idioplasm becomes separated from the disintegrating embryonic idioplasm in the earliest stage of development of the entire organ. According to our assumption, the individual determinants are present *singly* in the germ-plasm, and their multiplication increases the further ontogeny advances. As only those determinants which correspond to parts to be formed subsequently are required in the accessory idioplasm, the material for the latter is always present; and we need only assume that in each division of the primary cell of any bone, a portion of the determinants required for the formation of the subsequent parts becomes split off as secondary idioplasm, and remains inactive within the cell until a cause for regeneration arises.

I shall speak of this group of determinants as *accessory idioplasm* ('Neben-Idioplasma'), and its component determinants as *supplementary determinants* ('Ersatz-Determinanten'). We may imagine that these form a special and minute group enclosed within the id in the neighbourhood of the determinants which control the cell in question. A similar assumption may be made as regards the individual bones of the entire limb. The regeneration of the bisected humerus can be explained by supposing that each cell capable of regeneration possesses an accessory idioplasm, containing the determinants of the cells which will subsequently be formed in a distal direction; this formation will be possible because the necessary 'determinant-material' is present. The process only depends on the fact that in each differential cell-division a certain number of determinants, which ripen later on, become split off from the rest, and are retained in the cell as accessory idioplasm. The mechanism for regeneration is certainly a very complicated one, for each separate bone is controlled by a number of different determinants, and not by a single one; and all these special determinants are contained in the accessory idioplasm. As far as we can judge from the investigations made hitherto, the bones are at any rate regenerated in detail fairly exactly. The complexity of the mechanism accounts, in my opinion, for the fact that the fore-limb, which has such a marked power of regeneration in the salamander, has lost this power completely in the higher Vertebrates, for in them the mechanism would have become too complex.

A simpler mechanism than that which we have supposed to exist can only be conceived, if, with Herbert Spencer,* we attribute to each of the units composing the body the power combining to form any necessary organ just when it is wanted. We might then compare the entire animal to a large crystal, in the individual parts of which 'there dwells the intrinsic aptitude to aggregate into the form of that species; just as in the atoms of a salt there dwells the intrinsic aptitude to crystallise in a particular way.' The only difference between the particles of the crystal and those of the organism would be that the former are all permanently alike; and that the latter, in order that regeneration may be possible, are arranged in many different ways, according to whether an entire limb, a tail, a gill, or a single toe, fore-arm, or finger is to be replaced. How are the 'units' shown in each individual case what part is missing, and what form their arrangement is to take in order to produce the part anew? We are thus once more brought back to Blumenbach's '*nisus formativus*.' Spencer himself says:—'If in the case of the crystal we say that the whole aggregate exerts over its parts a force which constrains the newly-integrated atoms to take a certain definite form, we must, in the case of the organism, assume an analogous force.' This force would correspond to what was formerly spoken of as the '*spiritus rector*' or '*nisus formativus*;' and even supposing it to exist, it does not in the least help us in the attempt to explain the mechanism of the phenom-

* Herbert Spencer, 'The Principles of Biology,' Vol. 1, p. 181.

ena. Spencer adds that his view 'in truth is not a hypothesis,' but only 'a generalised expression of facts;' and remarks in another passage, that although it is 'difficult' to imagine regeneration as a sort of process of crystallisation, '*we see that it is so.*' It is just this point that I must object to. We see that it is so, or rather appears to be so, *sometimes*, but we also see that it is often *not* so. If the units of the body were capable of becoming modified at will under the influence of the whole, and of crystallising into the missing part, they must be able to do so in all species and in all organisms. This, however, is not the case. The limb of a salamander can be regenerated, and that of a lizard cannot. In a special section of this chapter I shall be able to show in greater detail that regeneration depends on special adaptation, and not on a *general* capacity of the animal-body.

It will be unnecessary to give a special diagram illustrating the regeneration of a single bone, such as that of the upper arm, and showing the supplementary determinants of each of the cells composing the bone which are necessary in order that regeneration may set in at any point. The diagram given for the entire limb is sufficient to make the general principle clear: an approach to an explanation of the actual details is out of the question, as is evident if we compare the number of cells given in the diagram with that of which the bones actually consist. For this reason I have not attempted to enter into minute histological details, or to define the quality of the cells which are capable of regeneration,—that is, to state whether they belong to the periosteum or to the bone itself, and whether all or only certain cells take part in the process. We only require a diagram which can be adapted to the actual details of the processes when these are known. It is sufficient at present to show that regeneration may be understood by considering the activity of the cells themselves, without having recourse to the assumption of an unknown directive agency. The '*nisus formativus*' descends from its previous position as a single force directing the whole, and breaks up into an unlimited number of material particles which are situated in the individual cells, and each of which prescribes the course of life of the cell. These particles are determined as regards their kind, and are distributed to their proper places so accurately, that by their united effect they give rise to a composite whole, such as, for instance, a series of bones, together with their articular capsules and ligaments, and the muscles, nerves, blood-vessels, connective tissue, and integument which come into relation with them. The diagram I have given to illustrate the regeneration of a bone can obviously be adapted to represent any other part or tissue. We must not look upon the bone as something quite disconnected from the rest of the limb, as we may very likely be inclined to do if we are specialists. The bone is in reality connected most intimately along its entire surface with the surrounding tissues, — the periosteum and loose connective tissue external to the latter, the numerous blood-vessels which penetrate into the substance of the bone, the nerves, and so on. The first rudiment of the limb consists, in fact, of a mass of mesodermic cells, which give no indication of the various structures which will later be developed from them. Nevertheless, their differentiation does not, in my opinion, depend on their accidental position within the limb, or in fact on any other external influences, but is primarily due to their individual nature, that is, *to the constitution of their idioplasm*. The determinants composing the id control the subsequent development of the cell and of its successors. The further changes which the id undergoes in the course of cell-division, and the manner in which the determinants undergo disintegration in the ids of the daughter-cells of all the subsequent generations, is decided by the composition of the id.

We can thus understand, at least to some extent, how it is possible that such a complicated part as a limb, the structure of which is so accurately prescribed, can arise by degrees from a mass of cells which are apparently all similar to one another. The harmony of the whole is primarily brought about by the variation and increase of the cells, the kind and rhythm of which respectively, is prescribed by the idioplasm of each individual cell, rather than by the mutual influence of the cells during their gradual differentiation. A muscle becomes developed at any definite spot, because one particular cell amongst all the apparently similar cells in the first rudiment of the limb contained the determinants which are capable of giving to a large number of the successors of this cell the special character of muscle-cells; and because, again, the id of this particular cell caused a rhythm of multiplication to set in, which, on mechanical grounds, rendered it necessary that certain successors of this cell which contained muscle-determinants should take up their position in the precise region of the limb in which this particular muscle is situated.

We must not, however, imply from what has been said above, that external influences are of no importance whatever in ontogeny, but merely that they certainly only play a secondary part in the process. A limb will certainly grow crooked if a corresponding external pressure is brought to bear upon it. Growing cells do not cease to multiply directly they are subjected to abnormal external influences, for they can accommodate themselves to circumstances. It is such cases as the regeneration of broken bones and the formation of new joints under abnormal external conditions, which prove that the cells continue to perform their functions of growing and of giving rise to organs under circumstances which deviate very markedly from the normal. These false joints also show what a considerable power of adaptation is possessed by the cells, and how efficient may be the parts which these cells are able to produce under abnormal conditions. But although the principle formulated by Roux * of the *struggle of the parts*, or as it might well be called '*intrabiontic selection*,' is certainly a very important one, I think it would be a great mistake to refer the normal process of ontogeny mainly to this principle. The groups and masses of cells must certainly press upon one another during the process of differentiation: in the process of the formation of a joint, for instance, proliferating connective-tissue cells do actually force themselves amongst the cartilage cells in one part of the rudimentary bone, in order to separate them from one another. But this proliferation and pressure are taken account of, just as much as are the processes of dissolution or absorption that occur in those cells in the primordial cartilage which are situated in the region of the joint. It might be supposed that the existence of so-called 'identical' human twins contradict my conception of ontogeny; for although they are undoubtedly derived from a single ovum and sperm-cell, and hence possess the same kind of germ-plasm, they are never really identical, but only very similar to one another. But apart from the fact that the *absolute* identity of the germ-plasm has not been proved in these cases, the very close resemblance between these twins shows

* W. Roux, 'Der Kampf der Theile im Organismus,' Leipzig, 1881.

how slightly the diversity of external influences affects the development of an organism. How wonderfully accurately the course of ontogeny must be prescribed, if it can be kept to so closely, through thousands of generations of cells, that 'identical' twins result! We may compare the process of development of such twins with the course taken by two ships, which, starting from the same place, proceed along the same devious route which has been carefully mapped out beforehand in all its thousands of definite changes in direction, until each finally reaches the same distant shore independently, within a mile of the other.

A careful consideration of such a case as this leaves no doubt that a very exact and definite course is mapped out for the egg-cell by its idioplasm, which, again, directs the special course to be taken by each of the innumerable generations of cells, in the direction of which course external influences can only play a very subordinate part. If this consideration be borne in mind, it will be less likely that the objection may be made that a much too complicated structure has been attributed to the idioplasm. *Its structure must be far more complex than we can possibly imagine*; and in this respect, its construction, as we have represented it theoretically, must certainly be far simpler than is the case in reality. For the same reason, it is less probable that similar objections may be made to the theory of regeneration as here stated. Complicated phenomena cannot possibly depend on a simple mechanism. The machines in a cotton factory cannot be constructed of a few simple levers, nor can a phonograph be manufactured from two lucifer matches.

That form of regeneration which has been considered above may be described as *palingenetic*, for it pursues the course taken by the primary or embryonic development; but as soon as it leaves this course and takes a shorter one, it may be distinguished as *cœnogenetic*.

Cœnogenetic variations of the primary process of development probably always occur in cases of regeneration of complex structures; and even the reconstruction of the extremities, which we have chosen above as an example, will hardly take place in exactly the same way as occurs in the primary development of these parts, although it may resemble the latter in its principal phases.

Even if mere abbreviation of the development of a part can

be easily conceived by supposing an aggregation and redistribution of the determinants to occur in the idioplasm, the process of idic division becomes very complicated when the primary and secondary development take place along different lines; for in the latter process the combinations of supplementary determinants in the id of the cell-generations must be different from those which occur in the former. But this difference is evidently due merely to a greater complication of the process, and it does not stand in the way of the theory. In all cases of regeneration, the mode in which the supplementary determinants become split off must be previously arranged for in the id. The assumption of a mere increase in the power of multiplication of certain determinants might seem sufficient in the case of palingenetic regeneration, for this would lead to one portion of a certain group of determinants becoming separated off as accessory idioplasm at a particular ontogenetic stage. In cœnogenetic regeneration, however, we can only assume that a double or still greater number of determinants are present in the germ-plasm, one set of which are destined for embryonic development and the others for regeneration; and these are previously arranged with reference to their internal forces, particularly that of multiplication, so that at a certain stage of development they become split off as 'accessory idioplasm,' either alone or together with the adjacent 'regenerative determinants.'

It seems to me, however, that palingenetic regeneration cannot be satisfactorily accounted for unless we assume the existence of special regenerative determinants, for it would otherwise be impossible to explain the phyletic origin of the cœnogenetic variations in the process of regeneration. These latter must, indeed, depend on variations in a determinant of the germ-plasm. If however the latter contained only the *one* determinant destined for embryogeny, variations must occur in the latter process at the same time. But this is not the case, and consequently a kind of double determinant must be contained in the germ for those hereditary parts (determinants) which are capable of becoming regenerated: — that is to say, two originally identical determinants must be present, one of which becomes functional in embryogeny and the other in regeneration. This will be made apparent if we take some examples.

In most existing amphibians the caudal region of the vertebral column may undergo regeneration, although its embryonic foundation, the notochord, is never formed anew. The cartilaginous sheath of the notochord has an important share in the primary formation of the vertebral column, but it disappears to a greater or less extent at a later stage. If it became possible for the vertebræ to undergo regeneration after a portion of the tail had been lost without a renewal of the notochord taking place, the result would be a useful abbreviation of the process of regeneration. Such an abbreviation has occurred, and everything supports the assumption that *at an earlier stage of phyletic development the notochord was capable of undergoing regeneration*, and that it has only lost this capacity secondarily. In the case of frog-tadpoles, the power has been retained of regenerating the tail when it is cut off *together with the notochord*. We must not assume that the notochord does not become restored in other amphibians because it no longer persists in the full-grown animal; for it is entirely absent only in a few of them (*e.g.*, *Salmandrina*), and the notochord of the larval salamander cannot be regenerated any more than that of the adult. Thus the capacity for regenerating the notochord has been lost by most amphibians in the course of phylogeny. Such a process of degeneration is certainly to be explained most easily by assuming the existence of special regenerative determinants, which may gradually disappear without in the least affecting their embryogenetic partners.

The necessity of this assumption is shown still more conclusively in the case, for instance, of the restoration of the solid axis of the tail in reptiles. The tail of a lizard quickly becomes restored after it has been cut off, but its structure is then different from that of the original tail; for, according to the statements of Leydig and Fraisse, the spinal cord and vertebral column are not renewed. The former is, however, represented by an epithelial tube, but gives off no nerves; and the latter is replaced by an unsegmented cartilaginous tube. As Fraisse points out, this tube does not correspond to the regenerated notochord, but is a new structure which is substituted for it.

A *phyletic development*, tending essentially towards a simplification of the parts, has taken place in this case *as regards the processes of regeneration*. A gradual degeneration has occurred, just as may take place in the tail or any other organ of an animal in the course of phylogeny. The caudal region of the vertebral column has undergone a reduction, which does not influence its primary (embryonic) ontogeny, but only its secondary formation by regeneration. A vertebral column is formed primarily; but if the re-formation of a part of it becomes necessary, in consequence of the loss of the tail, the secondary reduced process for the development of the axis comes into play, and a simple cartilaginous tube is formed. This process recalls the phenomena of 'dichogeny' which take place so frequently in plants, and in which the same group of cells may develop in either of two different ways, according to the nature of the external stimulus which is applied to them. Thus a shoot of ivy will produce roots on a certain side if it is shaded, and leaves if it is exposed to light. The determination of the sex of an animal may perhaps be referred to similar causes, — if, at least, we may assume that the sex is not always universally decided by the act of fertilisation, and that influences exerted upon the organism subsequently may have an effect in this determination. In the case of certain parasitic Crustaceans, the *Cymathoidæ*, the male sexual organs are developed first; and when the animal has fulfilled its function as a male, the female organs become developed, and give the animal the character of a female. The two developmental tendencies here come into operation temporarily, one after the other; just as in the case of the lizard's tail, in which the tendency to form the vertebral column first comes into play, and then that to form the secondary cartilaginous tube. The necessity for the formation of this tube certainly need not arise at all; just as that side of the shoot of ivy from which the roots arise need not necessarily be subsequently exposed to the light, and give rise to leaves; the possibility of such an occurrence is, however, foreseen by Nature. It might be urged that there is an important difference between the regeneration of a lizard's tail and the successive development of the two kinds of sexual organs in the Cymathoids, since in the latter case the rudiments of these organs are present in the embryo, and it is only their final development which takes place successively. This is certainly a difference, but it is just such a one as to indicate to us how these cases of supplementary substitution may be explained theoretically. The cells in the tail of a lizard which give rise to the secondary cartilaginous tube must contain determinants which differ from those of the embryonic formative cells of the caudal vertebræ, just as the idioplasm of the formative cells must contain different determinants for the testes and ovaries. *The supplementary determinants with which the idioplasm of certain cells of the vertebral column was provided for the purposes of regeneration, must have become changed in the course of phylogeny.*

A transmissible variation of this kind must, however, also have had some effect on embryogeny, if only one and the same determinant were present in the germ-plasm for the two modes of development. *Hence each determinant of these caudal vertebræ must be doubled in the germ-plasm.*

It would be premature to go beyond this assumption, and to attempt to decide anything about the manner in which the various supplementary determinants which are required for the restoration of one of the larger parts—such as, for instance, the caudal vertebræ—come together, and how and when they become separated from the primary determinants. The processes of regeneration have not as yet been examined from the point of view which I have here suggested; and in many cases it is not even known for certain from what cells regeneration proceeds.

Hitherto we have not discussed in detail the question as to *the kind of cells which contain the supplementary determinants, and from which regeneration thus takes place.* May these determinants be present in any kind of cell belonging to any tissue, or is their distribution always limited to young and apparently undifferentiated cells of the so-called 'embryonic type'?

If we only consider Man and the higher Vertebrates, we shall be disposed to look upon the latter of these two alternatives as the one which is in general correct. Even recently, in fact, many authors seemed to be in favour of this view: 'embryonic cells' were supposed to be contained in all those tissues which are capable of regeneration, and it was, indeed, believed by many that the leucocytes are cells of this nature. The latest investigations, however, lead us to the conclusion that this is not the case, and that although the white blood corpuscles are extremely important as conveyers of nutriment in the process of regeneration, they do not serve as formative elements in the construction of a tissue. In his text-book on Pathological Anatomy, Ziegler speaks of a formal 'law of the specific character of the tissues,' which he explains as follows:—'the successors of the various germinal layers which separate from one another at an early embryonic stage, can only give rise to those tissues which belong to the germinal layer from which they were developed.' But this statement can only be true in the case of the highest Vertebrates, for, as the brothers Hertwig have shown, the germinal layers of the Metazoa are not primitive organs in the histological sense; and moreover, in the lower animals, several, if not all of the tissues, can be formed from each of the germinal layers. In lower animals, not only all the varieties of tissue, but under certain circumstances even rows of cells of one primary germinal layer and even indeed the entire animal, may arise from young cells belonging to the other germinal layer. In the chapters on multiplication by fission and gemmation, this process will be traced to its origin in the idioplasm. At present we have only to deal with the question as to whether the determinants of the various kinds of cells which are required for regeneration are contained within young cells only, or whether they are also present in those which have become differentiated histologically.

Although the supplementary determinants are certainly in many cases contained in young cells without any specially marked histological character, their distribution can nevertheless hardly be limited to these cells exclusively. It may happen — as will be shown in greater detail subsequently — that cells, which are fully developed histologically, both in plants and in the lower animals, contain all the determinants of the species; that is to say, they may contain germ-plasm as supplementary idioplasm. Hence there is no reason to assume that smaller groups of determinants may not have been supplied to specific tissue cells wherever they were required, although I am unable to give a definite example of such a case.

Although regeneration may originate in most cases in young, or so-called 'embryonic' cells, it is nevertheless quite a mistake to connect the idea of the undifferentiated state of these cells with this fact, as is so often done. These 'embryonic cells' are not 'capable of giving rise to anything and everything,' for each of them can only develop into that kind of cell the determinant of which it contains. Under certain circumstances such a cell may contain several different determinants at the same time, which are only distributed amongst the individual cells in subsequent cell-generations; but the structure which can and will become developed from it always depends on the cell itself, and its fate is determined by the idioplasm it contains, and can only be affected secondarily by external influences. Cells moreover exist, the idioplasm of which *permanently* retains the possibility of development along one of two lines. 'Dichogeny' in plants, which has already been mentioned, is likewise determined by the idioplasm, inasmuch as the latter must contain two kinds of determinants, one or the other of which either remains inactive owing to the nature of the external influences acting upon the cell, or else becomes active and determines the cell.

There are, however, no such things as 'embryonic cells' in the sense in which this term is used by authors. In the freshwater polype (*Hydra*), for instance, cells which are young and histologically undifferentiated — the so-called 'interstitial cells' — are present in the deeper part of the ectoderm: these can certainly give rise to various structures, viz., to ordinary ectoderm-cells, nettle-cells, muscle-cells, sexual-cells, and in all probability to nerve-cells also. It would nevertheless be absurd to suppose that any particular interstitial cell is capable of developing into any one of these structures. It obviously contains either germ-plasm, *i.e.*, the whole of the determinants, — in which case it can develop into a sexual cell, — or only the determinants of a thread cell or of one of the other kinds of cells, and then it can only give rise to one of the corresponding structures, and can never develop into a sexual cell.

2. The Phylogeny of Regeneration

It may, I believe, be deduced with certainty from those phenomena of regeneration with which we are acquainted, that the *capacity for regeneration is not a primary quality of the organism, but that it is a phenomenon of adaptation.*

The power of regeneration has hitherto been practically always regarded as a primary quality of the organism, — that is to say, as a direct result of its organisation: it has been looked upon as a faculty for which no special arrangements are required, but which naturally results as an unintentional secondary effect of the organisation which exists independently of it.

This view is based on the idea, which is in general a correct one, that the regenerative power of an animal is inversely proportional to its degree of organisation.* If this were univer-

* Cf. Herbert Spencer (*loc. cit.*, p. 175), who, however, expresses himself very cautiously with regard to this difficult subject as follows: — 'so

sally true, it would nevertheless not be a convincing argument for the above view, although it would certainly support it. But a closer examination into the facts shows that this statement is not absolutely correct. Although the capacity for regeneration is never so far-reaching in the highest animals as it is in the case of the lower ones, — and this must be due to some cause, — the regenerative power may nevertheless even vary widely in animals of the same degree of organisation, and may in fact be far greater in one of the higher than in one of the lower forms. Thus fishes are unable to regenerate a lost pectoral or pelvic fin, while the much more highly organised salamander has been known to regenerate a limb six times in succession (Spallanzani).

The regenerative power often varies in degree even within the same group of animals. In *Triton* and *Salamandra* the entire limb grows again after amputation, but apparently, so far as I have been able to observe, this does not occur in *Proteus*. The tail, too, is only replaced slowly and imperfectly in the latter animal, whereas it easily becomes restored in the salamander. In the year 1878 I received a living *Siren lacertina*, the fore-limb of which had been torn off, so that only the stump of the upper arm remained, and the entire limb did not grow again in the course of the ten years during which I kept the voracious animal, and gave it abundant food. In this case again the power of regenerating the extremities seems to be less than in that of salamanders, which are much younger phyletically, and much more highly organised.

It is well known also that the limbs of a frog do not grow again when they have been cut off, even when the animal is in the larval condition. The fact that the regenerative power can vary so considerably within the same genus is still more remarkable. Schreiber observed that the power of regeneration in *Triton marmoratus* is relatively very slight as compared with that which is possessed by all other species of Triton which have been examined for this purpose. In captivity, at any rate, even slight injuries in such parts as the crest are never re-

that the power of reproducing lost parts is greatest where the organisation is lowest, and almost disappears where the organisation is highest. And though we cannot say that between these extremes there is a constant inverse relation between reparative power and degree of organisation, yet we may say that there is some approach to such a relation.'

placed while the animal invariably succumbs to greater injuries.' Fraisse gives similar instances. Thus 'an amputated extremity never grew again to its normal size: merely a somewhat deformed protuberance was formed on the stump. The tail also was only reproduced to a very slight extent.' *

With regard to reptiles, Fraisse points out that the regenerative capacity obtains to a much slighter extent in some groups than in others. Chelonians, crocodiles, and snakes are unable to regenerate lost parts to any extent, while lizards and geckos possess this capacity in a high degree.

The dissimilarity, moreover, as regards the power of regeneration which exists in *various members of the same species*, also indicates that adaptation is an important factor in this process. In *Proteus*, which in other respects possesses so slight a capacity for regeneration, the gills grow again rapidly when they have been cut off. In lizards, again, this power is confined to the tail, and the limbs cannot become restored: in these animals, however, the tail is obviously far more likely to become mutilated than are the limbs, which as a matter of fact are seldom lost, although individuals with stumps of limbs are occasionally met with. The physiological importance of the tail of a lizard consists in the fact that it preserves the animal from total destruction; for pursuers will generally aim at the long trailing tail, and thus the animal often escapes, as the tail breaks off when it is firmly seized. It is, in fact, as Leydig was the first to point out, specially adapted for breaking off, the bodies of the caudal vertebræ from the seventh onward being provided with a special plane of fracture, so that they easily break into two transversely. Now if this capability of fracture is provided for by a special arrangement and modification of the parts of the tail, we shall not be making too daring an inference if we regard the regenerative power of the tail as *a special adaptation, produced by selection, of this particular part of the body, the frequent loss of which is in a certain measure provided for*, and not as the outcome of an unknown 'regenerative power' possessed by the entire animal. This arrangement would not have been provided if the part had been of no, or only of slight, physiological importance, as is the case in snakes and chelonians, although these animals are as highly organised as lizards. The reason

* *loc. cit.*, p. 152.

that the limbs of lizards are not replaced is, I believe, due to the fact that these animals are seldom seized by the leg, owing to their extremely rapid movements. But if a lizard does happen to be caught by one of its limbs, it must fall a prey to its pursuer, and the capacity for regenerating the limb would be useless. The case is very different with regard to such animals as Tritons. Their movements are much less rapid, and their assailants, being too small to swallow the whole animal, frequently bite off a limb. They are often attacked by members of the same species, which gnaw off a gill, limb, or the tail of a weaker comrade, bit by bit. If a considerable power of regeneration were possible at all, it would certainly be provided in this case. This power is possessed in a much smaller degree by Proteus; but these animals are only found in the caves of Carniola, where enemies larger than themselves do not exist, and in which there is no great competition for food, and therefore, at least as far as my observations extend, they do not bite one another.

Spallanzani has stated that nature does not reproduce every part that is cut off; expressed in theoretical terms, this simply means that *the various organs of an animal possess the power of regeneration in different degrees*. If we inquire further into the question, we shall find that those parts which are most frequently exposed to injury or loss must possess the power of regeneration in the highest degree. So far as I can judge from the facts with which we are at present acquainted, this remark appears to me to be a perfectly correct one. Unfortunately Spallanzani gives no instance in support of the above statement, so that we do not know what parts he referred to. I have myself, however, made some investigations in order to ascertain whether the degree of regeneration of a part bears any relation to its liability to injury.

If regeneration is a phenomenon of adaptation, the internal organs — which are not exposed to injury in the natural life of the animal — cannot possess any regenerative power, even in those animals in which the external parts — which are exposed to the attacks of enemies — possess it in a high degree.

The following experiment bears upon this point: — I cut open a large newt (*Triton cristatus*), removed about half of the right lung, and sewed the skin together again. The animal soon recovered from the effects of the chloroform, and its wounds healed: it was then well cared for for fourteen months, and afterwards killed. An examination showed that the right lung had not become restored: it was only half as long as the left one, and its end was blunt, and not pointed as in the normal lung. Four other similar experiments yielded like results: in one of these it was doubtful whether a growth of the lung had not taken place, but even in this case it had not recovered its long, pointed form.

These experiments are still being continued, but we may already deduce from them that a striking disproportion exists between the regenerative power of the external parts of a newt and that of its lungs. This difference seems even more marked if we bear in mind that in the case of a limb the process of regeneration is a very complex one, for complicated parts, consisting of many entirely different portions, have to be reproduced; whereas a lung is a simple hollow sac, which has no joints, and the histological structure of which is relatively simple.

We therefore infer that the internal parts, which are not exposed to injuries of an ordinary kind, do not possess a greater capacity for regeneration in these species than they do in the highest Vertebrates, which are so exceptionally inferior to them as regards the regenerative power of the external parts. Hence *there is no such thing as a general power of regeneration: in each kind of animal this power is graduated according to the need of regeneration in the part under consideration:* that is to say, the degree in which it is present is mainly in proportion to the liability of the part to injury.

This conclusion is closely connected with the fact that the restoration of a part which possesses the power of regeneration in a high degree, can only take place as the result of definite injuries which are in a manner provided for, and not from *any* kind of injury. Philippeaux was the first to discover that the limb of a Triton does not grow again when it has been removed at the joint, and that, in fact, it only does so when it is cut or torn off, so that the bone is injured. This fact has been explained by referring it to the law of the specific nature of the tissues, according to which bone can only be formed from bone, and the bone of the limb must be injured before it can become capable of being formed anew. It seems to me that this explanation is insufficient, although it is founded on a correct principle, according to which the injury to the bone causes the stimulus by which the cells of the stump are incited to proliferate. This is certainly correct, and may be expressed according to our theory by saying that the supplementary determinants which are present in a passive condition in the cells, are prompted to become active by the stimulus. But if an articular cavity is exposed, a stimulus is likewise produced, which must affect the cells of the articular cartilage, and doubtless also those of the underlying bone or periosteum. If, therefore, all the cells in this region were capable of reproducing the missing bones, and if the exposure of the articulation were the ordinary form of injury, these cells would certainly be just as much adapted for and capable of responding to this stimulus, by a formative growth, as would those situated at the broken ends of a bone. *But the disarticulation of a limb, or of a part of a limb, hardly ever takes place in the natural conditions of life, and therefore could not have been provided for by the organism; the respective cells of the exposed articular cavity could not consequently have been supplied with the supplementary determinants necessary for regeneration.* Hence these cells are incapable of reacting in an adequate manner to the stimulus due to the disarticulation.

In spite of all the facts already mentioned, it might still appear doubtful whether regeneration really depends on a special adaptation of the part in question, and whether it does not result from the degree of organisation of the animal, or at any rate from a *general* regenerative force possessed by the entire organism. The following considerations must, however, I think, set aside all doubts on the question. Physiological and pathological regeneration obviously depend on the same causes, and often pass one into the other, so that no real line of demarcation can be drawn between them. We nevertheless find that in those animals in which the power of regeneration is extremely great physiologically, it is very slight pathologically. This proves that a slight power of pathological regeneration cannot possibly depend on a general regenerative force present within the organism, but rather that this power can be provided in those parts of the body which require a continual or periodic regeneration: in other words, *the regenerative power of a part depends on adaptation*. Let us take a few examples. It was mentioned above that fishes are said to possess a very slight 'general regenerative power,' because they are unable to replace lost external parts, especially such structures as fins. Nevertheless many fishes are provided with teeth which are very liable to become worn out, and consequently they possess the power of constantly producing new teeth to replace the old ones. In the mouth of a ray or dog-fish the teeth are arranged in several rows along the edges of the jaws, the outer rows containing those which are worn out, and the inner the younger teeth which take their place. Birds, again, possess a very slight power of repairing defects which have arisen accidentally, and hence they are considered to have a very slight capacity for regeneration. But their power of physiological regeneration with respect to certain parts is nevertheless extraordinarily great: — all the feathers are cast off and renewed once a year. Pathological regeneration occurs to a very slight extent in mammals; defects in the superficial epithelium, the epithelium of the ducts of glands, the various supporting tissues, including bone, and in nerve-fibres, can be repaired from the elements of the respective tissues; but in no mammal does a segment of a finger or an eyelid grow again when once it has been cut off. In certain mammals, however, the power of physiological regeneration with respect to certain parts is unusually marked. Male stags shed their antlers annually, and new ones are formed in four or five months. If we take into consideration the mass of organic tissue which is thus formed in such a short time, this feat outstrips even the regenerative performances of the full-grown salamander. For according to Spallanzani, it takes a salamander more than a year to restore an amputated limb to its normal size and strength. Young individuals can, however, certainly reproduce a limb in a few days; and this gifted experimenter observed in the case of a young Triton that the four limbs and tail when they were cut off grew again six times in the space of three summer months!

In one respect, however, viz., as regards *the complexity of the part replaced*, this remarkable regenerative power in stags and birds is far inferior to that which obtains in the Triton. Although a bird's feather is a very wonderful structure, it is formed merely from epidermic cells, and a stag's antler is only a dermal bone covered over by the epidermis. But the limb of a Triton, on the other hand, consists of every kind of tissue with the exception of endodermal epithelium, — viz., of skin, muscles, a large number of skeletal parts, connective tissue, blood-vessels, nerves, and so forth; and all these have a very definite arrangement, number, and form. There is no doubt therefore that the regeneration of a limb is a greater feat than the renewal of feathers or antlers; and the fact has been long recognised, that *the more complex organs are regenerated less easily than those which have a simpler structure.* A series of carefully performed experiments, made with the view of testing this somewhat vague statement, would be of great value theoretically. We may predict that in one sense it would be confirmed, and that we should find that under similar conditions the simpler organs are on the whole regenerated much more easily than the more complex ones in any particular species. Even in the human race, many simple tissues — such as the connective substances, epithelia and nerves — can be repaired, and it is only the cells of the glands and ganglia, which are the most highly differentiated histologically, which are not replaced at all, or at most only to a very slight extent. We can see from a theoretical point of view that a far less complex apparatus is required in these cases than in those which concern a regeneration of entire parts of the body, such as the tail or limbs; for it is only necessary that the respective tissues should contain cells which are capable of multiplying, in response to the stimulus produced by the loss of substance in their immediate neighbourhood, and which continue to do so until the loss is made good. When, however, several kinds of cells take part in the restoration, and a strict regulation as to their arrangement in groups, their direction of growth, and rate of reproduction is required, it becomes necessary for the individual cells from which the restoration takes place to be accurately provided with supplementary determinants of various kinds; and it is clear that this will gradually become more difficult and complex, the greater the complexity of the part to be regenerated, and the more accurately all the details of its structure have to be preserved.

If, however, we review the facts known to us concerning regeneration in animals of various degrees of organisation, we meet with such marked differences even as regards the regenerative power of homologous parts, that we cannot help receiving the impression, which has affected all writers on this subject, that in general *the regenerative power is greater in less highly organised animals than in those of a more complex structure.* The question thus arises as to how this view is to be interpreted and presented in a scientific form.

Even in Vertebrates, certain facts seem to indicate that the 'lower' forms, as such, always possess the power of replacing lost parts in a greater degree than do the higher ones. It is true that the capacity for regeneration is certainly much slighter in fishes than in the more highly organised amphibians; but although the limb of a Triton becomes restored, and the fin of a fish does not, it must not be forgotten that the physiological importance of the two organs is somewhat unequal. On the other hand, the fore limb of a Triton and the arm of a man are not only homologous structures, but are also of almost equal physiological importance, and yet their power of regeneration is very unequal. We must therefore inquire into the causes of this dissimilarity. The power of regeneration in any particular part cannot depend only on the conditions which exist as regards the species under consideration: it must also be due to arrangements for regeneration which have been transmitted by the series of ancestors of this species. Leaving this question aside, and regarding the power of regeneration as merely depending in each individual case on adaptation, we should arrive at some such conclusion as the following:—the provision of the cells of a certain part with supplementary determinants for the purposes of regeneration, depends primarily on the liability of this part to frequent injury—that is to say, on the degree of *probability of injury*. Precautions are not taken for injuries which seldom occur, even though these may be very disadvantageous to the individual; for the loss thereby resulting to the species as regards the number of individuals would be extremely small and unimportant, and therefore processes of selection would not take place in order to counterbalance this loss.

In the second place, the *physiological or biological importance of the organ* itself must be taken into consideration. A useless or almost useless rudimentary part may often be injured or torn off without causing processes of selection to occur which would produce in it a capacity for regeneration. Thus, so far as is known, the minute limbs of *Siren* and *Proteus*, which are often bitten off, or not replaced; while the gills of these animals and of the Axolotl, which are equally liable to similar injuries, become regenerated:—in the latter case the organs are *physiologically* valuable, while in the former they are not. The tail of a lizard, again, which is very liable to injury, becomes regenerated, because, as we have seen, it is of great importance to the individual, and if lost its owner is placed at a disadvantage.

Finally, *the complexity of the individual parts* constitute the third factor which is concerned in regulating the regenerative power of the part in question; for the more complex the structure is, the longer and more energetically the process of selection must act in order to provide the mechanism for regeneration, which consists in the equipment of a large number of different kinds of cells with supplementary determinants, which are accurately graduated, and regulated as regards their power of multiplication. Thus we can understand, for instance, why the fore-limb of a Triton becomes regenerated, while that of a bird does not, although the wing is of far greater importance and is much more indispensable to its owner than is the fore-limb in the case of the Triton. Although there are fewer bones in a bird's wing than in a Triton's limb, the former is by far the more complicated structure; for it is covered with feathers, and as each quill has a special size, form, and coloration, the wing must contain a large number of special determinants in its formative cells. These determinants must all be contained and arranged in the germ-plasm, so that they can be passed on during embryonic development through a certain series of cells, — first into the outer germinal layers, then into the epidermis of the fore-limb, and finally, by the agency of further series of cells arising in the course of growth, to the region to which they specially belong. It is difficult enough to imagine how the distribution of the determinants can possibly take place in so accurate and certain a manner as must be the case in reality, so that not only the shape of the feather but even every speck of colour on it is accurately repeated in every individual of the species; and it might well, indeed, be considered impossible that the whole of this complex mechanism should also be capable of becoming modified in such a manner, that the entire wing, with all its feathers and patches of colour, could be regenerated from a cut surface in any part. Did this occur, the cells of any section of the wing would, according to our theory, have to contain the whole of the determinants of all the cells required for the construction of the portion of the wing distal to the cut surface as supplementary determinants, in addition to their own special idioplasm; and moreover, these determinants must then be distributed proportionately among the cells of the radial and ulnar, and of the upper and under surfaces of the wing, and the power of multiplication of each cell and its successors would have to be accurately adjusted. Although we cannot easily judge as to what is possible in nature, and are so often impressed by the discouraging conviction that many vital processes are still incomprehensible to us, we may perhaps in this case feel justified in *inferring the impossibility of such an occurrence from the fact that it does not take place*; that is, to infer that the regeneration of a bird's wing is impossible on account of the complexity of the mechanism required for it, because it does not actually occur.

We cannot, however, regard this as a formal proof of the fact that regeneration does not take place in this case. This would be inadmissible, if only because the first of the three factors which, as we have assumed, produces the mechanism of regeneration — that is, the probability of loss — is not present. In the state of nature, at any rate, a bird's wing is seldom injured without loss of life ensuing at the same time. For this reason alone, selective processes in connection with a regenerative mechanism could not be introduced. I have not brought forward the above example for the purpose of proving the case for this instance in particular, but because it seemed to me to be specially fitted to show how extraordinarily the complexity of the regenerative mechanism must increase along with the greater complexity of the part. But this brings us back to the consideration of *the general power of regeneration possessed by the lower, in contrast to the higher, animals.*

The supposition that this power exists, may, I believe, be conceded in a certain sense: that is to say, in consequence of the slighter complexity in structure of *all* the parts in one of the lower groups of animals, any particular part may also become capable of regeneration more easily than in the case of the higher groups. We must, however, always presuppose that the two other factors — the probability of injury, and the physiological importance of the organ — are present in the required degree; so that in speaking of the greater power of regeneration possessed by animals of a lower type, we are only using another expression for the third factor which takes part in the process, viz., the complexity of the organ to be regenerated.

The question, however, arises as to whether the capacity of each part for regeneration results from special processes of adaptation, or whether regeneration occurs as the mere outcome — which is to some extent unforeseen — of the physical nature of an animal. Some statements which have been made on this subject seem hardly to admit of any but the latter explanation. Thus, according to Spallanzani, the jaw of a Triton may become regenerated along with its bones and teeth. Bonnet states that even the eye of this animal is replaced after it has been extirpated. It has never come before my notice that in the natural state Tritons frequently lose the lower jaw in combat; but some of these animals which I had put for a short time in a small vessel attacked each other vigorously, and several times one of them seized another by the lower jaw, and tugged and bit at it so violently that it would have been torn off if I had not separated the animals. The loss of part of the jaw or eye may therefore occur not infrequently in the natural state, and we may thus perhaps assume that these parts are adapted for regeneration. Kennel, moreover, gives an account of a stork, the upper beak of which had accidentally been broken off in the middle, the lower one then being sawn off to the same length, and both were subsequently regenerated. Such cases, the accuracy of which can scarcely be doubted, indicate that the capacity for regeneration does not depend only on the special adaptation of a particular organ, but that a general power also exists which belongs to the whole organism, and to a certain extent affects many, and perhaps even all, parts. By virtue of this power, moreover, simpler organs can be replaced even when they are not specially adapted for regeneration.

From our point of view, such cases are not incomprehensible in principle. We need only assume that in all, or at any rate in many, of the nuclear divisions in the embryo, some of the earlier determinants remain associated with later generations of cells as accessory idioplasm. It only remains to trace this arrangement — which is a more or less universal one, and affects the whole body — to its origin; for no arrangement can be produced which is not useful, especially when it concerns such a complicated mechanism as that for supplying the idioplasm with accessory determinants. We are therefore led to infer that *the general capacity of all parts for regeneration may have been acquired by selection in the lower and simpler forms, and that it gradually decreased in the course of phylogeny in correspondence with the increase in complexity of organisation*; but that it may, on the other hand, *be increased by special selective processes in each stage of its degeneration, in the case of certain parts which are physiologically important and are at the same time frequently exposed to loss.* In all probability this view is the correct one.

3. Facultative or Polygenetic Regeneration

The tail of a lizard or the limb of a Triton grows again when it has been cut off, but the part amputated does not reproduce the entire animal. In some segmented worms, on the other hand, such as *Nais* and *Lumbriculus*, not only does the amputated tail-end become restored, but this end itself reproduces the anterior part of the body, so that two animals are formed from one.

This fact evidently cannot be deduced merely from the assumption we have made with regard to supplementary determinants; for were this the case, determinants of one kind only — viz., those which are necessary for the construction of the lost part — would be present in the cells. But in the above instances the *same* cells give rise to entirely different parts, according to whether they are situated on the surface which is anterior or posterior to the plane of amputation: in the former case they reproduce the tail-end, and in the latter the head-end. The fact that both parts grow again when the worm is cut into two through any region of the body, proves that regeneration in either direction may proceed from the *same* cells; it therefore follows that the cells situated in any particular transverse plane of the body are not merely provided with the supplementary determinants for the formation of the head- or tail-end only, but every cell can react in either way, according to whether it is situated anteriorly or posteriorly to this plane. In order therefore to explain the twofold action of these cells in accordance with our fundamental view, — which presupposes that the cells taking part in regeneration are arranged and controlled by the forces situated within them, and not by an external agency, — it seems necessary to assume that each cell possesses two different supplementary determinants, one for the construction of the head-end, and one for that of the tail-end; and that the one or the other becomes active according to whether the stimulus, due to the exposure of the cell, is applied to its anterior or to its posterior surface.

Before attempting to verify this assumption, I must mention certain cases in which the regenerative activity of the cells may even be threefold.

It appears to me that the regenerative processes which have been observed in the fresh-water polype Hydra and in the sea-anemones (*Actiniæ*) are instances of this kind of regeneration. If a worm is cut through in the median or any other longitudinal plane, neither part grows again, and each soon dies. The case is different in Hydra. If this animal is cut through longitudinally, the two parts grow again into entire individuals, irrespective of the plane of section. As the transverse section of the animal at any point is likewise followed by the restoration of each part, it follows that Hydra, in every part of the body, must be capable of a threefold regeneration, *i.e.*, of regeneration in the three directions of space. And as the body is differently constructed in these three directions, we are compelled to assume that each cell contains groups of determinants of three different kinds, viz., those which are concerned in the formation of the proximal and distal ends, and in the completion of the body-wall. An individual cell * must therefore be capable of dividing in three different planes, and of giving rise to a part of one of three different regions of the body; and, moreover, the plane in which division actually occurs, and consequently the kind of determinants which become active and control the cell, is decided not by the quality, but by the kind of division resulting from the stimulus produced by the injury.

The processes of regeneration in Hydra can, I think, to a certain extent be understood on this assumption. If, for instance, the group of supplementary determinants of the proximal end of the body becomes active, it will cause the development of linear rows of cells extending in the direction of the axis of the body and united laterally so as to give rise to a tube; these cells, moreover, will have the tendency to close in towards the centre as soon as possible, so as to form the disc or foot, and will also cause the differentiation of the ectoderm cells of the foot into glandular cells which secrete slime: the determinants for the formation of tentacles are wanting in this group. If, again, the group of supplementary determinants of the distal end becomes active, rows of cells arise which will tend to close in to form the oral disc, leaving a large space in the centre for the mouth. Tentacles will then grow out from certain points around

* I shall not refer to the histological details with regard to the process of regeneration in Hydra, as the necessary data appear to be too uncertain and incomplete.

the mouth, and it is certainly not easy to explain why the determinants which cause their formation become active at these points only. It will, however, be shown later on that the cells of Hydra — and probably those of all animal tissues — are in a certain sense polarised; that is to say, they are differently constituted in the three directions of space. The fact that the determinants of the tentacles — which we must suppose to exist in all regions of the body — only become active in certain cells around the margin of the mouth, may be due to the polarisation of the cells as well as to the peculiar conditions of pressure within the cellular dome of the oral disc.

What has just been said can certainly not be looked upon as anything more than the merest provisional explanation of the facts, but it appears to me to be impossible to give a better one at present. It nevertheless, I think, penetrates somewhat further into the problem than does Herbert Spencer's hypothesis, in which regeneration is compared in general to crystallisation, and the capacity of arranging itself on every occasion under the influence of the whole aggregate in the manner required for the renewal of the missing part, is attributed to *every* ultimate particle. If we take the fresh-water polypes alone into consideration, one of these explanations seems just as good as the other; but if other groups of animals are included, it is at once apparent that this capacity is not by any means always possessed by the particles, but that even the cell may give rise by regeneration sometimes to various parts of the whole aggregate, at other times only to one certain part, and at others again only to those similar to itself, and that it must therefore contain something which makes it specially capable of one or of the other kind of regeneration. This something is the group of supplementary determinants.

If a polype or worm is cut through transversely, or if a loss of substance is caused artificially in any organism, the conditions of pressure previously existing in the cell in the region of the injury become changed, the pressure previously exerted by the lost part suddenly ceasing. This induces a change in the vital conditions of the cells thus affected, which must have a definite morphological and physiological result. We are unable at present to state more precisely what this change is; but as we know that such losses of substance are followed by the multiplication of the cells, we may safely assume that it exerts a stimulus on the cell, and more especially on its idioplasm, which forces the latter to undergo multiplication. This view is maintained by those who have the greatest opportunity of investigating the details of such processes, — I refer to the pathological anatomists. The proliferation which ensues in the surrounding tissue after a loss of substance, is not explained by them as being due indeed to a stimulus — in the ordinary sense of the word — exerted on the surrounding cells, but rather to a cessation of the '*resistance to growth*,' and this may in one sense also be described as a 'stimulus,' inasmuch as it is an 'incitement' to growth.

If the cells were constituted alike in the three directions of space, the effect on the idioplasm would be the same whether the stimulus due to the loss of substance acts from before, from behind, or from the side. *One* of the three groups of determinants could not possibly be alone effected by the stimulus and thus rendered active in one case, the second only in another, and the third only in a third instance. We have, however, every reason to suppose that the structure of one of these tissue-cells is not the same in the three directions of space, and that they are, in fact, variously differentiated according to each of these, and consequently respond to stimuli in different ways according to the direction in which the latter act upon them. Vöchting * has proved that at any rate in higher plants, 'a different upper and lower, anterior and posterior, and right and left half, can be distinguished in each living cell in the root and stem.' Portions of the root of the poplar transplanted on to the stem, or portions of the stem transplanted on to the root, only grew and flourished when they were fixed in a certain position; in the reverse position they sometimes indeed grew, but soon showed phenomena of degeneration. Vöchting infers from this observation that the cells are 'polarised,' this term being taken merely in an analogous sense to that in which it is generally used. The root and stem behave in a certain sense like a cylindrical magnet, which is composed of sections equally magnetised in the radial and longitudinal directions. Such a magnet, like the stem and root, may be separated into pieces. If the smooth adjoining surfaces of the portions of the magnet are

* H. Vöchting, 'Uber Transplantation am Planzenkorper,' Tubingen, 1889, p. 400.

placed with their opposite poles as close together as possible, the entire magnet is once more formed. Similarly, if the root of a poplar is cut in half transversely, each half produces buds and roots at the corresponding poles; but if, on the other hand, the two portions are joined together in the same relative position as that which they occupied originally, they become united together, so that a single piece of root, with its two poles, results, quite similar to the original piece.

These important results which Vöchting has obtained by his experiments on transplantation, are mentioned in this place because they can be utilised in considering the phenomena of regeneration in animals, which have just been discussed. We may in this respect compare a fresh-water polype with a poplar root. After a Hydra has been cut in half transversely, the distal portion gives rise to a new foot at its proximal end, and the proximal portion produces an oral region at its distal end. We might therefore in this case speak of pedal- and oral-poles, instead of root- and stem-poles, as in the case of the poplar. And, in fact, if a Hydra is cut transversely into three portions, the distal part or oral pole of the middle piece develops a new oral region, and its proximal part or pedal pole gives rise to a new foot. It might not be impossible for a clever experimenter to cause this middle piece to unite with the two terminal portions of the body before the former had had time to develop into a complete animal, by joining the three portions together with bristles. This would result in a union just as in the case of the poplar.

It would be a mistake to try to deduce that one of the poles of the poplar root must grow shoots and the other roots merely from the fact of its polarisation: one might as well try to deduce it from the fact of the polarisation of a real magnet. Something more is required before this can take place:—*the cells of the poplar root must contain the primary constituents for the formation of shoots and roots*; that state of the cells which Vöchting describes as polarisation only produces the conditions under which one or other of the primary constituents becomes active, and thus undergoes development. The hypothesis of the polarisation of the cells does not, therefore, relieve us in the least from the necessity of making a theoretical assumption to explain how it comes about that the primary constituents of different kinds of structures are present in one and the same cell. According to my view, we must assume in the case of the poplar root that the cells are provided with two different kinds of idioplasm, which remain inactive until the adequate stimulus arises and causes the idioplasm of either the root or of the stem to become active. In both cases the loss of substance must be regarded as the stimulus, and the direction in which it acts must decide the quality of the reaction.

If the idioplasm of the tissue-cells were capable in itself of responding to the effect of this stimulus by causing a regeneration of the missing parts of the body, worms possessing the regenerative power in a high degree, such as *Nais* and *Lumbriculus*, would be capable of regeneration in a lateral as well as in the anterior and posterior directions. This, however, as Bonnet has previously proved, is not the case: when cut in half longitudinally, the missing right or left half is not reproduced, and the cells of these animals must therefore be wanting in that substance — viz., in the *antimeral* supplementary determinants — which renders this kind of reproduction possible.

From our point of view, it is not surprising that these determinants are absent in worms; for in the natural state these animals are never torn in half longitudinally, and there was therefore no need for Nature to provide for such a contingency.

If we consider that the groups of supplementary determinants must become more complicated in proportion as the organism and the part to which they give rise increase in complexity, we can understand why *facultative* regeneration only occurs in relatively simple organisms, and that it apparently takes place in three dimensions in Polypes and Flat-worms only, in two dimensions in Annelids and Starfishes, and merely in one dimension in Arthropods, Molluscs, and Vertebrates.

It must not be supposed that other factors do not also take part in limiting the capacity for regeneration, — such as, in particular, the vulnerability of the higher organisms, and the fact that they are dependent on the circulation and temperature of the blood, even apart from the influence of the nervous system, of which we are practically still very ignorant. The relatively small quantity of substance in the part removed, as compared with that of the rest of the body, would also prevent the amputated limb of a salamander, for instance, from becoming regenerated into an entire animal. All these considerations help to explain why bi-dimensional regeneration —

that is, regeneration in two directions — cannot take place in the higher animals.

If, then, regeneration depends on the distribution of supplementary determinants to certain cells, which occurs whenever it is necessary or possible, the process must be primarily traceable in the case of the Metazoa to the doubling of the ids in a certain ontogenetic stage. And since a division and doubling of the idants takes place in every mitotic nuclear division, this hypothesis is supported by actual fact, even although we are still far from being acquainted even with the general details of the processes of growth and doubling of the ids and determinants, not to mention the systematic transference of such inactive determinants to definite cells and series of cells. Here again, however, Nature will have caused an advance from the simple to the more complex; and it therefore follows that, just as complicated organisms could only arise in the course of innumerable series of generations and species, so also the complex apparatus for regeneration in the tail or limb of a newt could not have been developed suddenly, but must have arisen in consequence of similar modifications in innumerable ancestors.

It might be possible to picture to one's self approximately the series of modifications which the apparatus for regeneration has gradually undergone, beginning at the lowest multicellular forms, and passing upwards to those animals in which the power of regeneration is the most highly developed and complex. I shall not, however, attempt to do so. At some future date it may perhaps be found that differences occur as regards the number of ids contained in the cells of those which have, and in those which have not, a marked capacity for regeneration: it will not be worth while to trace in detail the courses which the development of the power of regeneration has taken, until our knowledge of the idioplasm is sufficiently complete to furnish a basis for the theory in fact.

4. Regeneration in Plants

The process which may be described as regeneration in the case of the lower plants — the algæ, fungi, and mosses — will be treated of in greater detail subsequently. In this place, I merely wish to point out that true regeneration only occurs in a very slight degree in all the higher plants which are regarded as cormophytes or plant-stocks. If a piece is cut out of a leaf of a tree or of any other Phanerogam, the leaf does not become regenerated. If, again, an anther or a stigma is cut off from a flower, the corresponding filament or style will not give rise to a new anther or stigma. The cells of these organs are therefore not adapted for regeneration, and do not contain 'supplementary determinants.'

Botanists might be inclined to explain this fact by supposing it to be due to the cells having already reached their full size, and having therefore lost their power of multiplication. This is certainly the case, but it does not explain matters in the sense I mean: the question still remains as to why these cells have not been provided with supplementary determinants. The large number of cases in which adult cells of leaves or other parts, which have reached their full size, may under certain circumstances begin to multiply, and form buds from which entire plants arise (*e.g.*, *Begonia*), proves that such a provision is possible.

The solution of the above problem is to be sought for in the fact that it would have been of far too slight importance to the plant to be able to restore such defects in its leaves, as it possesses the power of producing *new* leaves. Buds can be formed and undergo further development in many parts, and thus the plant gains much more than it could possibly do by mere regeneration. *Regeneration can be dispensed with, as the far more important power of budding is possessed by the plant.*

The fact that the higher plants are unable to restore such parts as portions of leaves, furnishes an additional important proof that regeneration is dependent on external circumstances, and that it is a phenomenon of adaptation. True regeneration, however, occurs in those cases in which the losses or injuries would be harmful to the plant, and cannot be made good by the development of buds. Thus a loss of substance in the bark of a tree becomes replaced by the formation of callus, which arises from the edges of the wound, and grows over it, and thus the underlying wood is protected from injury. The cut or broken surface of a branch, even in the case of many herbaceous stems, becomes covered over in a similar manner by a mass of proliferating callus, which may even give rise to new growing points of shoots and roots, and thus become the place of origin of new individuals.* The stimulus to proliferation, as in the case of regeneration in animals, is due to the removal of the opposition to growth; the cells must, however, be adapted for this reaction, otherwise the proliferation cannot take place; the stems as well as the roots and veins of herbaceous plants do not by any means always respond to an injury by the formation of callus. This process is therefore not a primary quality of the plant, but an adaptation, due, in my opinion, to the association of certain supplementary determinants with the active idioplasm of certain kinds of cells.

The formation of callus is probably the only process in plants which can be regarded as an actual regeneration.

5. Regeneration in Animal Embryos, and the Principles of Ontogeny

The theory of heredity which has now been formulated, — and more especially that portion of it which concerns the composition of the germ-plasm out of determinants, and the gradual disintegration of the mass of determinants in the germ-plasm during the course of ontogeny, — is based on the assumption that the *cells control themselves:* that is to say, the fate of the cells is determined by forces situated within them, and not by external influences. The primary cells of the ectoderm and of the endoderm arise by the division of the fertilised egg-cell and its contained germ-plasm, because the determinants of the ectoderm are passed into one cell and those of the endoderm into the other, and not because some external influence, such as the force of gravity, affects the cells in a different manner. Similarly a certain cell in a subsequent embryonic stage does not give rise to a nerve-, a muscle-, or an epithelial-cell because it happens to be so situated as to be influenced by certain other cells in one way or another, but because it contains special determinants for nerve-, muscle-, or epithelial-cells.

This conception of the predestination of the individual cells, the fate of which, together with that of their successors, is determined by the idioplasm they contain, was first imperfectly expressed in the theory formerly propounded by His,† in which he

* J. Sachs, 'Lectures on the Physiology of Plants,' Leipzig, 1882, p. 709. (English edition, translated by H. Marshall Ward, Oxford, 1887.)

† Wilhelm His, 'Unsre Korperform u. das physiologische Problem ihrer Entstehung,' Leipzig, 1874.

formulated the existence of 'special regions in the germ, which give rise to special organs.' His imagined that the 'primary constituents of the organs of a chick were present in superficial extension in the germinal disc,' *i.e.* in the cell-body of the ovum, and that each organ is therefore represented by a definite part of the body of the egg. As has already been mentioned in the historical introduction, subsequent investigations, made in the course of the following ten years, proved that the 'primary constituents' of the various structures are to be found in the nuclear substance. The special form in which His expressed his views was thus certainly contradicted, although the fundamental principle of his theory was not thereby affected in its general sense, which indicates that the differentiating principle of ontogeny is to be looked for in the cells themselves, and not in external influences. Wilhelm Roux * was the first to prove definitely that the differentiation of the egg into the embryo is certainly not caused by influences existing apart from the egg, but that it is due to causes originating in the egg itself. Pflüger † showed with regard to the ovum of the frog, that whatever position the egg is forced to take up the upper side always gives rise to the animal pole of the embryo, and it was thought that this must be due to the force of gravity. Roux, however, proved that frogs' eggs which are rotated slowly in a vertical direction, develop just as well as those on which the force of gravity is not interfered with. It has further been proved by Born ‡ that, although when an egg undergoes development in a fixed position the substance of the cell-body does not become displaced at first, the nucleus nevertheless changes its position, for it very soon passes to the upper pole of the egg, at which point development then begins. These observations undoubtedly proved that the formative forces are situated in the egg itself; but they still left it undecided whether the differentiation of the ovum is due essentially to the action of the individual cells alone,—that is to say, whether differentiation occurs independently in each individual cell, so that it would, if necessary, be capable of passing through

* Wilhelm Roux, 'Beitrage zur Entwicklungsmechanik des Embryo,' München, 1885.

† Pfluger, 'Ueber den Einfluss der Schwerkraft auf die Thielung der Zellen u. auf die Entwicklung des Embryo,' Arch. f. Physiol., Bd. xxxii., 1883, p. 68.

‡ Born, 'Biologische Untersuchungen,' (I.) Arch. f. mikr. Anat. Bd. 24.

its prescribed course of development apart from the rest of the embryonic cells, — or whether the various cells of the embryo become differentiated by their mutual interaction: or, in other words, whether a determinating influence is to a certain extent exerted by *the whole on its parts* and thus prescribes the fate of the various cells.

The experimental proof of the self-differentiation or predisposition of the individual cells was, I believe, furnished by Roux,* whose ingenious experiments are always accompanied by keen deductions. Roux destroyed a single segmentation-cell in each of a series of frogs' eggs by means of a hot needle, and then observed that eggs treated in this manner developed into 'half or three-quarter embryos,' that part being absent which corresponded to the cell thus destroyed. When one of the first two segmentation-cells was demolished, half of the embryo was formed, and this corresponded either to a lateral or to the anterior or posterior half, according to whether the first segmentation had resulted in a division of the hereditary substance into portions belonging to the right and left, or to the anterior and posterior halves. The process of segmentation in the frog is known to vary in this respect. When one of the first four segmentation-cells was destroyed, three-quarters of the embryo was formed.

These experiments must be regarded as affording a proof of the self-differentiation of the cells. Observations have since been made which seem to contradict this deduction; and although these are still incomplete, and can only be regarded as the preliminaries to more detailed investigations, they must not be passed over in silence, especially as I am convinced that they do not really contradict the hypothesis of the self-determination of the cells.

Chabry's † experiments on the eggs of Ascidians must be mentioned first. By means of a special apparatus, he destroyed one of the first two segmentation-cells, and then observed that the remaining cell continued to develop, and eventually gave rise, not indeed to half an embryo, but to an *entire* one of half the normal size. Such embryos were certainly not quite

* Wilhelm Roux, 'Beiträge zur Entwicklungsmechanik des Embryo,' (V.) Virch. Arch. Bd. 94.

† L. Chabry, 'Embryologie normale et tératologique des Ascidies,' Paris, 1887.

perfect, but only organs of slight importance were wanting in them. Chabry himself has drawn no theoretical conclusions from his observations; Driesch,* however, has made certain deductions from a series of similar experiments on the eggs of Sea-urchins. By continued shaking, Driesch effected a mechanical separation of the two first segmentation-cells, and observed that at first each of them continued to undergo further segmentation just as would occur in the entire egg, but that later on the resulting *hemi*-blastula became completed to form an entire one. In some of these hemi-blastulæ development proceeded still further, the invagination taking place to form the primary digestive cavity of the gastrula, so that eventually a rudimentary pluteus-larva — which, though small, was in other respects normal — could be recognised.

Driesch sums up his results in the following words: — 'These experiments therefore show that under certain circumstances each of the two first segmentation-cells of *Echinus micro-tuberculatus* can give rise to a larva of the normal form, which is *entire* as regards its shape; and that a *partial formation*, and not a *semi-formation*, occurs in this case.' The author concludes that his results 'fundamentally disprove the existence of special regions in the germ which give rise to special organs,' and adopts the following view stated by Hallez †: — 'Il n'est pas dès lors permis de croire que chaque sphère de segmentation doit occuper une place et jouer un rôle, qui lui sont assignés à l'avance.'

Although I am far from wishing to assert that we are at present in a position to give a perfectly reliable and detailed explanation of the extremely interesting and important results of the experiments just described, I nevertheless cannot help thinking that they do not in the least necessitate the giving up of the view which entails a predestination of the individual segmentation-cells, and, in fact, of cells in general. Other than experimental methods may lead us to fundamental views, and an experiment may not always be the safest guide, although it may at first appear perfectly conclusive. Even Driesch himself doubts whether the above-mentioned experiments made by Roux are really conclusive, though, in my opinion, he is wrong

---

* H. Driesch, 'Entwicklungsmechanische Studien, Zeitschrift f. wiss. Zoologie,' Bd. 53, 1891.

† Hallez, 'Recherches sur l'embryologie des Nématodes,' Paris, 1885.

in doing so: he asks, in fact, whether the uninjured segmentation-sphere of the frog would not behave exactly in the same manner as that of the sea-urchin if it could be *actually isolated*, instead of remaining in close connection with the other injured sphere. Thus even the apparently incontrovertible result of this experiment may be doubted.

It seems to me that careful conclusions, drawn from the general facts of heredity, are far more reliable in this case than are the results of experiments, which, though extremely valuable and worthy of careful consideration, are never perfectly definite and unquestionable. If what was said in support of the theory of determinants in the first chapter of this book be borne in mind, the conviction that ontogeny can only be explained by evolution, and not by epigenesis, seems to force itself upon us. It would be impossible for any small portion of the skin of a human being to undergo a hereditary and independent change from the germ onwards, unless a small vital element corresponding to this particular part of the skin existed in the germ-substance, a variation in this element causing a corresponding variation in the part concerned. Were this not the case, 'birth-marks' would not exist. If, however, determinants are contained in the germ-plasm, these can only take part in controlling the formation of the body if, in the course of embryogeny, they reach those particular cells which they have to control,—that is to say, if the differentiation of a cell depends primarily *on itself*, and not on any external factor.

If therefore ontogeny is not, as Roux aptly expresses it, a '*new formation*' of multiplicity, or an epigenesis, but is merely the *unfolding* of multiplicity, *i.e.* an evolution,—or, as it might also be called, *the appearance of a previously invisible multiplicity*,—the principle of self-determination is certainly only established with regard to the egg *as a whole:* the self-determination of each cell, and its control of ontogeny, do not necessarily follow from this conclusion. We can only thereby arrive at the very simple assumptions, that the primary constituents of the germ-plasm are distributed by means of the processes which can actually be observed in the nuclear divisions, so that they come to be situated in those regions which correspond to the various parts of the body, and that those primary constituents are present in each cell which correspond to the parts arising from it.

As has just been shown, it is also possible to make the reverse hypothesis, and to suppose that although the whole of the idioplasm is contained in each cell, only that particular primary constituent which properly concerns the individual cell has any effect upon it. The activity of a primary constituent would thus depend not on the idioplasm of the cell, but on the influences arising from all the cells of the organism as a whole. We should thus have to suppose that each region of the body is controlled by all the other regions, and should therefore practically be brought back to Spencer's conception of the organism as a complex crystal. This simply means giving up the attempt to explain the problem at all, for we cannot form any conception of such a controlling influence exerted by the whole on the millions of different parts of which it consists, nor can we bring forward any analogy to support such a view, the acceptance of which would render a great number of observations on the phenomena of heredity totally incomprehensible. What explanation, for instance, could be given of the fact that a certain human birthmark is always inherited on the left side only? According to this hypothesis, the germ-plasm contained in the cells of this region would be present on the right side just as much as on the left: as the two halves of the body are alike in other respects, we cannot suppose that the whole aggregate exerts different influences as regards this region on the left and on the right sides.

It seems to me, therefore, that we must not give up the hypothesis of the self-determination of the cells, in spite of its apparent refutation by the facts described by Chabry and Driesch. Moreover, I think these facts can be explained—in principle at any rate—in another manner, viz., *by attributing the processes observed to regeneration*, the arrangement for which, however, has not been provided for the first stages of segmentation, but for a later period of ontogeny.

It is hardly to be expected that the first stages of segmentation should be in a sense *purposely* arranged for regeneration. Both in Ascidians and sea-urchins the number of eggs produced is so large, that it probably matters very little whether a segmenting ovum perishes or becomes regenerated when one half of it has been eaten by a small enemy. I do not, however, wish to do away entirely with the idea that the eggs of certain animals may conceivably be protected in this manner from numerous enemies, but in this place I must refrain from including such a possible occurrence in the argument.

The following explanation of the phenomena, however, still remains. The first division of the ovum separates the group of determinants into two, viz , that for the right and that for the left half of the body; each of these groups does not constitute a perfect germ-plasm, as each determinant it contains is not doubled; but it is very probable that the ids are capable under certain circumstances of dividing in such a way that each becomes doubled. Such a germ-plasm could not contain *in potentia* a birthmark, or any other asymmetrical peculiarity of the other side of the body, but it would be able to give rise to a complete animal. The destruction or mechanical removal of one segmentation-cell in the first stage of segmentation may be the primary cause of the doubling of the ids in the other cell.

The capability of becoming doubled, which the undivided germ-plasm possesses in certain cases, may be mentioned in support of this view of the regeneration of an isolated cell in the first stage of segmentation. The fact that in each integral division of the cell and nucleus, a longitudinal splitting of the nuclear rods and their contained macrosomes occurs, shows that the ids are as a rule capable of growth and of doubling their number by division. The assumption of a doubling of the ids of germ-plasm must be made in dealing with the origin of *identical* twins, *i.e.* those twins in which we must suppose that the division of the nucleus of the ovum from which they arise occurs *after* and not *before* fertilisation; for otherwise the embryos could not be identical, as two spermatozoa would then take part in the process. In the case of facultative parthenogenesis, a doubling probably also occurs in the ids and idants of the ovum, half of them having previously been removed by the 'reducing divisions.'

The formation of an entire embryo by the regeneration of one of the two first blastomeres admits, however, of another interpretation. Ascidians multiply very freely by budding, and not only by sexual reproduction. It is true that this is not the case with sea-urchins, but the power of regeneration which these animals possess is unusually great. This fact was explained in the present chapter by assuming that certain idic stages of ontogeny are provided with an 'accessory idioplasm,' consisting of the determinants required for regeneration. In a subsequent chapter I shall have occasion to show that we must make a similar assumption in the case of budding. Such assumptions are indispensable if we accept the hypothesis of the germ-plasm and determinants. The accessory idioplasm required for budding causes the reproduction of the entire animal, and must therefore contain all the determinants of the germ-plasm, and must exist in the ovum before segmentation, remaining in a latent condition in a definite series of cells during all the stages of development. If now this accessory idioplasm were capable of becoming active under certain abnormal influences, — such as that produced by the destruction of the other blastomere, — a regeneration of the whole embryo might thus result.

These explanations are, however, only possible ones, and I should not have been sorry to leave them out of consideration altogether, for I am fully aware of their incompleteness and unreliability: I merely wish to show that the observations mentioned above do not render an explanation impossible, even although we are not able at present to state that any particular interpretation of the phenomena is the correct one, because the observations themselves are far too incomplete and deficient. For this reason I shall not attempt to give a more precise explanation of the peculiar development of these embryos.

I must, however, draw attention to the different behaviour of the eggs in the case of the frog and in that of Ascidians and sea-urchins. Leaving aside the question of 'post-generation,' we have seen that only half an embryo arises from *one* blastomere of a frog's ovum, while an *entire* animal becomes developed from one blastomere in the case of either of the other two animals mentioned. However imperfect the explanation I have offered may be, the fundamental assumption on which it is based must in general be a correct one, — viz., that the first blastomeres of the egg of an Ascidian or sea-urchin must possess a capacity which is absent in the case of the frog's egg. As, however, forces are dependent on substances, it is probable that the blastomeres of an Ascidian and of a sea-urchin contain an excess of substance — *the accessory idioplasm* — which gives them the power of regeneration, and that this substance is wanting in the blastomeres of the frog. Driesch, as already mentioned, expresses a doubt as to whether the blastomere of a frog would not behave in a similar manner to that of a sea-urchin, if, like the latter, it could be completely separated and isolated from its injured fellow blastomere. This doubt seems, however, to be hardly justified, as such an isolation was not effected in Chabry's experiments on the ascidian ovum, but nevertheless the development into a complete animal ensued just as in the case of the egg of the sea-urchin.

Although the half of a frog's egg develops into half an embryo only in the first place, the latter may subsequently become completed by a very peculiar regenerative process, which was first observed by Roux in 'half' and 'three-quarter embryos,' and which he designated as 'post-generation.'

Roux observed that a segmentation-cell of a frog's egg may be 're-animated' after it has been deprived of its capacity for development. A considerable number of nuclei pass into the vitellus of the injured part from the normally developed half of the egg, and there increase and give rise to cells. 'The post-generative formation of the germinal layers takes place from the cell-material subsequently formed, while the process of differentiation continues to advance in the quiescent cell-material.' Roux thought he observed that a complete restoration of the embryo may take place in this manner, so that it can continue to live; and, in fact, he actually succeeded in keeping such an embryo alive for some time.

Considerable attention has naturally been drawn to these observations, which are certainly of the greatest interest; but I doubt whether in their present state they are sufficiently complete to form the basis of fundamental theoretical conclusions. With all respect for Roux's accuracy of observation and skill in research, I cannot help thinking that the half embryos which were subsequently 'post-generated' to entire animals, were possibly those in which the thrust with the hot needle had not affected the nucleus of the segmentation-cell.

In any case, it was only possible to observe the actual effect of the operation and its result on the whole series of processes which followed, and which led to the restoration of individuals *other* than those which ultimately became complete. To pierce a segmentation-cell with a hot needle must be a tolerably rough operation, and something different may be destroyed each time it is performed: not only the nuclear matter as a whole, but also the *individual idants*, might possibly remain uninjured. The idants, again, might subsequently increase to the normal number by doubling, and so bring about the development of the half of the egg. Roux certainly states that 'post-generation' does not occur in the same manner as does the normal development of the two primary halves, — that is to say, the germinal layers are not formed independently in each; but the processes which take place in the interior of the ovum can only be followed out by means of sections, the preparation of which necessitates the killing of the embryo.

In such experiments, moreover, no two cases are alike, and it would be necessary to examine a very large amount of material before stating with any degree of certainty that the egg which has been cut into sections, and that in which the development and post-generation have been followed out, have a precisely similar internal structure.

Roux observed a 're-animation' of three different kinds in the halves of the eggs operated upon, one of which consisted in a growth of the cells in the external layer of the living half around the dead half. In this instance, however, post-generation did not result: it only occurred in *certain*, but not *all*, of those cases mentioned above in which nuclei passed from the living half into the part which had been operated upon, *and in which only slight pathological changes had occurred in the yolk*. It is therefore natural to suppose that post-generation only occurred when the injury was a slight one, and when some nuclear matter remained and subsequently caused a formation of cells. This, however, does not imply that living 'nuclei' did not penetrate into the injured half of the egg; the segmentation-cells, even in normal development, have to undergo an enormous increase, and it is therefore not surprising that after the opposition to growth has been removed by the operation on the other half of the egg, they should increase at the expense of the latter. In those cases in which the other half of the embryo was subsequently completed, this completion must have resulted from a kind of infection of the cell, of such a nature that mere contact with ectoderm or mesoderm cells, for example, caused the undifferentiated cells of the injured half of the egg to become correspondingly differentiated into ectoderm and mesoderm cells. But I could only accept such a revolutionary hypothesis as this if it could be proved by incontestable facts.

Roux himself has, however, only looked upon his contributions to this subject as 'a first instalment of a large work,' and has led us to expect a continuation of his experiments. But as long as the processes which he describes admit of more than one interpretation, we cannot reject the hypothesis of the predestination of the cells by means of the distribution of certain determinants and groups of determinants to them, for this view is supported by so large a number of facts, and even by the earlier experiments of Roux himself. It would certainly, however, have to be rejected if we could prove that the cells of the germinal layers were really capable of being determined in their nature by the region which they accidentally reach, or by their accidental surroundings.

Further research along the line opened up by Roux will, I am convinced, show us the facts in another light, and will enable us to reconcile them to the rest of our conceptions as to the causes of ontogeny. But I do not consider it worth while at present to enumerate all the possible causes which must be taken into account in an attempt to explain 'post-generation.'

CHAPTER III

MULTIPLICATION BY FISSION

1. Preliminary Remarks

Until a short time ago the process of multiplication by gemmation was looked upon as having been derived phyletically from the corresponding process by fission, and the two were thought to be closely related, and connected by gradual transitions. Von Wagner * has, however, recently attempted to contest this opinion, and to show that the two processes should be more distinctly separated from one another than they have hitherto been, and that they are, in fact, genetically distinct. By the term (fission), Wagner means to indicate a process of multiplication which is preceded by a symmetrical growth of the parent organism, by means of which the individuality of the latter becomes changed and to a certain extent abolished: the term (gemmation), on the other hand, he takes to mean a process of multiplication which is preceded by an unsymmetrical (differential) growth of the parent organism, in which the individuality of the latter is not abolished and its place taken by a new individual.

This view I agree with in so far as I am convinced that in multicellular organisms the processes of multiplication by fission and budding have not arisen *genetically* from one another: these processes differ so essentially that it will be advisable to discuss them separately.

Following von Wagner's example, I shall include under this head of fission all the processes of asexual multiplication which occur in the flat-worms (*Turbellaria*, *Cestoda*), the annelids proper (*Syllidæ*, *Naidæ*, *Tubificidæ*, &c.), and also in the higher Medusæ (strobilation). In all these cases multiplication is effected by the division of the parent animal into two or more

---

* Franz von Wagner, 'Zur Kentniss der ungeschlechtlichen Fortpflanzung von Microstoma nebst allegemeinen Bemerkungen über Thielung u. Knospung im Thierreich,' Zool. Jahrbücher, Abth. f. Anat. u. Ontogenie, Bd. iv., Jena, 1890.

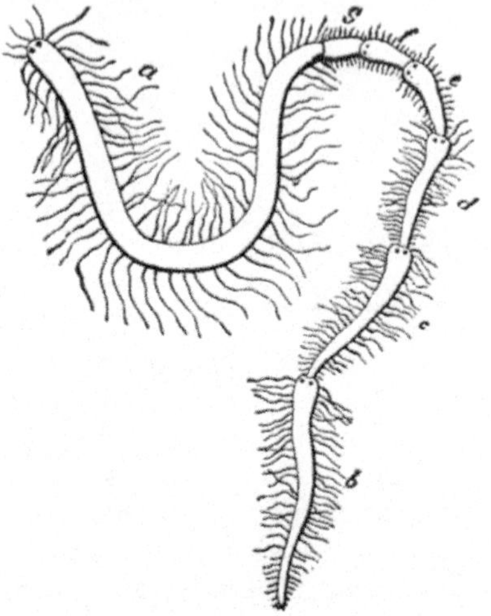

FIG. 4.—*Myrianida*, a marine worm which multiplies by fission (after Milne-Edwards, from Hatschek's 'Lehrbuch der Zoologie'). The letters *a*—*g* indicate the relative ages of the daughter-individuals resulting from the division of the parent.

parts: this therefore necessitates a regeneration of one or of the other end of the body, or even of both ends. This process may begin to take place either after (*Lumbriculus*) or before the division has taken place, and in the latter case is more or less complete before the fission begins. The actual process of the formation of the new organism is essentially the same in both cases, and important differences only occur as regards the various groups of animals.

We are particularly well acquainted with these processes of regeneration — which may either precede or succeed fission — in the case of various kinds of worms, and we will therefore first illustrate them in their main features by reference to these animals.

2. THE PROCESS OF FISSION IN THE NAIDÆ

The process of fission in these small fresh-water segmented worms has been very accurately followed out by Semper. An individual undergoes division into two, or usually into several, daughter-individuals at the same time, the fission being regularly preceded by a circular growth of cells taking place around the circumference of the body at one or more definite regions, each of such ring-like thickenings eventually giving rise to a new head- and tail-end respectively. These rings of cells have hitherto been spoken of as 'zones of gemmation;' but it would be better to call them 'zones of regeneration,' as they are not concerned with budding in the true sense of the word. Two of these rings are as a general rule formed in each animal, and when the anterior and posterior ends of each of the resulting three sections are fully developed, the separation into the corresponding three daughter-individuals takes place by a construction in the middle of each zone of regeneration.

In *Nais* the zones of regeneration are always formed at the boundary line of two segments: that is to say, they arise from the contiguous margins of two segments, in the following way. Cells of the epidermis first begin to multiply, and give rise to a circular layer of small stratified cells, which is thickest on the ventral side. The cells have at first no definite histological character. At the same time an increase in the length of the internal organs takes place: this is rendered necessary by the growing zone penetrating between the segments from which it arises and thus forcing them apart. The alimentary canal, however, is the only internal organ which becomes regenerated from its own cells: all the other new formations, including the ventral nerve cord, muscles, blood-vessels, 'liver'-cells, and excretory-organs, are developed from the ring of proliferating epidermic cells.

As Semper has pointed out, the process of the reconstruction of the anterior and posterior ends which prepares the way for fission, may in a sense be compared with the embryonic development of the animal subsequent to the gastrula stage, in which the two primary germinal layers are already distinct.

In these regenerative processes two layers of formative cells are likewise produced, owing to the proliferation of the ectoderm cells on the one hand, and those of the alimentary canal on the other; the epithelial lining of the latter only is formed from the internal layer, the outer layer giving rise to all the other organs, including the mesodermic structures as well as those which belong to the ectodermic part of the integument. In fact the resemblance between the processes which take place in embryogeny and in regeneration is so close, that in both cases the mesoderm becomes split off from the mass of formative ectoderm cells in the form of two longitudinal bands, from which the blood-vessels, muscles, &c., are then differentiated.

In order to explain these processes theoretically from our point of view, we must suppose that those cells of the epidermis from which the formative cells arise possess an '*accessory idioplasm*,' containing the determinants of those organs which are formed from them in regeneration in addition to their own specific idioplasm. The rate of division of each of these cells. as well as the manner in which the groups of determinants contained in them becomes disintegrated in the course of the subsequent divisions, is strictly definite, and determines the number of successors which each cell produces, as well as the relative position and combination into organs, and histological differentiation of the cells. When the process of proliferation begins, the newly-formed cells no longer retain the specific epidermic character, and their successors may, indeed, be said to possess an 'embryonic character,' in the sense in which that term has usually been used, — if it is not thereby understood that they must all contain similar primary constituents. Their further development shows that this cannot be so: the cells of one particular region give rise, for instance, to the dorsal vessel; those of another to the nerve-cord; those of a third region to certain muscles, and so on.

We must therefore suppose that the various epidermic cells of the parent animal are provided with active accessory idioplasm, somewhat in the manner I have indicated in the accompanying diagram. The cells marked *n*, for instance, would contain the groups of inactive determinants for the formation of the ventral nerve-cord; those of the epidermis marked *m*, groups of determinants for the mesoderm structures, in addition to their own proper idioplasm, — m^1 containing those for the lateral muscles, m^2 those for the ventral blood-vessel, m^3 those for the 'liver-cells' and mesodermic part of the intestine, m^4 those for the segmental organs and the adjacent system of muscles, m^5 those for the dorsal vessel, m^6 those for the dorsal muscles, and so on. For the sake of simplicity I have supposed that the epidermis consists of only *one* layer of cells, although in reality there are two layers in many parts: the diagram, in fact, is not by any means intended to represent the actual structure of the

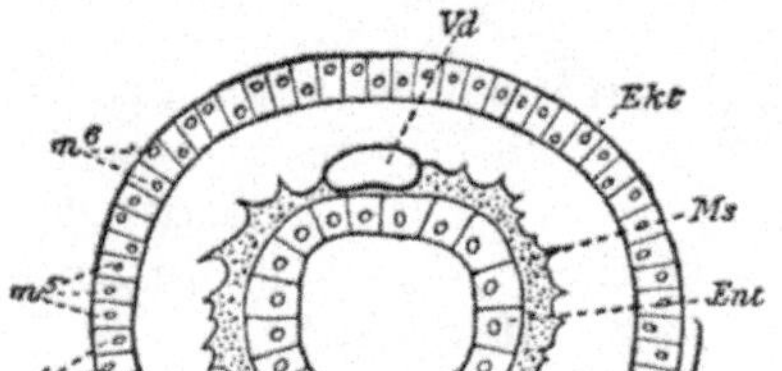

Fig. 5. — Transverse section through a Nais in the region of the zone of regeneration (modified from Semper): *Ekt*, Integument; *Ent*, Epithelium of the alimentary canal; *N*, Nerve-cord; *Ms*, Visceral mesoderm; *Vd*, Dorsal blood-vessel; *m*, Cells with accessory determinants for the mesoderm; *n*, Cells with accessory determinants for the nerve-cord.

animal in detail, or even to indicate accurately the part which the individual cells take in the process of budding.

The question of the *origin of the supplementary determinants* which we have assumed to exist in the cells of the epidermis, does not stand in the way of this explanation of the regenerative process; for, as already stated, a similar course is in the main followed both in embryogeny and regeneration. In both processes the primary mesoderm arises from the primary ectoderm. The definitive ectoderm cells have therefore an opportunity during their embryonic development of taking over certain primary mesoderm determinants as accessory idioplasm from the primary ectoderm cells: these can then become separated into several groups during the multiplication of the ectoderm cells, so that the epidermic cells around the circumference of the body are provided with an accessory idioplasm consisting of various mesoderm determinants.

It must, however, also be borne in mind that the growth in length of the worm only takes place at the posterior end of the body, just as occurs in the regeneration which prepares the way for division. In both cases a new body-segment is formed between the last segment and the last but one, in which process the epithelium of the intestine alone arises from the endoderm, the integument and all the mesodermic structures being formed from the ectoderm. Thus the accessory determinants which we have assumed to exist in the epidermic cells, and which render the subsequent regeneration possible, are not derived from the embryo directly, but from the zone of growth in the tail-end, into which again they have passed during embryogeny.

3. The Process of Fission in the Microstomidæ

It is not, then, in the nature of every ectoderm cell to give rise to all the possible kinds of cells and organs with the exception of the epithelium of the alimentary canal: each one must be specially equipped for the purpose. This is proved by the fact that the ectoderm by no means always performs this function in animals which multiply by fission: even in some worms this is not the case.

According to von Wagner's * excellent researches, the cells of the epidermis in a certain flat-worm, *Microstoma lineare*, take

* *Loc. cit.*

a very small share in the reconstruction of the anterior and posterior ends of the animal during the process of fission, the restoration being effected in this case chiefly by the mesoderm, or so-called 'connective tissue' cells, which 'are suspended in large numbers in the perivisceral fluid between the supporting trabeculæ.' These cells begin to increase in number when the animal is preparing for division, and by their multiplication they form a ventral mass of so-called 'embryonic' cells, which gives rise to the pharynx, the pharyngeal and prostomial glands, all the parts generally known as 'parenchymatous' or 'mesodermic structures,' and also apparently to certain parts of the nervous system. Kennel * found that a similar mode of development of these parts occurs in a Planarian. In such cases we must therefore suppose that the accessory determinants required for regeneration are supplied to the mesoderm cells, instead of to those of the ectoderm. We cannot at present determine whether this is effected by each of these cells being provided with all the supplementary determinants for the mesoderm, which only become disintegrated and distributed amongst the other cells when these begin to multiply, or — as we assumed in the case of the ectoderm cells of *Nais* — by the distribution of the different determinants to a number of these cells *before* proliferation occurs.

The regularity with which all organs are formed in the proper position and mutual relation, may perhaps be taken as a proof of the assumption that they contain latent determinants which are from the first separate, and which differ according to the topographical position of the organ. It is hardly possible that the contrary assumption can be the correct one, for this would render it necessary to suppose that although all the determinants are certainly present in every formative cell, only that one can undergo development which corresponds to the region in which the cell happened to be situated.

Here, again, we meet with no serious difficulty as regards the derivation of the required supplementary determinants in ontogeny: in fact there is less difficulty in this case than in that of *Nais*, for the cells of the different layers of the body contain the determinants for the *corresponding* organs.

---

* J. Kennel, 'Untersuchungen an neuen Turbellarien,' Zool. Jahrbucher, Bd. 3. Abth. f. Anat. u. Ontog. d Tie e, p. 447.

4. THE PHYLOGENY OF THE PROCESS OF MULTIPLICATION BY FISSION

There can be little doubt that the process of spontaneous division which occurs in flat-worms and in annelids is to be derived phylogenetically from regeneration, as Kennel* has recently attempted to prove. He has rightly, I believe, shown that multiplication by a spontaneous separation into parts, such as occurs regularly in the freshwater worm *Lumbriculus*, must be looked upon as a preliminary stage of that kind of fission, accompanied by regeneration, which occurs in the *Naidæ*, for instance. The difference between the two processes consists essentially in the fact that in *Nais* the separation into parts is preceded and prepared for by the formation of new head- and tail-ends, which appear between the old segments at the point at which the separation is to take place. Such a preparatory process does not occur in *Lumbriculus*: the region in which division will take place in this worm cannot previously be distinguished, and the new head- and tail-ends are formed subsequently, after the division has occurred.

The capacity for division of an individual into parts must naturally be looked upon as an adaptation, and it presupposes some kind of histological and physiological arrangement of which we are at present ignorant. It is, however, quite conceivable that when fission had once occurred in a species, it may have been advantageous for a more thorough preparation for the process to take place, and for the structures necessary for the completion of the individuals thus formed to become developed beforehand. Such a capacity for multiplication by spontaneous division necessitates, moreover, the previous possession of the power of regeneration. Hence the latter must have existed in the animal before spontaneous division could take place regularly in the species, and we must thus conclude that the capacity for regenerating portions of the body which had been accidentally torn asunder was first acquired very early in the philogeny of multicellular animals; and that the special arrangements for multiplication by fission subsequently originated from this capacity for regeneration, and was followed by the formation of new head- and tail-ends. The formation of the new parts pre-

* J. Kennel, 'Uber Theilung u. Knospung der Thiere,' Dorpat, 1888.

viously to the division must be looked upon as a still later modification of the process.

This conclusion receives further support from the fact that, as already shown, the capacity of regeneration is not by any means an inherent quality of the organism: that is to say, it is not a direct and inevitable result of a particular degree of organisation, but is due to an adaptation produced by natural selection, and constitutes a special arrangement which may exist in different degrees of perfection, or which may, again, be entirely absent. If an earthworm is cut into two, the anterior portion develops a new tail-end, but the posterior portion does not give rise to a new head-end: the arrangement existing in *Lumbriculus* and *Nais* is therefore absent in this case. This fact I should explain by assuming that in the last-named animals the determinants required for the formation of the head-end are supplied to the cells of the integument and alimentary canal as accessory idioplasm, while in the earthworm these cells only possess the determinants required for the formation of the tail-end.

It is very possible that the arrangement for the regeneration of the tail-end may have taken place more easily than that for the restoration of the head in the case of segmented worms, owing to the fact that the last segment possessed the power of giving rise to entire new segments. The growth of the animal is effected by the formation of new segments at the posterior end of the body, which would therefore be already provided with the requisite accessory determinants, and it would then only be necessary that these should be transferred to the corresponding cells of the other body-segments. This might have taken place in a relatively simple manner in the course of philogeny, by a portion of the accessory determinants being left in the cells of each new body-segment as it became formed. The determinants of the head-end, on the other hand, can only have been supplied to the respective cells as accessory idioplasm before or during embryonic development; we can therefore understand why the capacity for forming a new head-end was only acquired later, and that some worms are able to regenerate the posterior, but not the anterior end by the body when it is cut in half.

We can therefore trace a series of stages of gradually increasing complexity in the development of the process of regeneration in worms, beginning with the formation of segments at the growing tail-end, and then passing from the regeneration — first of the tail, and then of the head-end, to the actual fission of *Lumbriculus*, and finally to that of *Nais*. And according to our view, this course of development depends on the regular distribution of certain accessory determinants to particular tissue-cells, and the gradual increase in the complexity of this distribution.

The regenerative process which renders fission possible must be traced to the doubling of certain groups of determinants in the idioplasm, so that the half of them remain latent. I imagine that this doubling need not necessarily take place and be followed by the subsequent multiplication of the inactive groups of determinants in the germ-plasm itself: such a multiplication would, in fact, be a useless encumbrance to the germ-plasm. The latter need only contain the determinants for fission when this process leads to an alternation of generations; that is to say, when the animals formed by division have a different structure from those which arise directly from the ovum: for in the latter case the forms which arise by fission are independently variable hereditarily. This must be the case in the alternation of generations in certain marine annelids, such as the *Syllidæ*, and also in the strobilation of polypes, which will be discussed later on. In all such cases, two kinds of ids must be assumed to exist in the germ-plasm. In the ordinary kind of fission which occurs in the fresh-water annelids, on the other hand, the separation of the groups of determinants necessary for the regeneration of a part may take place during embryogeny: nothing definite, however, can be said on this point at present, but in any case the process of splitting off of the accessory determinants may conceivably be thrown back from the later to the earlier stages of ontogeny, until it finally takes place in the fertilised ovum, so that double ids are present in the germ-plasm. It will be assumed in the next section that certain forms of budding owe their origin to the presence of these double ids.

CHAPTER IV

MULTIPLICATION BY GEMMATION

1. The Process of Gemmation in Animals

If, with von Wagner, we look upon gemmation as 'a process in which entire individuals are formed anew,' and which depends 'exclusively on a special (differential) growth differing from the normal one,' we must include under this term the processes of asexual multiplication which occur in most of the *Cœlenterata*, the *Polyzoa*, and the *Tunicata*.

A.—Cœlenterata

Hitherto it has been considered that we were fully acquainted with the process of gemmation in the *Cœlentarata*, especially in the case of the *Hydrozoa*. It had been observed that the two layers of cells which form the body-wall of these animals are present even in very young buds of Medusæ and Hydroid-polypes. These layers surround the digestive-cavity just as they do in the parent animal, and since the body-wall as well as the cavity it encloses are in direct connection with those of the parent, nothing was more natural than to suppose that the bud arises as an *evagination of the body-wall of the parent, both layers of the latter taking part in its formation from the first*. A doubt as to the correctness of this statement was less likely to arise owing to the fact that even in the youngest buds of a Hydroid-polype, before they become hollow, the ectoderm and endoderm were seen to consist of a number of cells engaged in active multiplication. I myself made such a statement in connection with my investigations on the formation of the sexual cells in Hydroids,* and no doubt has yet been raised as to its correctness, or rather as to its interpretation.

The assumption that both germinal layers of the parent take part in the formation of the bud is nevertheless an incorrect one; for *the bud arises from the ectoderm only*, and the young cells

* 'Die Enstehung der Sexualzellen bei den Hydromedusen' (with 25 plates), Jena, 1883.

which form the endoderm in these buds are not derived from the endoderm of the parent, but have migrated from the ectoderm.

Purely theoretical considerations first led me to suppose that this must be so. The origin of the process of gemmation in the idioplasm can only be brought into agreement with the theory of the continuity of the germ-plasm, if the cells of the parent-organism from which buds arise collectively contain all the determinants of the species as accessory idioplasm. If this were not the case, an entire animal, capable of reproduction, could never arise from the bud. If, now, a certain cell of the ectoderm contained all the determinants for the outer, and one of the endodermal cells all those for the inner layer, a bud could only be formed when these two cells happened to lie exactly opposite to one another in the body-wall. As, however, the endoderm cells form a definite and continuous epithelial layer, and have a fixed relative position, and, moreover, the position of the ectoderm cells, although not quite so definite, is still on the whole a fixed one, I found it difficult to imagine how budding could take place at perfectly definite parts of the polype and of the stock in such a regular manner as actually occurs in many cases. The assumption that *all* the cells of the ectoderm and endoderm are equally provided with the necessary accessory idioplasm is excluded by the fact that budding occurs in such a regular manner. I am therefore led to suppose that the distribution of the 'blastogenic * germ-plasm' might possibly be confined to *one* germinal layer only; and since it is known that in Hydroids the germ-cells are always developed from the ectoderm, it is natural to conclude that the blastogenic idioplasm is contained in the cells of this layer.

This conclusion has now been confirmed by investigations carried out by Mr. Albert Lang, in the Zoological Institute at Freiburg. In various Hydroid-polypes (*Eudendrium*, *Plumularia*, and *Hydra*) the bud arises in the following way. The cells in a certain small circumscribed region of the ectoderm first begin to multiply, the 'supporting lamella,' which separates the two layers of the body-wall, gradually becoming thinner and softer at the same time, and then a few of the newly-formed cells penetrate into the endoderm through this membrane. Here they

* The term 'blastogenic idioplasm' is here used in the special sense of 'Knospungs-Idioplasma,' and not in the more general sense in which it is usually used by the author (cf., *e.g.*, the chapter on 'The Supposed Transmission of Acquired Characters'.—W. N. P.

form a layer of young actively dividing cells, such as I had formerly observed in very young buds: this layer forces the older endoderm cells away from the supporting membrane, in consequence of which they loosen their connection with the rest of the endoderm, undergo disintegration, and gradually become absorbed. The cells, however, which have migrated from the ectoderm then give rise to the endoderm of the bud.

Now that these facts have been proved by Lang's investigations,* it is easier to give a theoretical explanation of the process

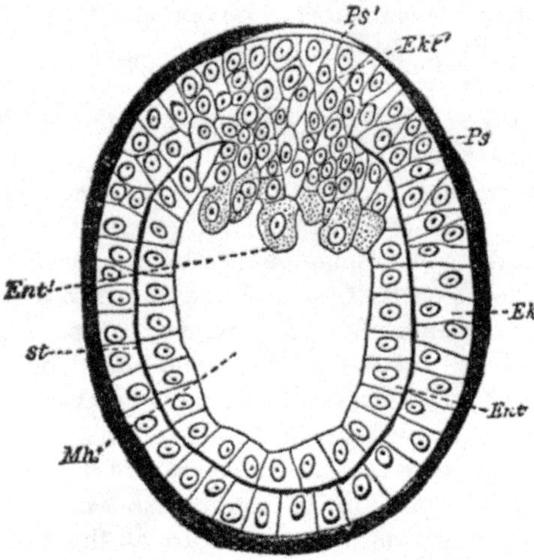

Fig. 6.—Diagrammatic section through the rudiment of a bud of *Eudendrium*. (Modified from a figure by Albert Lang.) *Ps*, the horny perisarc; *Ps'*, portion of the perisarc which has become very thin owing to the proliferation of the underlying ectoderm (*Ekt'*); *Ent'*, the region of the endoderm, in which a number of proliferating ectoderm cells have broken through the supporting lamella (*st*), migrated into the endoderm, and caused the latter to project into the gastric cavity.

of gemmation in Hydroids. We must, however, still assume that certain cells and series of cells in the ectoderm are provided with an accessory idioplasm, which contains all the determinants of the species, and which is therefore a kind of germ-plasm, though perhaps not quite identical with the germ-plasm proper:

* Albert Lang, 'Ueb. die Knospung bei Hydra u. einigen Hydropolypen,' Zeitschr. f. wiss. Zool., Bd. 54, 1892, p. 365.

I therefore speak of it as 'blastogenic idioplasm.' It cannot be stated with certainty *which* cells of the ectoderm contain this idioplasm; it seems, however, that the growth of the bud originates in the deeper layer, *i.e.*, in the 'interstitial' cells. We may therefore suppose that some of these interstitial cells contain inactive blastogenic idioplasm, which, after a certain series of cell-divisions necessary for the growth of the polype, obtains the control of one of the offspring of these cells, and so causes budding to take place. Each bud must originally arise from *one* cell only, although this fact has not as yet been actually proved; and in the first division, or at any rate in the early divisions of this cell, the group of determinants of the ectoderm must become separated from that of the endoderm, the 'bearers' of the latter group migrating into the old endoderm through the disintegrating supporting lamella. The remaining details of the process require no further explanation.

In the Hydromedusæ, then, each bud originates in a single cell, and the process of multiplication by gemmation therefore differs essentially from that of reproduction by fission. For gemmation owes its origin to the entire mass of the determinants of the species, which only undergo disintegration at a later stage; while the new structures which arise by fission originate simultaneously from numerous smaller groups of determinants, corresponding with those of the later stages of ontogeny.

We should nevertheless be mistaken in supposing that the essential difference between fission and gemmation is due altogether to this difference as regards the group of determinants concerned in the two processes. This is rendered evident by a comparison with the processes of budding in other groups of animals. It still remains to be shewn whether the process of gemmation in other Cœlenterates, viz., in the *Actinozoa*, the higher *Medusæ*, and the *Ctenophora*, also only apparently originates from both layers of the body-wall, or whether it actually arises from *one* layer only. As the possibility of the latter mode of origin has not till now been considered, it is very possible that the migration of cells may have been overlooked in this case also.

If we now turn our attention to the other groups of the animal kingdom in which gemmation occurs, viz., to the *Polyzoa* and *Tunicata*, we shall find that we possess the results of very excellent investigations on which our arguments can be based; and the histological structure of these animals is such as to render it unlikely that any oversight as regards the migration of cells can have occurred.

B.—Polyzoa

The small stocks or colonies formed in the *Polyzoa* arise by a process of gemmation; and even the small number of species which do not form stocks multiply vigorously by budding, but in these cases the buds become detached from the parent sooner or later.

The process of gemmation seems to be essentially similar in all Polyzoa. A proliferation, which primarily originates in *one* cell, takes place in a certain region of the epidermis; the masses of cells which are thus produced form a hollow invagination, which extends into the body-cavity of the animal and gives rise to the entire alimentary canal—including the fore-, mid-, and hind-guts, as well as to the preoral 'atrium' with the tentacles ('lophophore'). Certain 'free mesoderm cells' are then said to migrate from the body-cavity of the parent into the bud, in which they give rise to the muscles and sexual organs, and also in certain groups of Polyzoa to the outer (serous) layer of the intestine; while in others again they form a subcutaneous layer of cells. This is at any rate the case according to the recent observations made by Seeliger,* which are undoubtedly very accurate and trustworthy. But one point, however, still seems to be doubtful, viz., whether the sexual organs may not perhaps, after all, arise from the primary proliferation of the epidermic cells.

These processes of gemmation interfere very considerably with the ordinarily accepted and extremely conventional ideas of the germinal membranes; for the epithelium of the alimentary canal, which characteristically belongs to the inner germinal layer, here arises from the ectoderm. This, however, causes no difficulty from the point of view of the theory of the germ-plasm: we need only assume that the group of determinants for the endoderm is passed on to certain cells of the epidermis as accessory idioplasm. This transference must take place at an early stage in embryogeny, before the separation of the primary endoderm and ectoderm occurs.

Nitsche, whose means of observation were comparatively im-

* O. Seeliger, 'Die ungeschlechtliche Vermehrung der endoprokten Bryozoen,' and 'Bemerkungen zur Knospenentwicklung der Bryozoen,' Zeitschr. f. wiss. Zool., Bd. 49 and 50, 1889 and 1890.

perfect, but whose researches are nevertheless of great value, concluded that the *whole* bud was derived from the proliferation of the ectoderm. Had his statement proved correct, the explanation of the process of budding in the Polyzoa, based on the idioplasm, would be just as simple as in the case of Hydroids: it would then only have been necessary to suppose that the cell in which proliferation first began contained accessory idioplasm in the form of 'blastogenic idioplasm.' Seeliger was, however, unable to support Nitsche's statements, and the most recent observations of Oka,* Davenport,† and Braem,‡ prove beyond doubt that the 'mesoderm cells' of the parent take part in the formation of the buds. We must therefore suppose that certain mesoderm cells, provided with definite groups of determinants for muscles, endothelia, and sexual organs, migrate into the bud. It is quite conceivable that muscles, and more especially endothelia, should be developed in this manner, but it would be difficult to understand how free cells from the body-cavity of the parent could migrate into the bud, and there give rise to sexual organs at *perfectly definite regions*: were this so, we must suppose that in reality *certain* of the cells only, and not *any* of them, are concerned in the migration. Such an assumption is, however, contradicted by the abnormal processes of budding which occur, for instance, in *Pedicillina*. I therefore do not consider that the question of the origin of the sexual organs is yet decided, but I suspect, nevertheless, that one or two of the mesoderm cells of the bud are derived from the primary proliferation of the ectoderm. This view is supported by Seeliger's statement that he considers such a derivation of individual mesoderm cells of the bud possible, at any rate, in the case of *Loxosoma*.§

As, however, we are not specially concerned with the process of budding in the Polyzoa in particular, but are only making

---

* A. Oka, 'Observations on Freshwater Polyzoa,' Journ. of College of Science, Imperial University, Japan, Vol. iv., Pt. 1, 1890.

† C. B. Davenport, 'Observations on Budding in Paludicilla and some other Bryozoa,' Bull. of the Museum of Comp. Zool. at Harvard College, Vol. xxii., No. 1, 1891.

‡ F. Braem, 'Untersuch. uber d. Bryozoen des sussen Wassers,' Bibl. Zool., Cassel, 1890.

§ Seeliger, 'Bemerkungen zur Knospenentwicklung der Bryozoen,' Zeitschr. f. wiss. Zool., Bd. 50, p. 564.

use of it as an example of gemmation in which two germinal layers are primarily concerned, we can leave this question aside. It is at any rate true that in the Polyzoa parenchymatous cells, as well as a certain ectoderm cell, take part in each process of budding. We must therefore assume that the determinants of the species cannot be contained altogether in *one* cell as blastogenic idioplasm, as in the case of the Hydrozoa, but that a number of them — including those for the muscles, endothelia, blood-corpuscles, and perhaps those for the sexual organs also — are supplied to certain mesoderm cells of the parent. The development of sexual cells renders it necessary that those cells from which they arise shall also contain germ-plasm; and the formation of the epidermis of the bud, which results to some extent on purely mechanical grounds, presupposes the existence of determinants for the ectoderm in the epidermic cells of the parents.

The disintegration of the determinants, which is necessary before budding can take place, is obviously, however, of a very different kind from that which occurs in embryonic development. Seeliger, indeed, has called attention to the fact that the ontogeny which results from gemmation is a much shorter process than that which occurs when an embryo and larva are formed. In the former case, not only are the whole series of stages of segmentation and development of a free-swimming larva absent, but even in the later periods of development none of the stages in embryogeny and in gemmation exactly correspond to one another. Without following out these two processes in detail, I should be inclined to explain them in general by assuming that the groups of latent supplementary determinants, with which certain cells are provided in the course of embryogeny, contain combinations of determinants different from those which lead to the development of the embryo.

C. — *Tunicata*

The fixed Ascidians usually multiply very extensively by gemmation, and thus give rise to stocks, the individual persons of which are more or less closely connected with one another.

We owe our detailed knowledge of the process of budding in the genus *Clavelina* to the researches of Seeliger.* The

* O. Seeliger, 'Zur Entwicklungsgeschichte der Ascidien; Eibildung u. Knospung von *Clavelina lepadiformis*,' Sitzungsber. d. Wien. Akademie, Bd. 85 1882.

parent, which was developed from an ovum, produces long stalk-like processes or stolons, on which new animals are produced by budding. Each of these stolons is made up of three layers of cells—an outer ectoderm, an inner endoderm, and an intermediate layer of motile 'mesoderm cells.' The ectoderm layer gives rise merely to the epidermis of the bud: the epithelium of the alimentary canal and its accessory organs, the branchial sac ('peribranchial tube'), and the pericardial tube, being developed from the endoderm; and the muscles, ganglion (?), and sexual glands from the 'free mesoderm cells.'

The endodermal tube mainly determines the form of the animal in these processes: it becomes definitely segmented, and on it the growing ectodermal tube is moulded, so to speak. We may thus conclude that a series of homologous formative zones of structure are to be found in the endodermal tube of the stolon, each of which may consist originally of a *single* circular layer of cells. At the point where a bud will arise, the corresponding zone of cells grows out to form a bladder-shaped enlargement, which becomes detached from its point of origin on the endodermal tube of the stolon and regularly differentiated, so as to give rise to the peribranchial tube, the intestine, and so on. The cells of this endodermic vesicle cannot all be equivalent, nor can they contain exactly similar determinants: were that the case, such a differentiation could not occur, and the walls of the peribranchial chamber could not arise from one part, and the intestine from another. But even as regards the primitive intestinal vesicle itself, one cell must contain the determinants of the stomach, another those of the hind-gut, and so on. In short, we must assume that—just as occurs in principle, if not as regards actual details in the case of embryogeny—*a disintegration of the idioplasm and a distribution of the groups of determinants among the different cells takes place during development.* The determinants of all parts arising in connection with each endodermal vesicle, must be collectively contained in each zone of cells of the endodermal tube from which such a vesicle is developed.

The formation of those organs which arise from the 'free mesoderm cells' of the stolon is the most difficult to understand. There is certainly no reason why we should not suppose that these cells contain very different kinds of idioplasm: one, for instance, might contain 'muscle-determinants,' another 'nerve-determinants,' and a third 'blood-corpuscle determinants.' Various kinds of these cells may easily be distinguished while they still float freely in the blood of the stolon. The difficulty only consists in ascertaining the *exact* part they play in the formation of the developing bud. Those which are to give rise to the longitudinal muscles become arranged in rows, which, diverging obliquely from one or two definite points, extend over the animal from behind forwards, and are attached at more or less definitely fixed points anteriorly. The ganglion and the sexual glands have also perfectly definite positions in the animal. In embryogeny, as well as in the development of the endodermal vesicle of the bud, the position of every cell is assigned to it mechanically, in consequence of its origin from previous cell-generations,—that is, by the rhythm of the cell-divisions. In the case, however, of the ganglion for instance, the cells of which it is composed must come together at the right place by means of their power of locomotion. A similar process is known to occur in embryogeny in the case of several groups of animals, such as the *Echinodermata*, for instance; and until we know more of the actual facts concerned, we can only—however unsatisfactory such an assumption may be—attribute to the cells a tendency to become attached at definite points according to the manner in which they have previously been determined. The reverse assumption—that these cells develop into muscle-, nerve-, or sexual-cells according to their point of attachment—seems to me at any rate a less likely one.

If we compare the processes of gemmation and embryogeny in Ascidians, important differences are seen to exist between them. In the former, all the stages of segmentation of the egg and gastrulation, together with the formation of the mesoderm, are omitted; and many parts, again, arise from the ectoderm in the embryo and from the mesoderm in the bud. These differences are perhaps still more marked in the free-swimming *Salpæ*. These animals also multiply by buds produced on a kind of stolon; and, as in the other Ascidians referred to, the ectoderm forms practically nothing except the epidermis, and the endoderm gives rise to only a few structures, by far the greater number of parts arising from the 'mesoderm-cells.' Seeliger * explains this by supposing that 'the mesoderm

* Seeliger, 'Die Knospung der Salpen,' Jena, 1885.

of the mother-animal which passes into the buds, practically corresponds only to the future sexual apparatus.' This, however, can merely be taken as an explanation of the fact in so far as it indicates the possibility of the 'mesoderm cells' of the stolon and bud containing the groups of determinants required for these different structures. For *all* the determinants must be present in the sexual cells, and, owing to their disintegration during cell-division, they may become arranged in very varied groups, so that certain mesoderm cells may become furnished with one group and others with another. This certainly presupposes that the process of the distribution of the determinants in this case is entirely different from that which takes place during embryogeny, and this difference, again, can only depend on a difference in the original architecture of the idioplasm. In discussing the process of alternation of generations I shall once more return to this point, which, from a theoretical point of view, is a very fundamental one.

2. The Process of Gemmation in Plants

Our conception of the process of gemmation has been in the first instance derived from the vegetable kingdom: all the higher plants correspond to stocks or corms which arise by copious and regular budding, much as occurs in the case of such animal-stocks as those of the Hydrozoa, for instance. Although the physiological individuality of separate 'persons' in a plant is often less defined than in the case of many animal colonies, there can nevertheless be no doubt as to the morphological value of a shoot as a 'person,' in the sense in which Haeckel uses the term.

Although as regards animal colonies, it has not yet in all instances been possible to ascertain with absolute certainty the actual origin of the processes of budding in connection with the cell-generations of the first person of the colony, this has been done very accurately in the case of plants; a theory of heredity can therefore be much more safely applied to the process of gemmation in plants than to that in animals.

In many plants, at any rate, budding originates from a *single* cell, situated at the apex of the growing shoot, and known as the '*apical cell*.' This cell grows and undergoes a series of divisions, much as occurs in the development of the ovum, and thus gives rise to a group of cells, the number, form, and arrangement of which is perfectly definite. The primary constituents of the entire new shoot are contained in this group, and it is possible to predict what parts of the shoot will be formed from each of its cells. The successors of this group of cells continue to multiply up to a certain limit, and have then only to become elongated in one or more directions, and more highly differentiated, in order to give rise to a fully developed 'person' of the stock. This person does not undergo any further essential changes, but it is capable of giving rise to a new person from its apical cell;

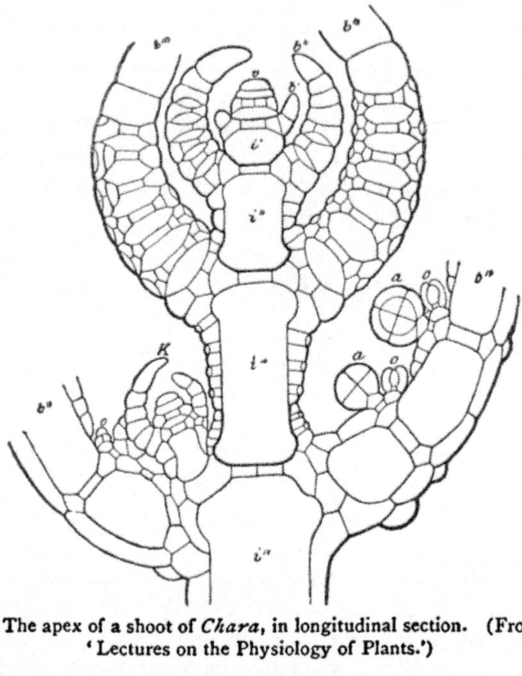

FIG. 7.—The apex of a shoot of *Chara*, in longitudinal section. (From Sachs' 'Lectures on the Physiology of Plants.')

for the latter is always being renewed, or, in other words, it always remains the same.

Let us take as an example the alga *Chara*. A glance at Fig. 7 will at once make it apparent that the idioplasm of the apical cell (*v*) cannot undergo separation into different groups of determinants in the first division, because one of the resulting two cells remains as the apical cell, while the other, or 'segmental cell,' gives rise to an entire shoot,—that is to say, to that very structure which the apical cell is capable of producing. The next division of the lower of the two daughter-cells, however, separates the determinants into two dissimilar groups, for it results in the production of an upper biconcave 'nodal cell,' from which the leaves (b^{I}, b^{II}, b^{III}, b^{IV}), the lateral shoot (*k*), and the sexual organs (*a* and *o*), will subsequently arise; and of a lower biconvex cell, which does not undergo further division, but only grows considerably in length, so as to form a segment of the main axis (i^{I}, i^{II}, i^{III}, i^{IV}). The idioplasm of this 'internodal cell' does not therefore undergo further disintegration; the nodal cell, however, divides vertically, so as to form cells which, since they give rise to other parts of the shoot, must contain various groups of determinants. Thus a comparison of the younger with the older segments of the shoot, shows that the outer of the five nodal cells in the figure gives rise to a whole leaf, together with the sexual organs, the inner ones forming the actual node. The division of the outer cell is accompanied by constant though usually unimportant changes as regards its idioplasm: a glance at the structure of the leaf, in which similar segments are repeated many times over, will make this evident. If we now leave out of consideration the accessory idioplasm which is present in the cells along with the primary idioplasm, it will be seen that the distribution of the group of determinants derived from the apical cell must simply take place so as to result in each cell, as it is formed, receiving that group of determinants only, the individual constituents of which are required by its successors for the control of the individual cells. We must therefore suppose that the internodal cells of the stem only contain their own specific idioplasm, composed of 'internodal determinants,' for they do not give rise to any other structures. The primary nodal cell, on the other hand, must contain an entire group of determinants, as it gives rise to a number of cells which have various forms and perform various functions.

Although the cells of plants are often apparently very much alike, and no essential difference can be observed between them, such a difference must exist if the origin of the specific leaf, stem, and reproductive organs can be proved theoretically at all. For the origin of these structures can only be explained, at any rate in principle, by supposing that each of these centres of vitality is controlled by a specific idioplasm; that is, by a determinant which differs in some way or other from those in the other cells.

3. Comparison of the Process of Gemmation in Animals and Plants

Various stages may be recognised in the different kinds of gemmation with regard to the kind of idioplasm concerned in the process. The simplest form of budding is seen in those plants in which the production of a new 'person' by budding always originates from a single cell. We must therefore assume that the idioplasm of this cell contains all the determinants of the shoot, and very probably those of the root also. For most of the shoots of a plant, when they have been cut off from the stem, are capable of giving rise to roots under favourable circumstances. This does not as a rule occur under normal conditions, — that is to say, while the shoot is still connected with the parent-plant. The 'blastogenic idioplasm' cannot be quite identical with germ-plasm proper; for although precisely the same parts may arise from it as from the fertilised egg-cell, the different succession of cells which results in embryogeny and in gemmation indicates that the determinants must at any rate be differently arranged in the idioplasm, and that possibly their proportional number is also different. *'Blastogenic idioplasm' and germ-plasm may in a sense be regarded as 'isomeric' idioplasms*, using the term in an analogous sense to that of isomeric chemical compounds.

The same would be true as regards such animals as Hydroids, in which the formation of a bud originates from a *single* cell. In this case, again, the resemblance between embryonic development and the process of gemmation, although to a certain extent approximate, is not a complete one; and it must again be assumed that the *whole* of the determinants of the species are contained in the blastogenic idioplasm, — not only those which as a rule undergo development, but also those required for the formation of the attached ends in the case of Hydroids, or of roots in the case of plants. This conclusion is supported by the phenomena of budding in polypes like *Hydra*, in which the buds regularly become detached, and carry on an independent existence. In such cases the daughter-polypes do not develop a 'foot' until they become detached from the parent.

The next stage in the process of budding is seen in the Polyzoa. All the determinants of the species from which the bud is formed are no longer contained in a *single* cell, but are arranged into main groups, one of which is supplied as accessory idioplasm to one cell of the ectoderm, and the other to one or more cells of the mesoderm. The *single* ectoderm cell gives rise to the entire endoderm, but it must nevertheless not be considered equivalent to those cells of the embryo which give rise to the endoderm by invagination, for it forms parts which are either not developed at all in the embryo, or else arise from other ectoderm cells. Without entering into details here, the facts may be expressed in terms of the idioplasm by supposing that the ectoderm-cell from which the bud arises is provided with an idioplasm which contains the whole of the determinants for the endoderm, as well as a number of others, and that this combination of determinants does not occur in embryogeny. The mesoderm-cells of the parent which gives rise to the endothelia, muscles, &c., of the bud, must also contain a peculiar combination of determinants which is not exactly similar to that which occurs in embryogeny. The gemmation must therefore be prepared for in embryogeny by certain series of cells in the ectoderm and mesoderm being provided with these groups of determinants in the form of accessory idioplasm.

A third stage is represented by the gemmation of fixed Ascidians and Salps. In these the bud originates—in the fully-formed animals, or in those which are still undergoing development—from three kinds of cells, viz., those of the ectoderm, mesoderm, and endoderm. And here again those groups of determinants which must be supposed to exist in the three kinds of cells do not correspond exactly to those which must be contained in the primary ectoderm, endoderm, and mesoderm cells. In fact, no group of cells which occurs in embryogeny can contain precisely the same group of determinants as does the endoderm cell of the bud. A collection of determinants especially adapted for budding must therefore be provided on a large scale during embryogeny in this case, so that eventually certain cells may receive their supply from it in the form of accessory idioplasm.

This last kind of budding resembles regeneration very closely as regards the idioplasm concerned in the process. It must not, however, be therefore implied that the former process has been derived phylogenetically from the latter. The resemblance only consists in the formation of a new 'person,' which in both cases originates in several cells provided with different groups of determinants, these eventually completing one another, and interacting in such a manner that a fully-formed person must result.

4. The Phylogeny of the Process of Multiplication by Gemmation

In all probability the phylogeny of gemmation has taken place along different lines. The process most likely arose independently in animals and in plants, and perhaps even in different groups of animals it has had a different origin.

In many of the lower plants, the cells and organs of which are only slightly differentiated, all, or at any rate many, of the cells can individually give rise to a new plant under certain circumstances. In such cases we might be inclined to suppose that each cell contained originally, *i.e.* at the time of its phyletic origin, the entire mass of determinants of the species, or, in other words, contained germ-plasm. The various differentiations of the cells on the upper and lower surfaces, for instance, would consequently depend on the different determinants becoming active in response to different external stimuli: some, for instance, might be stimulated by a bright light, and others by a dim light.

This explanation would hardly suffice in the case of the higher plants, the differentiation of which is far too complicated to be due to the effect of external causes. A large number of the cells must nevertheless contain germ-plasm, which, however, is in the unalterable ('gebundenen') state,—that is to say, it is not merely inactive, but is incapable at the time of undergoing disintegration. This stage in the phylogeny of gemmation may be derived from the earliest stage. As the plant underwent an increasing differentiation, cells appeared which only contained special determinants, in addition to those with germ-plasm proper; and this may have led to the condition which we now find in the highest plants, and which is distinguished by the fact that many cells only contain specific determinants, while a large number of others possess in addition germ-plasm in the unalterable condition, which only becomes active under certain influences. I shall have occasion to return to this subject later on.

In the case, again, of the various groups of the lower animals which multiply by gemmation, we cannot assume that this process has a common origin. But although it may have arisen independently in the various subdivisions of the animal kingdom, the history of its origin will have been essentially the same in all cases, for 'blastogenic' idioplasm must have become differentiated from the germ-plasm even in the egg-cell, as all the determinants of the species are contained only in the latter. Even at the present day the blastogenic idioplasm must be present as such in the germ-plasm, for otherwise it could not have undergone independent and hereditary variation: the formation of medusæ from polypes by gemmation, and many other cases of alternation of generations, prove that this has actually occurred.

Balfour attempted to derive the process of budding from a division of the fertilised ovum into two separate parts, such as has been observed in certain animal forms, and which leads to the formation of two individuals. He imagined that if this process of doubling were transferred to a later ontogenetic stage, budding would result, and expressed his views as follows:—'While it is next to impossible to understand how production of a bud could commence for the first time in the adult of a highly organised form, it is not difficult to form a picture of the steps by which the fission of the germ might eventually lead to the formation of buds in the adult state.'* Unfortunately this gifted observer did not work out this idea in detail: it seems to me, however, that the derivation of budding from the doubling of the fertilised ovum by division is not so simple or self-evident as we might expect at first sight.

Let us suppose that a fertilised ovum became capable of dividing into two parts: these two first segmentation-cells would not then be blastomeres, but would correspond to egg-cells, each of which could give rise to an entire animal. But this could not be called gemmation, nor would the latter process occur if the doubling were transferred to a later stage:—this would only cause a multiplication of the egg-cell, which would result in the formation of four, eight, sixteen, &c., ova, instead of two.

If, however, we suppose that the division of the egg is of such a kind that the two halves at first remain together so as to form

* F. M. Balfour, 'Comparative Embryology,' Vol. I., Introduction, p. 13.

only one embryo, the condition observed in a certain earthworm (*Lumbricus trapezoides*) by Kleinenberg would result. In this animal the development is *apparently* single up to the gastrula stage, at which the separation of the two embryos first occurs. Did this separation take place at a much later stage, perhaps not until the two individuals are fully developed, the process would not be one of budding, but only of a doubling of the embryo.

An essential modification of this process is indispensable if gemmation is to result from it, and this consists in the *postponement of the development of one half of the egg*. Let us suppose that one of the two equivalent blastomeres of an ovum did not at once undergo development at the same time as the other, but remained in a unicellular condition enclosed within the embryo formed from the active blastomere, and subsequently began to develop when the latter had already given rise to a full-grown animal: this would be a true process of gemmation. I do not wish to assert definitely that the phylogeny of budding might not have taken place on similar lines. A postponement and subsequent transference to a later stage of ontogeny of the development of one of the blastomeres is not actually inconceivable. But such a transference must have undergone a still further modification, before even the simplest form of budding with which we are acquainted could arise. The shifting must have occurred in a backward as well as in a forward direction; that is to say, *the division of the egg into two separate ones must have been suppressed, and represented by the mere division of the germ-plasm*.

Thus in Hydroids and other animals which multiply by budding, we see, in fact, that one of the two blastomeres into which the egg-cell divides does not serve, so to speak, as a reserve cell for subsequent gemmation; both blastomeres, on the contrary, continue to divide, and together give rise to the embryo: and even in the latter none of the cells can be distinguished as 'blastogenic-cells': the cells which take part in the formation of the buds only appear at a much later stage, when the polype is fully formed. If therefore gemmation has in this case originated from the doubling of the egg, the latter process must itself have become degenerated, only the essential part of it remaining: the germ-plasm concerned in it must have remained associated with that of the egg-cell in the form of '*unalterable*' germ-plasm, and must then have been passed into certain series of cells in the course of ontogeny.

Whether the process of budding has actually been derived from that of the doubling of the egg or not, it seems to me to be certain at any rate that the first process undergone by the idioplasm *must have been that of the doubling of the ids of the germ-plasm in the fertilised egg-cell*, and that this was not connected with the division of the egg-cell; one half of the germ-plasm consequently remained in an unalterable and inactive condition, in which, however, it was capable of development. This blastogenic germ-plasm was then supplied to one of the first segmentation-cells in the form of accessory idioplasm; and from these it was passed on through certain series of cells in an unalterable condition, only becoming active when it had reached certain parts in the fully formed animal, in which it then caused gemmation to occur.

It does not seem to be inconceivable that the process of budding owes its origin phyletically to such a spontaneous division and doubling of the germ-plasm, and that this was originally connected with the inactivity of half the germ-plasm: its connection with the doubling of the ovum was consequently not such as was indicated above,—that is to say, gemmation did not owe its origin to the doubling of the egg, but *both processes originated primarily in the division and doubling of the germ-plasm of the egg-cell*, to which in any case the doubling of the egg must be due. The difference between the two processes would then consist in the fact that in budding one-half of the germ-plasm would pass into the inactive condition, while in the doubling of the egg both halves would at once become active.

The modifications of the idioplasm which result in gemmation must become more complex as soon as two, or all three, of the germinal layers take part in the process, instead of one only. In such cases the blastogenic germ-plasm must undergo disintegration at certain ontogenetic stages, *e.g.*, at the separation of the ectoderm from the endoderm, and again at the separation of the mesoderm from one of the two primary germinal layers. Precisely the same combination of determinants need not necessarily be produced by the disintegration of the accessory germ-plasm into two or three groups of unalterable accessory idioplasm, such as are formed in embryogeny. We can thus explain the origin of endodermal organs from the ectoderm cells of the bud, as occurs in the Polyzoa for instance, and also even the co-operation of three germinal layers in the formation of the bud.

It seems to me to be improbable that the phylogeny of gemmation in animals has taken place in the reverse manner. We might assume that in the lowest Metazoa, which no longer exist at the present day, all or many of the cells also contained germ-plasm proper, just as in the case of the lower multicellular plants. Under certain circumstances a perfect animal might have been produced from each of these cells. But this assumption would only suffice as long as the individual formed by budding was exactly similar to that arising from the egg. Even the slightest difference between these would necessitate the presence of special ids in the germ-plasm. For such a difference can only depend on the fact that the two kinds of individuals are capable of independent variation from the germ onwards. We should therefore have to assume further, that in the course of phylogeny the germ-plasm of these somatic cells from which the buds originated became doubled in the earlier stages of ontogeny, and that it was consequently present in the germ-plasm of the egg-cell in the form of a special group of ids. But this, to say the least, is a very involved assumption, and can hardly be considered very probable: that which presupposes a primary doubling of the ids of germ-plasm is certainly far preferable to it. The following chapter will make this point still more evident.

CHAPTER V

ALTERNATION OF GENERATIONS IN ITS RELATION TO THE IDIOPLASM

STARTING from the germs specially adapted for amphimixis (sexual intermingling), we have designated as *germ-plasm* the definitely arranged group of determinants which must be contained in the sexual cells. By this term is meant an idioplasm which contains all the determinants of the species. At the same time a large number of species exist in which the sexual cells are not the only ones which contain all the determinants, and in which the *development takes place, for the second time during the course of the life-cycle, from a single cell*; the idioplasm of this cell must therefore also be composed of all the determinants of the species. This is the case in most of the lower plants, such as mosses, horse-tails, and ferns, — in all of which sexual reproduction alternates with the formation of asexual 'spores,' — as well as in those groups of animals in which that form of alternation of generations which is known as *heterogeny* occurs. But even in the case of alternation of generations in the more restricted sense, — *i.e.*, the alternation of sexual reproduction and gemmation, the development of an individual may take place twice successively from a *single* cell, as was mentioned above with regard to the stocks or colonies of plants and of Hydroid-polypes. In all these a cell, the idioplasm of which contains all the specific determinants, occurs twice in the course of the life-cycle from one fertilised ovum to the next one. The question therefore arises as to whether the idioplasm in each case is to be considered identical, and may merely be described as germ-plasm.

This question has already been discussed in the section on the process of gemmation in plants; and it was there concluded that the idioplasm of the apical cell and that of the fertilised ovum cannot be assumed to be perfectly identical, owing to the fact that the course taken by embryogeny — in which process the first shoot and roots are formed — is different from that followed by the cell-divisions which result in the apical cell producing a new shoot. The same is true as regards the formation of a new polype from a blastogenic cell and from an ovum. In both cases the final result is the same, or at any rate very similar, though the method by which it is attained is different. Although a precisely similar organism might be produced by either of the two methods of development, and the primary cells would therefore contain the same determinants in both cases, the grouping of the latter in the two idioplasms at any rate must be different, for they must pass through different groups during ontogeny before their ultimate disintegration into single determinants. Even in this very simple case it is necessary to distinguish the 'germ-plasm proper' of the egg- and sperm-cells from the 'apical-plasm' or 'blastogenic germ-plasm.' It is convenient, however, to speak of every kind of idioplasm which contains all the determinants of the species as *germ-plasm in the wider sense*, and to distinguish its various subdivisions as 'blastogenic' and 'sporogenic' germ-plasm, and so on; these latter may all be included under the term *accessory germ-plasms* or *para-germ-plasms*, in contrast to the *primary* or *ancestral germ-plasm*.

Wherever two kinds of germ-plasm occur in the life-cycle of a species, we might be inclined to assume that they change into one another in the course of life. But this view is untenable, as has been shown above, and we are on the contrary forced to assume that *both kinds of germ-plasm continually pass simultaneously along the germ-tracks, and that each of them becomes active in turn*.

This assumption is unavoidable, for the phyletic development of the species shows that *the individual generations in cases of alternation of generations can vary independently and hereditarily*. It, however, presupposes that *special* determinants are present in the germ-plasm in each generation, for otherwise both generations would be affected at the same time by a variation in the germ. A similar assumption must be made in the case of metamorphosis. The wings of a butterfly must be represented in the germ-plasm by a group of determinants. If the wings were formed by the transformation of some of the determinants of the caterpillar, they could never vary without at the same time producing a variation in some parts of the caterpillar, and *vice versâ*.

It will not be uninteresting to give some examples by way of illustration.

We will first take a case of *heterogeny*, or alternation of generations in which the two generations do not differ at all from one another in the full-grown condition. As a rule, the difference between the two successive generations in the *Daphnidæ* or water-fleas, for example, consists in the fact that one generation is developed from summer-eggs, which contain a small amount of yolk, while the other arises from winter-eggs, in which the yolk is very abundant. From both of these two kinds of eggs similar females are developed: — the complication arising from the periodic appearance of the male may be neglected for the present. The summer-eggs are nourished by the blood of the mother, while the winter-eggs are not; for the amount of yolk in the latter necessitates a different kind of

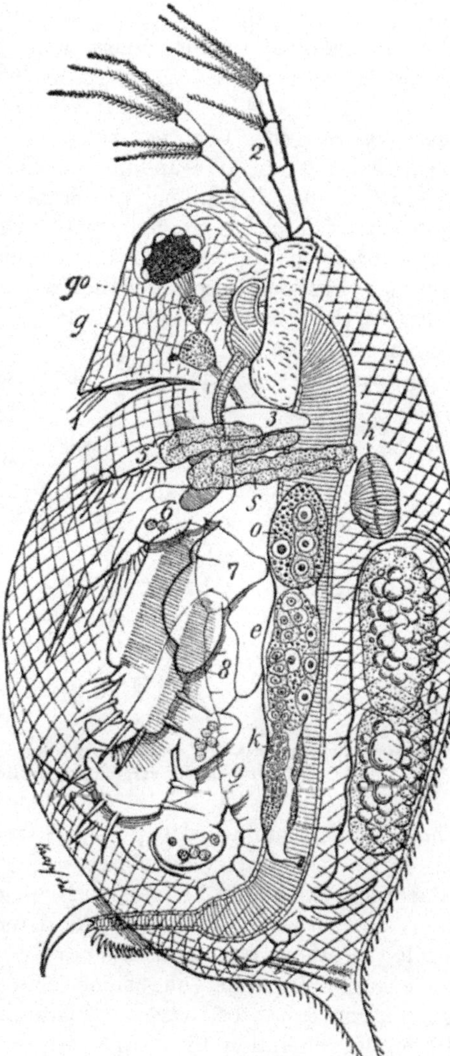

Fig. 8. — A female of *Daphnia pulex*, with two parthenogenetic eggs in the brood-chamber (*b*). (After R. Hertwig.)

ontogeny, and this presupposes not only a difference in the arrangement of the determinants in the germ-plasm as compared with the meta-germ-plasm ('Nach-Keimplasma'), but also the presence of *different determinants for some of the embryonic stages.* The case becomes still clearer if we take one particular species of Daphnid (*Leptodora hyalina*) into consideration. In this animal the embryogeny of the winter-eggs only extends as far as to the formation of the primitive crustacean larva, or nauplius, which possesses three pairs of limbs: the summer-eggs, on the other hand, develop at once into the adult form of the animal, in which all the limbs are present. The summer-eggs certainly also pass through the stages from the ripe ovum to the nauplius, but these are abbreviated, and

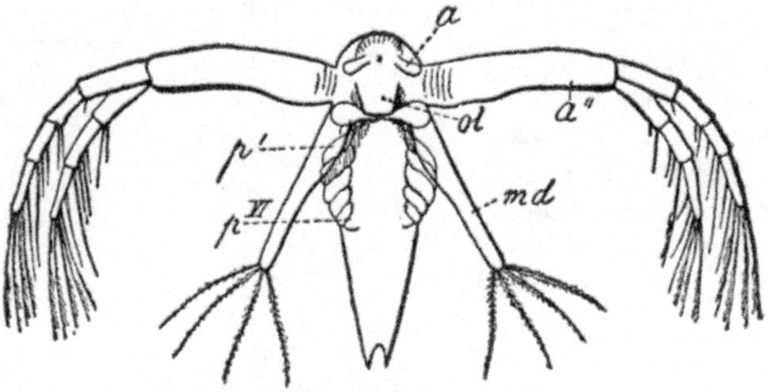

Fig. 9.—Nauplius larva of *Leptodora hyalina*. (After Sars, from Korschelt and Heider's 'Lehrbuch der vergleichenden Entwicklungsgeschichte.')

though this nauplius also possesses three pairs of limbs, these are only rudimentary, and are useless as swimming organs. There must therefore be two kinds of germ-plasm in *Leptodora*, one of which still contains all the determinants of the nauplius. while the other only contains a portion of them, and even these have probably undergone some change. The two kinds of germ-plasm must be passed separately along the germ-tracks from one generation to another, so that each must always contain the other, which is, so to speak, stored away in it in an inactive condition. It seems to me impossible to explain the facts in any other way, for it is inconceivable that the germ-plasm of the summer-eggs, which has undergone reduction, and possesses comparatively few determinants, *should be able to develop the lost determinants out of its own substance.*

The *phyletic development* of these two kinds of germ-plasm would be very enigmatical if we were compelled to assume that only a single unit of the germ-plasm is present in the nucleus of the germ-cell. We have, however, made the reverse assumption from the first, and it will be shown later on that a consideration of sexual reproduction, or amphimixis, leads us to assume that several, and in fact probably a large number of units or ids must be contained in the germ-plasm of every species which multiplies sexually. If now a reduction of the determinants for the nauplius in the summer generation of *Leptodora* were advantageous, it would have appeared, increased, and become fixed in the course of generations by selection, and an abbreviation of embryogeny would thus have resulted. This would only have occurred gradually, so that at first the summer-eggs would contain more reduced than unreduced ids only in the case of a few individuals; and if the original unabbreviated form of embryogeny were of greater advantage to the winter generation, the determinants for the nauplius would not become lost or modified in all the ids, but only in certain of them. A balance of the two kinds of ids would finally take place from the struggle between the modified ids, which were more advantageous in summer, and the unmodified ones, which were of greater advantage in winter, and this would result in the germ-plasm of the species being composed of an equal number of modified and unmodified ids; these would *alternately* control the cell, so that each would remain inactive and unalterable during a certain number of generations, and become active during certain others.

This regular alternation between definite periods of activity and inactivity in the two kinds of germ-plasm can be directly observed, for we can determine how many generations occur which give rise to summer-eggs before one again appears in which winter-eggs are produced. As I was able to prove a considerable time ago, this regularity varies very much in different species of Daphnidæ, and stands in close relation to the mode of life of the species. In those species which live in very small bodies of water which are liable to become rapidly dried up, the formation of the two kinds of eggs alternates very frequently: this is due to the fact that the extermination of the animals by the sudden drying up of the pond is only prevented by the thick shells by which the winter-eggs are surrounded. On the other hand, all the species which live in large bodies of water, such as pools and lakes which never become dried up, produce summer-eggs alone for a large number of generations, and only give rise to eggs of the other kind on the approach of winter; and these, on the death of the animals which produced them, ensure the continuance of the species in the following spring.

The occurrence of changes in the *final stages* of ontogeny must be accounted for in a similar way.

In plant-lice belonging to the genus *Aphis*, the fertilised egg gives rise to females, which are, however, incapable of being fertilised, for the receptaculum seminis is wanting, and this is essential in the process. Their eggs are, however, capable of undergoing development in the ovary parthenogenetically. The resulting offspring give rise to similar females possessing no arrangements for fertilisation, and these again produce others of the same kind. Ultimately, however, one of these gives rise parthenogenetically to females which are capable of fertilisation, as well as to males. The two sexes as a rule differ as regards the shape and colour of the body, apart from the structure of the reproductive organs and sexual products, but the embryogeny of these sexual animals is similar to that of the others.

In this case, therefore, the determinants of the mature animal become modified in the parthenogenetic generations, for sexual reproduction is the more primitive of the two forms of the process. If therefore we make the assumption, which, however, is not a strictly correct one, that the sexual generations have remained quite unaltered since the introduction of alternation of generations in these animals, we should have to represent the phyletic change in the germ-plasm as being of such a nature as to cause the degeneration of the determinants of the seminal vesicle in one half of the ids, and to produce a change in other determinants, such as those which control the colour of the integument, for instance. The modified, as well as the unmodified ids, must be contained within the same germ-plasm, but they control the egg *alternately*, and never become active at the same time.

In this instance the generations which have been interpolated have only suffered a slight change as regards the structure of the whole body. But in many cases of alternation of generations very important differences of structure occur, so that not infrequently one might easily believe that the different generations belong to two entirely different groups of animals.

This is the case in the *alternation of generations in medusæ.* The polype is the original form, and even at the present day the fertilised ovum of the medusa gives rise to a polype in most species. By the budding of this polype, or at any rate of the offspring which have been produced by gemmation, medusæ are again developed. If, for the sake of simplicity, we neglect the slight differences which may exist between the germ-plasm of

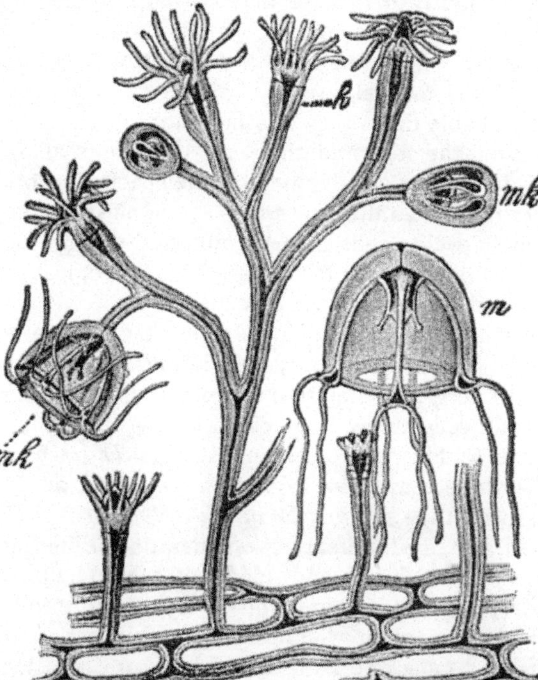

FIG. 10.—*Bougainvillea ramosa.* (After Allman.) Polype stock with gastrozooids (*h*) and medusoid buds (*mk*); *m*, young Medusa (*Margelis ramosa*), which has become free. (From A. Lang's 'Lehrbuch der vergleichenden Anatomie.')

the egg and that of the bud, it is evident that two germ-plasms take part in the cycle of development of the species, and these must differ as regards very many, if not almost all, of their determinants, for the medusa is provided with a number of parts and organs which the simple polype does not possess. Thus we must assume that there are two different kinds of ids of which the germ-plasm is composed in equal numbers, the periods of activity of which alternate with one another. The ids of the accessory germ-plasm, which arose subsequently, must be larger than those of the ancestral germ-plasm, because they contain more numerous determinants. If at some future time it should be definitely ascertained that those granules or microsomes, which are arranged like beads in a necklace in the nuclear rods, really correspond to ids, we may possibly, perhaps, be able to prove by the aid of the microscope that such differences in size actually exist. A knowledge of the entire number of nuclear rods or idants may also possibly help to confirm the theory, for it is probable that in species in which alternation of generations occurs, the ids, and therefore the idants also, have been doubled during the development of the species. For if my view is correct that a definite amount of germ-plasm is necessary for the normal development of a certain kind of egg, the periodical inactivity of half the ids must have been accompanied by a corresponding doubling of these structures.

The mechanism of the idioplasm in alternation of generations becomes somewhat different, and rather more complicated, as soon as the second generation arises, not from a single cell, but from several cells at the same time, derived from different layers of the body. This is the case as regards *the strobilation of the higher medusæ and that of tape-worms*, and an intermediate stage is seen in the *gemmation of the Salpæ*.

In the last-named animals, two generations differing as regards the form of the body and mode of reproduction follow one another. A number of individuals are united into a 'chain' in the first generation, in which sexual multiplication occurs; and in the second generation the individuals are separate, and multiply by budding. It has already been pointed out in the chapter on gemmation that this budding is produced by a co-operation of the ectoderm and mesoderm cells. We must imagine that in this case, again, the germ-plasm of the egg- and sperm-cells is composed of two kinds of ids, which alternately become active, one of which contains the determinants for gemmation, and the other those for embryogeny. In the case of the Hydroid-polypes and medusæ, the determinants of the 'blastogenic' ids remain together in *one* cell, but in the single form of Salpa they must be separated into groups during embryogeny, and these groups would be supplied — in part to the ectoderm, and in part to the mesoderm and endoderm—as inactive and unalterable accessory idioplasm. These only become active, and cause budding, when they have reached some definite part, such as the ovary or proliferating stolon.

The development of the higher medusæ or *Acalephæ* by strobilation can easily be traced from the above processes. In these animals the sexual forms arise asexually. the polype becomes divided into disc-shaped portions, and so comes to resemble a pile of saucers, each disc eventually being trans-

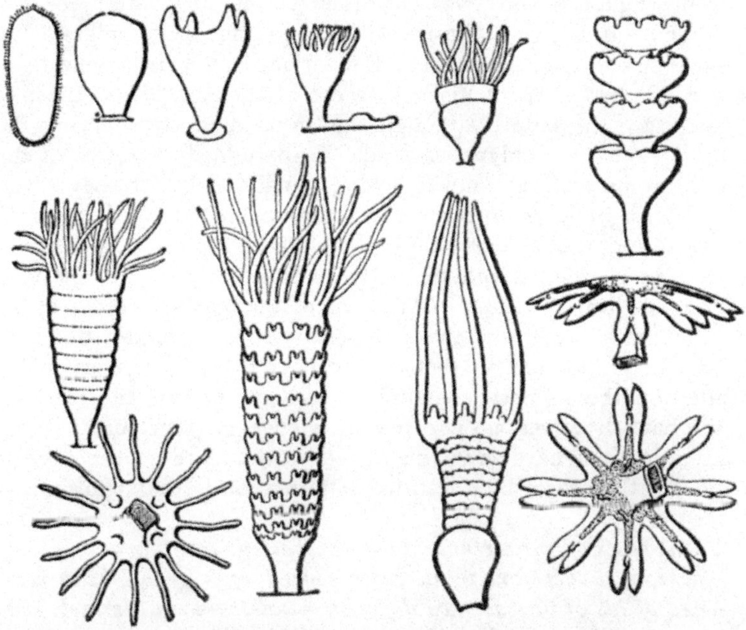

Fig. 11.—***Development of Medusæ by strobilation***—**1, the young larva; 2–5, its development into a polype; 7, a polype viewed from the oral pole; 6, 8, and 9, transverse division of a polype into disc-shaped portions; 10, the constriction of these portions into young Medusæ; 11 and 12, a young Medusa. (From Hatschek's 'Lehrbuch der Zoologie.')**

formed into a medusa. If the medusa underwent division, the process would be one of simple regeneration : the differentiation of one of these discs into a medusa depends on a mechanism in the idioplasm exactly similar to that which gives rise to the process of regeneration in a worm the anterior end of which has been cut off, or which has undergone spontaneous division. The various cells in the body of the polype must be provided with different groups of determinants of the medusæ in the form of inactive accessory idioplasm, and these must become active in the process of strobilation, and cause the development of highly complex medusæ with eight or more radii, and provided with eyes, auditory organs, and olfactory pits. The difference between this process and that of simple division followed by regeneration, consists in the fact that in the latter the supplementary determinants of the cells of the body are of the same kind as those from which the body was constructed: in strobilation, on the other hand, the germ-plasm of the egg- and sperm-cells, which gives rise to the sexual generation or medusa, must contain not one, but two kinds of ids, viz., those of the polype and those of medusa; the latter, although they remain inactive during the ontogeny of the polype, and take no part in the control of the cell, are nevertheless not absolutely unalterable, for they break up during ontogeny into many different groups of determinants, and at the same time become distributed among different cells in a regular and perfectly definite manner. It is very probable, however, that *all* the cells of the polype — those of the ectoderm as well as of the endoderm — are provided with accessory determinants, so that each cell of the polype contains in addition the primary constituents of some cell of the medusa. We have, however, no positive knowledge on this point, for no investigations have as yet been made with regard to the succession of the cells which lead to the formation of the medusa from the polype.

The basis of the alternation of generations as regards the idioplasm must therefore in all cases consist of a germ-plasm composed of ids of at least two different kinds, which ultimately take over the control of the organism to which they give rise.

CHAPTER VI

THE FORMATION OF GERM-CELLS

1. The Continuity of the Germ-plasm

If heredity depends on the presence of a substance, the germ-plasm, which causes the production of the new individual by directing the process of division in ontogeny, in the course of which it becomes changed in a regular manner, the question arises as to how unaltered germ-plasm can nevertheless reappear in the germ-cells of the new individual. The transmission of characters from parent to child can only depend on the germ-cell from which the offspring arises containing ids of germ-plasm precisely similar to those of the germ-cell from which the parent was developed. The germ-plasm, however, undergoes an enormous number of changes during the development of the ovum into the parent: how is it possible therefore that this substance can reappear in the germ-cells of this parent?

There are obviously two possible solutions of this problem. The changes which the germ-plasm undergoes during the construction of the body must either be of such a kind that they can take place in the reverse order when the idioplasm of all, or at least of a portion of, the somatic cells is re-transformed into the germ-plasm from which it was, in fact, *indirectly* derived; or, if such a reversal is impossible, the germ-plasm of the germ-cells must be handed on *directly* from parent to offspring. This latter hypothesis was suggested by me some years ago under the name of the *continuity of the germ-plasm*.* A third solution of the problem is impossible, for it is quite out of the question that the germ-plasm can be entirely formed anew.

The hypothesis of the continuity of the germ-plasm depends on the assumption of a contrast between the *somatic* and the *reproductive* cells, such as can be observed, in fact, in all multicellular plants and animals, from the most highly differentiated forms to the lowest heteroplastids amongst the colonial Algæ.

* 'Die Continuität des Keimplasma's als Grundlage einer Theorie der Vererbung,' Jena, 1885 (English translation, 2nd ed., p. 163).

I assume that germ-cells can only be formed in those parts of the body in which germ-plasm is present, and that the latter is derived directly, without undergoing any change, from that which existed in the parental germ-cell. Hence, according to my view, a portion of the germ-plasm contained in the nucleus of the egg-cell must remain unchanged during each ontogeny, and be supplied, as such, to certain series of cells in the developing body. This germ-plasm is in an *inactive* condition, so that it does not prevent the active idioplasm of each cell from impressing a specific character on the latter in a greater or less degree. It must, moreover, differ from ordinary idioplasm, inasmuch as the determinants it contains are kept closely together, and are not distributed in groups among the daughter-cells. This accessory germ-plasm is thus passed on in an unalterable condition through longer or shorter series of cells, until it ceases to be inactive in a certain group of cells, more or less remote from the egg-cell, and then impresses upon the particular cell the character of a germ-cell. The transmission of the germ-plasm from the ovum to the place of origin of the reproductive cells ('Keimstätte') takes place in a regular manner, through perfectly definite series of cells which I call *germ-tracks*. These are not actually recognisable, but if the pedigree of the cells in the embryogeny is known, they may be traced from their termination in the germ-cells backwards to the ovum.

This assumption is supported by the fact that a direct, or at any rate a very close, connection can be proved to exist, although only in rare instances, between the germ-cells of two consecutive generations. In the *Diptera* the first division of the egg-cell separates the nuclear material of the subsequent germ-cells of the embryo from that of the somatic-cells, so that in this case a direct continuity can be traced between the germ-plasm in the germ-cells of the parent and offspring.

The process in this case must certainly, however, be looked upon, not as a primary one which has been passed on unchanged from very ancient times, but as a special arrangement peculiar to this order of insects. It nevertheless proves the possibility of each generation of germ-cells being derived directly from the preceding one, and also that the germ-plasm which has been prevented from taking part in the construction of the somatic portion of the embryo is not required in this process.

We may next take the case of the embryogeny of the

Daphnidæ. In these animals the primary germ-cells become separated from the somatic cells in the first stages of the segmentation of the egg; and in *Sagitta* again, this separation takes place at the gastrula-stage. In Vertebrates this process occurs much later, although it always takes place within the first half of embryogeny; while in Hydroids — both in colonial and solitary forms — the germ-cells do not appear in the 'person' which is developed from the ovum at all, and only arise in a much later generation, which is produced from the first by continued budding. The same is true as regards the higher plants, in which the first shoot arising from the seed never contains germ-cells, or even cells which subsequently become differentiated into germ-cells. In all these last-mentioned cases the germ-cells are not present in the first person arising by embryogeny as special cells, but are only formed in much later cell-generations from the offspring of certain cells of which this first person was composed. These ancestors of the germ-cells cannot be recognised as such: they are somatic cells, — that is to say, they, like the numerous other somatic cells, take part in the construction of the body, and may be histologically differentiated in various degrees.

A series of organic species might therefore be formed in which the formation of the germ-cells begins at very different degrees of remoteness from the egg-cell. This would admit of the interpretation that the fertilised egg-cell of the earliest Metazoa first divided into two cells, one serving for the formation of the body (soma), and the other for that of the germ-cells; and that a shifting had occurred subsequently, owing to a separation of the material for both parts in the germ-plasm, so that the portion of the germ-plasm which remained unchanged was supplied in an inactive condition, in the form of accessory idioplasm, to one of the somatic halves of the egg-cell, and was transmitted by the latter to a somatic cell of the second, third, or fourth generation. The shifting of the process of separation into germ-cells and somatic cells finally reached its extreme limit, as in the case of the Hydroids, and the unchanged germ-plasm of the fertilised ovum then only led to the formation of germ-cells after passing through a long series of somatic cells.

These facts do not, however, as yet constitute an actual proof of the correctness of this interpretation: they might be taken as indicating that the series has been developed in the reverse direction, the late differentiation of the germ-cells being the primary condition and the earlier separation of the two parts then having arisen gradually in individual cases. There can hardly be any doubt, indeed, that the early differentiation of the germ-cells ot the Diptera and Daphnidæ is of a secondary nature; and it will presently be shown that in the case of Hydroids such a shifting of the formative areas ('Bildungsstätte') of the germ-cells — *i.e.*, the fact of their earlier differentiation — can be actually proved. But the facts which have been stated still support the interpretation of them given above, in so far as they show that the germ-cells are by no means formed at the time and in the place where they are actually wanted, and that the time of their formation, in fact, varies very much, and must have been changed in the course of phylogeny. The direction in which this shifting originally took place — that is to say, whether it proceeded from the egg to the close of ontogeny or in the reverse direction — must be decided when our knowledge of the facts is more complete.

We might here lay stress on the fact that the destruction of the sexual glands in an animal, however low in the scale, is not followed by the formation of sexual cells in any other part of the body. Castration might be expected to have this effect if germ-cells could be formed from any young cells of the body. But just as in the case of any other highly specialised organs, such as the liver, kidneys, and central portions of the nervous system in Vertebrates, such a replacement never occurs. This fact is to be explained according to our present view by supposing that the formation of these latter organs anew is impossible, because the determinants necessary for such a development are not present in any other cells of the body. The same conclusion will, it seems to me, be inevitable in the case of the germ-cells; *the idioplasm necessary for the formation of germ-cells — i.e., germ-plasm — must be absent in these cases, and germ-plasm at any rate cannot be formed from somatic idioplasm.*

The case of the Hydroids * is probably still more convincing, for here a natural shifting of the place of origin of the germ-cells has actually taken place. As has already been mentioned, the germ-cells of Hydroids first arise very late in the life-cycle,

* Weismann, 'Die Enstehung der Sexualzellen bei den Hydromedusen,' Jena, 1883.

and hundreds or even thousands of cell-generations are passed through from the fertilised egg-cell onwards before they appear. In species which exhibit a complete alternation of generations, they are first formed in particular parts of the medusæ which have arisen from the polype-stock by budding—usually in the ectoderm of the manubrium. No trace of them is to be seen in

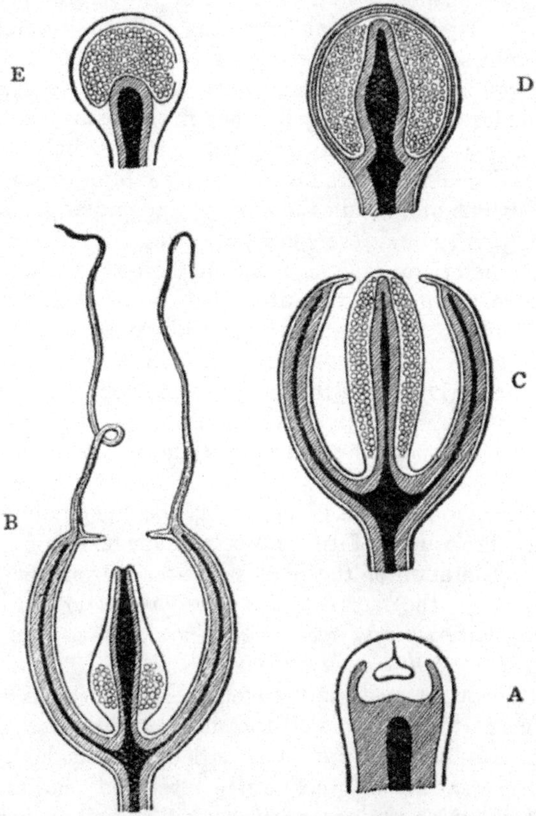

FIG. 12.—*Diagram of the degeneration of the Medusa into a mere gonophore.* **A, Medusoid bud; B, a Medusa shortly before it is set free; C, degenerated Medusa, in which the manubrium is present, but the mouth and tentacles are wanting; D and E, further stages in degeneration. (From Hatschek's 'Lehrbuch der Zoologie.')**

the young bud, and in many cases they only become differentiated from the other ectoderm-cells after the medusa has become detached from the stock, and has developed into an independent, free-swimming animal. Some of the ectoderm cells of the part in question then become transformed into egg- or sperm-cells.

In the case of certain species of polypes, free sexual medusæ were produced in the earlier period of the development of the species, but at the present day these do not become detached, but always remain attached to the stock: they thus no longer serve for the dispersal and ripening of the sexual products but only for their production and ripening. These species illustrate the different stages in the process of degeneration of the medusæ to mere gonophores, or sexual sacs. In some species the form of the medusa is completely retained in the sexual persons of the stock, only the eyes and marginal tentacles being absent; in others, the bell has become degenerated into a closed sac, the walls of which still retain the circular and radial canals; and in other species again, these canals have also disappeared, only the three characteristic layers of the medusa remaining, and even these have become so thin that their presence can only be detected in microscopic sections. Finally, these three layers also undergo degeneration, the wall then consisting of a single layer, so that the derivation of the sac from the bell of the medusa can only be proved indirectly. Throughout all these stages of degeneration, however, the ova or spermatozoa always ripen in the gonophores.

The behaviour of the germ-cells is the chief point of interest to us in the course of this process of degeneration. For the entire degeneration of the medusa proceeds from its germ-cells, and is due to the fact that *the development of the latter has gradually to be thrown back to earlier stages, so that the sexual elements are ripened more quickly*.

It will not be necessary to enter into the reasons for this hastening of the sexual maturity; it is sufficient to know that in some species in which the medusæ become detached, *e.g.*, *Podocoryne cornea*, the egg-cells are developed earlier than the medusæ in which they subsequently ripen, and in proportion as the degeneration of the medusa advances the place of origin of the germ-cells recedes more and more into the older and earlier formed parts of the stock. The advantage of this is, that the germ-cells develop earlier, and afterwards enter the germ-sacs in a riper stage: they thus reach maturity much more quickly.

The remarkable thing about this process is the fact that

active migrations of the germ-cells take part in it. Originating in the ectoderm, these cells wander into the endoderm, and subsequently back again into the ectoderm: and this remarkable process occurs in a definitely prescribed and regular manner. In spite of the relegation of their place of origin to earlier persons of the stock, the germ-cells always originate from *the same layer of cells* as that from which they arose in the ancestors of the species. It may thus be said that they are developed ontogenetically from the ancestors of those cells from which they would have arisen if the polype stock still produced free medusæ; or, in other words, *they arise lower down on the germ-track at present than they did formerly.* Thus in *Hydractinia echinata*, for instance, the youngest egg-cells first become visible in the endoderm of certain polypes in the same regions from which gonophores (degenerate medusæ) subsequently bud out. The egg-cells then migrate into the latter, and enter the ectoderm of the manubrium as soon as it is formed; and in this way *they return to the old place of ripening, which in earlier times was also the place in which they were formed.* At the present day, however, the egg-cells only apparently originate in the endoderm of the polype: it can indeed be proved that they are derived from the ectoderm, but migrate into the endoderm while still in a very young condition, before they exhibit the definite character of egg-cells. They therefore originate in the same region in which at an earlier phyletic period the ectodermal layer of the manubrium of the medusa was developed; or, in other words, *the same ontogenetic series of cells which produce the egg-cell at the present day did so in former times.* This fact probably only admits of one interpretation, and this is, that *only certain series of cells contain the primary constituents of the germ-cells*, and wherever it became useful in the course of phylogeny for the germ-cells to be situated in another position and in another layer of the body-wall, this change of position could only be effected by the cells of the germ-track becoming transformed into germ-cells at an *earlier stage*, and at the same time migrating into the other layer of the body-wall. If any — I will not say all — of the cells could give rise to germcells, this complicated mode of procedure would be quite inexplicable, for Nature always takes the shortest possible course.

If this reasoning is correct, the hypothesis of the *germ-tracks*, as I have formerly stated it, is inevitable; and the fact that the cells lying in these tracks are alone capable of giving rise to germ-cells, can hardly be explained otherwise than by assuming that these cells alone contain germ-plasm along with their special idioplasm. If germ-plasm could be produced from the idioplasm of ordinary somatic cells, it would be impossible to see why germ-cells should not be formed in Hydroids in case of need by the transformation of young ectoderm cells: but this never happens. And even if we wished to assume that the endoderm cells, as such, possessed an idioplasm which could not be transformed into germ-plasm, while the nature of the ectoderm cells rendeied such a transformation possible, this assumption would be contradicted by other facts; for, as far as we know, the germ-cells arise exclusively from the endoderm cells in the higher medusæ, and in the polypes nearly related to them. In this case therefore the germ-tracks are situated in the endoderm, — that is to say, the germ-plasm is only passed into certain series of cells in the endoderm, and the reserve material of unalterable germ-plasm, which will serve for the formation of the germ-cells, is passed into the primary endoderm cell only in the process of segmentation of the ovum, and is handed on by it. In the Vertebrata the germ-cells become differentiated from certain groups of mesoderm cells, and they are never found in any other part of the body. In this case the germ-track passes from the fertilised egg-cell into those segmentation cells from which the primary cells of the whole mesoderm are formed, in which latter it follows a closely confined course.

All these facts support the assumption that somatic idioplasm is never transformed into germ-plasm, and this conclusion forms the basis of the theory of the composition of the germ-plasm as propounded here. It is obvious that its composition out of determinants which gradually split up into smaller and smaller groups in the course of ontogeny, cannot be brought into agreement with the conception of the re-transformation of somatic idioplasm into germ-plasm. If, as we have assumed, each cell in the body only contains *one* determinant, the germ-plasm — which is composed of hundreds of thousands of determinants — could only be produced from somatic idioplasm if cells containing all the different kinds of determinants which are present in the body were to become fused together into *one* cell, their contained idioplasm likewise combining to form *one* nucleus. And, strictly speaking, even this assumption would be by no means sufficient, for it does not account for the architecture of the germ-plasm: the material only would be provided. Such a complex structure can obviously only arise historically.

The fact that somatic idioplasm cannot again give rise to germ-plasm serves as an additional support for the theory of the germ-plasm as here developed. Invisible, or at any rate unrecognisable, masses of unalterable germ-plasm must have been contained in the body-cells in all cases in which such a transformation has apparently occurred.

These masses need not necessarily be invisible, for they cannot be smaller than ids: and if it should subsequently be proved that the microsomes of the nuclear rods do actually correspond to ids, we may hope to ascertain the exact number of these ids in the individual species. An extensive field would then be opened out for further investigation, for it would be possible to decide by direct investigation whether the cell-series of the germ-tracks carry along with them a larger or a smaller number of ids than is contained in the fertilised egg-cell, and also the relative proportion of the number of ids in the somatic cells in the germ-track. We may thus hope that facts will come to light which can be utilised in connection with this theory.

Observations of this kind have already been made which indicate an actual continuity of the germ-plasm. Boveri * observed that the differentiation of the somatic cells from the primary sexual cell in the segmenting egg of *Ascaris megalocephala* is accompanied by a peculiar diversity in the nuclear structure. The nuclei of the somatic cells throw off a large part of their chromatin, in which process each idant loses a similar amount of substance. Further facts and illustrations of the process are still wanting, but even did we possess them it would be necessary to postpone the detailed theoretical explanation of such observations until we were able to judge as to the universal occurrence of the process. Observations which my assistant, Dr. V. Häcker,† has made on the segmenting ovum of a crus-

* Theodor Boveri, 'Anatom. Anzeiger II. Jahrgang,' No. 22 1887; and 'Zellen-Studien,' Heft 3, Jena, 1890, p. 70.

† Valentin Hacker, Zool. Jahrbücher, Abth. f. Anat. und Ontog., Bd. v., 1892; and Archiv. f. mik. Anat., Bd. 40, 1892.

tacean (*Cyclops*), have indeed also proved that the behaviour of the somatic segmentation cells is different from that of the primary sexual cell, but the process differs essentially from that which occurs in *Ascaris*. When we are in possession of similar observations on various types of animals, we shall be able to recognise the essential parts of the process, and shall then be in a position to offer an explanation of them.

From a theoretical point of view, we must expect that the ids of germ-plasm become doubled in the nucleus of the fertilised egg-cell or even previously, one half of such a double id being in the active condition in which it can undergo disintegration, and the other being in the inactive and unalterable condition. The former direct ontogeny, and the latter are passed on in a passive condition to the primary sexual cells. As these, however, behave at first like somatic cells,—that is to say, they multiply in a regular manner, and are distributed amongst definite series of cells to definite parts of the body,—they must possess active idioplasm in addition to unalterable germ-plasm. They must therefore contain *more* ids in their nuclear matter than do the somatic cells. The above-mentioned observations on *Ascaris* can thus be explained in accordance with our theory up to this point, but more than this cannot be stated at present.

2. The Germ-tracks

Taking sexual reproduction only into consideration for the present, the course of the germ-tracks in existing Metazoa apparently varies both as regards its length and the direction it takes. The germ-track is shortest in the *Diptera*, in which the primary germ-cell becomes separated off in the first division of the ovum, so that in this case we might speak of a division of the ovum into a primary germ-cell and a primary somatic-cell. In the *Daphnidæ* the germ-track is longer; for, counting from the ovum, five successive divisions occur before the primary germ-cell is formed. In the free-swimming marine worm *Sagitta* it is longer still, two primary germ-cells only appearing after ten or more successive divisions have occurred, and the mass of embryonic cells has already given rise to a gastrula-larva. In other worms, such as the Nematodes, the primary germ-cells become separated from the somatic cells in a still later generation of cells, which has so far not been actually determined; and in most of the higher Metazoa this only occurs after the formation of hundreds or thousands of cell-generations.

The position of the germ-track may also vary. In the *Diptera* it is quite distinct from the somatic cell-tracks, and the genealogical trees of these two kinds of cells separate at the root. In the *Daphnidæ* the germ-track passes through each of the first four segmentation-cells, and then branches off from the somatic

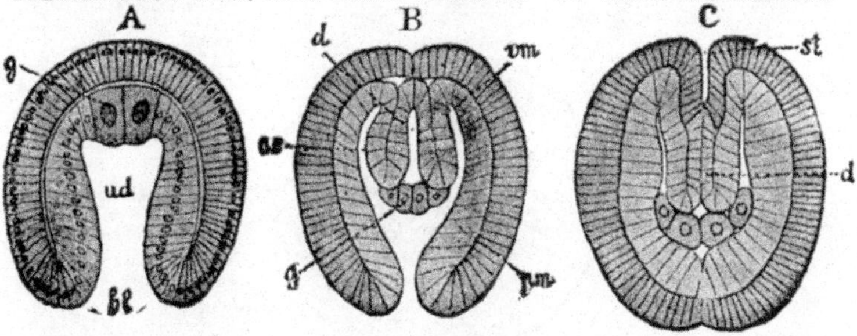

Fig. 13. — ***Three early stages in the development of Sagitta.*** — **(After O. Hertwig.) In *A* the differentiation of two primitive germ-cells (g) from the endoderm, and in *B* and *C* the multiplication and separation of these cells is shown. (From Lang's 'Lehrbuch der vergleichenden Anatomie.')**

tracks. In *Sagitta* it passes through the primary endoderm cell, and the primary germ-cells separate from the primary endoderm before the definitive endoderm of the alimentary canal has been formed from the latter. In *Rhabditis nigrovenosa* the germ-track extends through three generations of endoderm cells, passes into the primary mesoderm, and after several generations branches off from two of the mesoderm cells. In most of the Metazoa, however, the formation of the primary germ-cells is postponed to a still later period, so that the separation of the germ-branch from the somatic branch takes place at a much higher level on the genealogical tree of the cells, and often first occurs in the younger and smaller lateral branches. The primary germ-cells do not always branch off from the track of the endoderm, but may just as often diverge from that of the ectoderm. In the lower Medusæ, for instance, in which the development is a direct one, the germ-cells become differentiated at a very late stage from the ectoderm cells of the body, which is already fully formed and often independent and self-supporting; while in the higher Medusæ and Ctenophora the primary germ-cells are derived from the endoderm. We thus see that the germ-tracks or series of cells which lead from the egg-cell to the primary germ-cells frequently take very different courses: they are in some cases very short, and in others longer—sometimes so long that they pass through very different embryonic cells; in some instances they branch off from the primary endoderm-cells, and in others from those of the mesoderm, and they may even arise from later generations of the mesoderm, ectoderm, or endoderm-cells.

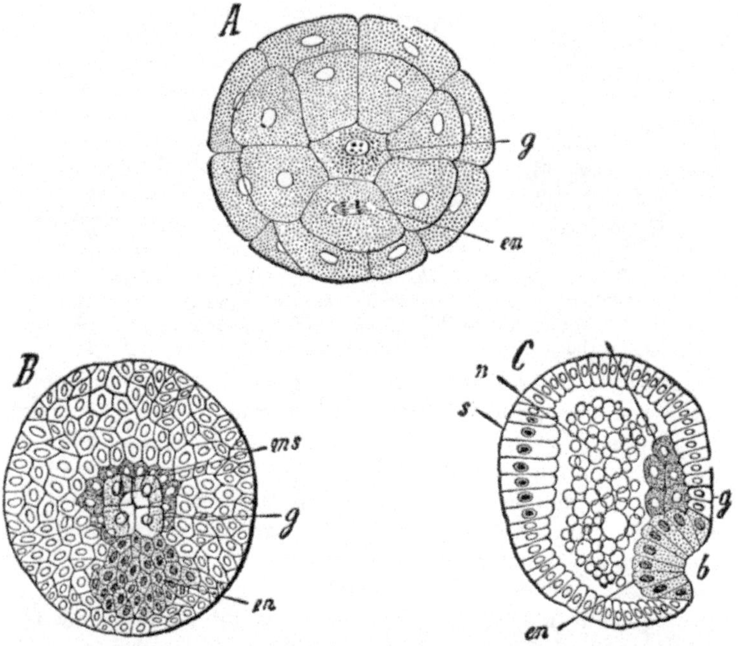

Fig. 14.—*Three Stages in the development of the summer eggs of Moina.*—(After Grobben.) *A*, Stage of segmentation viewed from the vegetative pole, in which thirty-two cells are present; *B*, Blastula stage, from the vegetative pole; *C*, Gastrula stage, in longitudinal section; *g*, the primitive germ-cells. (From Korshelt and Heider's 'Lehrbuch der vergleichenden Entwickelungsgeschichte.')

The same germ-track is always strictly followed in each of these cases respectively, no deviation ever taking place: thus the primary germ-cells never arise from endoderm-cells in a group in which the normal germ-track lies in the ectoderm, and *vice versâ*. We consequently cannot help arriving at the conclusion *that the cells in the germ-track must have some advantage over the rest of the cell-tracks in ontogeny, for they alone are capable of giving rise to the primary germs-cells.* Moreover, if we remember that in the case of the Hydroid polypes the period of the separation of the primary germ-cells can be relegated to earlier or later stages, it will be clear that not only the cells at the terminations of the germ-track in which this separation actually occurs in individual cases, but also the entire preceding series of cells, possess qualities which are absent in the other cells of the organism, and which, sooner or later, render the cells of the germ-track capable of giving rise to primary germ-cells.

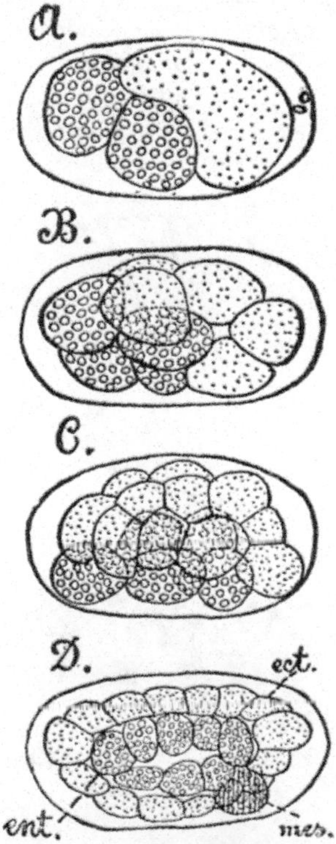

FIG. 15.—*Stages in the segmentation of the ovum and formation of the germinal layers in Rhabditis nigrovenosa.*—(After Götte.) *ect*, Ectoderm; *ent*, Endoderm; *mes*, Mesoderm.

The cells of the germ-track do not themselves correspond to primary germ-cells, the character of which latter is not as yet apparent; they are cells of a mixed character,—that is to say, they contain different primary constituents, which are gradually separated off until eventually only two of them remain, and these then separate from one another by means of a final cell-division.

The embryogeny of a parasitic worm (*Rhabditis nigrovenosa*) from the frog's lung may serve to illustrate this point. In fig. 15, A to D represent the first four stages in segmentation up to the differentiation of the primary mesoderm-cell (*mes*). This and the following stages are represented diagrammatically in fig. 16, which shows the genealogical tree of the cells and the germ-track. The ovum (*Eiz*) must of course be considered as containing the whole of the primary constituents of the organism before its first division into a primary ectoderm (*urEkt*) and a primary endoderm cell (*urEnt*). The latter retains all the primary constituents of the mesoderm and primary germ-cells, in addition to those of the endoderm, and is therefore not *merely* a primary endoderm cell. This then divides again and forms two cells, of which the one marked 3 on the left side of the figure only contains primary constituents of the endoderm, and is therefore an endoderm cell proper; while that marked 3′ represents the first rudiment of the mesoderm and of certain portions of the endoderm, and *contains in addition the primary constituents of the primary germ-cells*. This cell (3′) divides into two (4′ and 4″), thus separating the above-named rudiments into those for the right and the left sides of the body; and finally, the

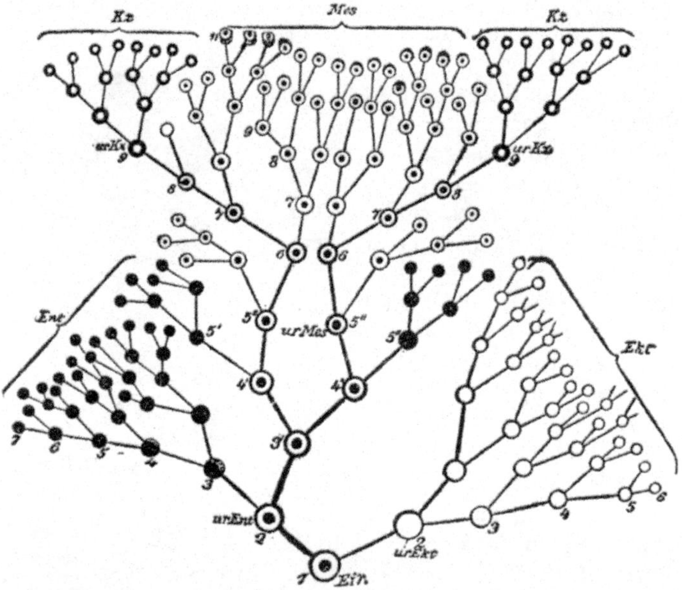

Fig. 16.—***Diagram of the germ-track of Rhabditis nigrovenosa.***—**The various generations of cells are indicated by Arabic numbers, the cells of the germ-track are connected by thick lines, and the chief kinds of cells are distinguished by various markings:—the cells of the germ-track by black nuclei, those of the mesoblast (*Mes*) by a dot in each, those of the ectoderm (*Ekt*) are white, those of the endoderm (*Ent*) black; in the primitive germ-cells (*ur Kz*) the nuclei are white. The cells are only indicated up to the twelfth generation.**

complete separation of the rudiments of the mesoderm and endoderm occurs, and results in one daughter-cell (5″), containing the primary constituents of the mesoderm and primary germ-cells, while the other gives rise to a cell of the endoderm proper. The primary constituents of the primary germ-cells remain connected with certain of those of the mesoderm during several generations of cells, and in each subsequent division certain of the latter pass out *alone* into one of the daughter-cells, the other retaining the primary constituents of the primary germ-cells in addition to those of the mesoderm. Finally, in the ninth series of cells in the diagram — in which the processes are represented as greatly abbreviated — the separation of these two sets of primary constituents occurs, and the first primary germ-cell (*ur Kz*) is formed.

So much is certain, and does not depend on any hypothesis. Opinions may differ as to whether the cells situated in the germ-track are to be described as real somatic cells. I have called them so, because in many cases the germ-tracks extend far beyond the period of embryogeny into the fully-developed functional tissues, and because it can be proved that *even cells which are histologically differentiated may produce germ-cells under certain circumstances.* This occurs amongst plants — in the prothallus of ferns, for instance — and also in the cells of certain Polyzoa from which gemmation may take place, and which must therefore contain inactive germ-plasm. In these cases it is certain that real somatic cells are situated along the germ-tracks; in all cases the cells of the germ-track *are not germ-cells from the first*, and they always take part in the construction of the body. And if we further consider that a large number of somatic cells must contain accessory idioplasm of some kind, — either that which will serve for simple regeneration, or for the regeneration of more complex parts, or again, for the formation of buds, — we can hardly assume that the character of a somatic cell is thereby abolished: I can see no advantage *in objecting to describe a cell of the germ-track as a somatic cell.*

The change which the idioplasm of the cells constituting the germ-track undergoes, can obviously only consist in its *active* portion gradually becoming separated off in the course of the ontogenetic cell-divisions, so that ultimately the cell contains germ-plasm only, which then stamps it as a germ-cell. Even then the germ-plasm remains unalterable as long as this first or primary germ-cell continues to produce others similar to itself. The cells only become differentiated into spermatozoa and ova when this multiplication ceases, and this presupposes the splitting off of special spermatogenetic or ontogenetic determinants The disintegration of the germ-plasm which results in a new embryogeny — provided that the necessary conditions have been fulfilled — can only begin when this has occurred.

3. Historical Account of the Theory of the Continuity of the Germ-plasm

When my essay on the 'Continuity of the Germ-plasm' appeared seven years ago,* I was under the impression that I was the first to give utterance to this conception. Since then, however, I have found that similar ideas had arisen, in a more or less distinct form, in other brains; and I gradually discovered that a number of authors had independently recognised more or less clearly the distinction between the body-cells and germ-cells and the direct connection between the germ-cells of different generations: some had merely made the assertion, and others had attempted to support it by facts. I shall here give an account of those theories which preceded mine, taking them in chronological order.

As early as 1849, Sir Richard Owen had indicated that differences may arise in the developing germ-cells between those which become greatly changed and form the body, and those which only undergo a slight change and form the reproductive organs.†

Francis Galton was the first to express certain ideas which bore some resemblance to the conception of the continuity of the germ-plasm. In a paper which appeared as early as 1872, the individual is conceived 'as consisting of two parts, one of which is latent, and only known to us by its effects on his posterity, while the other is patent, and constitutes the person manifest to our senses. The adjacent and, in a broad sense, separate lines of growth in which the patent and latent elements are situated, diverge from a common group, and converge to a common contribution, because they were both evolved out of elements contained in a structureless ovum, and they jointly contribute the elements which form the structureless ova of their offspring.' ‡

---

* 'Der Continuität des Keimplasma's,' Jena, 1855 (Essay iv., p. 163, in the second English edition).

† I quote this statement from Geddes and Thomson's 'Evolution of Sex' (London, 1889), p. 93, in which the original authority is not given.

‡ Proc. Roy. Soc., No. 136, p. 394.

A few years later, Galton changed his opinion and adopted Darwin's theory of pangenesis, which he modified considerably, and only used 'as a supplementary and subordinate part of a complete theory of heredity.' This theory has already been discussed in the Historical Introduction to this book. The 'gemmules' which are contained in the fertilised ovum together constitute the 'stirp' or stock, which by means of the egg-cell gives rise to a new individual. Each 'sort of gemmule' is represented by a number of gemmules which differ somewhat from and compete with one another; and since the successful ones in the competition for taking part in the construction of the body form the various parts of the body and are therefore contained in them, the rest remain unused, thus constituting the 'residual germs.' These, then, are 'the parents of the sexual elements and buds.' The 'dominant' gemmules may also take a part, though only a slight one, in the formation of the germ-cells, 'as they are the least fertile in the production of gemmules.' The germ-cells are therefore mostly formed from gemmules which have remained latent, and this accounts for the fact that the offspring usually do not exhibit the most marked peculiarities of the parent. As this hypothesis only accounts for the dissimilarity between parent and child, so far as it exists, and not for the far commoner resemblance between them, Galton assumes that the parts of the body can also give off gemmules which become distributed and extend beyond the boundaries of the cells in which they arose, and so may even penetrate into the sexual elements. He thus substitutes the idea of a locally restricted distribution of the gemmules for Darwin's view of their 'free circulation.' If we attempt to make this somewhat vague and unrealistic idea rather more comprehensible, by considering the 'residue of the stirp' as equivalent to the 'unalterable' reserve germ-plasm, Galton's hypothesis will be found to bear some resemblance to the theory of the continuity of the germ-plasm. But there is still a fundamental difference between them, for Galton's idea is only conceivable on the presupposition of the occurrence of sexual reproduction, while the theory of the continuity of the germ-plasm is entirely independent of any assumption as to whether each primary constituent is present in the germ *singly* or in numbers. According to my idea, the active and the reserve germ-plasm contain precisely similar primary constituents, gemmules, or determinants; and on this the resemblance of a child to its parent depends. The theory of the continuity of the germ-plasm, as I understand it, is not based on the fact that each 'gemmule' necessary for the construction of the soma is present many times over, so that a residue remains from which the germ-cells of the next generation may be formed: it is founded on the view of the existence of a special adaptation, which is inevitable in the case of multicellular organisms, and which consists in the germ-plasm of the fertilised egg-cell becoming doubled primarily, one of the resulting portions being reserved for the formation of germ-cells.

Gustav Jäger * was the first to express the idea that in the higher organisms the body consists of two kinds of cells, which he calls respectively 'ontogentic' and 'phylogentic;' and that the latter, or reproductive cells, are not a product of the former, or body-cells, but are derived directly from the germ-cell of the parent.† He took it for granted that the 'formation of reproductive substances occurs in an animal during the early embryonic stages,' and imagined that he had thus proved the existence of a connection between the germ-cells of the parent and those of the child. Although these opinions were not founded on fact or followed out in detail, they ought to have led to further ideas on the subject. They, however, together with the book in which they were contained, remained unnoticed.

A few casual remarks made by Rauber,‡ in a paper on 'Formbildung und Formstörung in der Entwicklung von Wirbelthieren,'

---

* Gustav Jager, 'Lehrbuch der allgemeinen Zoologie,' Leipzig, 1878, II. Abtheilung.

† The praiseworthy attempt to do justice to my predecessors in this particular subject has perhaps been carried too far. In Geddes and Thomson's 'Evolution of Sex' (p. 93), for instance, a quotation is given from Jager which seems to prove that he anticipated me with regard to the theory under consideration. The quotation in which this idea is expressed is, however, not taken from the book published in 1878, but from an essay written ten years later, and it concludes with the following words: —'This reservation of the phylogenetic material I described as *the continuity of the germ-plasm.*' But no mention is made by Jager of the continuity of the germ-plasm in his book which appeared in 1878, in which a connection between the germ-*cells* of different generations is supposed to exist: — and this is not the case. The entirely new statement of his ideas has been influenced by those contained in my essays which had appeared in the meanwhile.

‡ 'Morphol. Jahrbuch,' Bd. 6, 1880.

suffered the same fate. This author states that 'as regards the effect of fertilisation, it can only convert a *portion* of the egg, viz., the personal part, into the form of a person; the other portion does not experience this effect, for it has a stronger power of persistence.'

Finally, Nussbaum * was likewise led to the idea of the continuity of the germ-*cells*. He, too, assumed that 'the segmented ovum divides into the cell-material of the individual and the cells for the preservation of the species,' and he supports this statement by quoting the cases already mentioned of the very early differentiation of the sexual cells.

I will conclude this section with the words which appeared in the preface to a short paper intended as a defence against the accusation of plagiarism which had been made against me. 'A fertile scientific idea has rarely appeared without having been contested on the one hand, and set down as already known on the other. The former is certainly a perfectly justifiable and even necessary course of proceeding, for a clear and definite knowledge of the truth can only result from the contest of opinions; and even the latter is to some extent justifiable, for an idea of this kind probably very rarely arises without having been preceded by similar attempts directed towards the same object; and it is only natural that those who first made such attempts should overlook the difference between these struggles towards the desired object and its attainment.'

Others may decide the reason why no attention had been drawn to the suggestions mentioned above as having been made previously to my theory of the continuity of the germ-plasm, and why these did not exert any influence on scientific thought. This is certainly the case: and it practically follows from the fact, that all the objections which have been made have been directed against *me*. Some of these objections will be discussed in the following chapter. That I am far from desirous of detracting from the merit of others, has, I trust, been shown by the fact that as soon as I became aware of previous views on the subject I brought them forward Jäger's ideas, for instance, might have long remained unnoticed, had not I brought them to light. But an historical account of the various previous views

* M. Nussbaum, 'Die Differenzirung des Geschlechts im Thierreich,' Archiv. f. mikr. Anatomie, Bd. xviii., 1880.

on this subject * cannot be considered to be an impartial one, if *no mention is at the same time made of the fact that all these suggestions remained unnoticed, and had no effect on the progress of scientific thought.* That this is the case there can be no doubt. And although it may be a satisfaction to every one to have expressed a correct idea, no such idea can be considered as fertile, and as having an important influence on the progress of scientific thought, unless its meaning is so obvious that it results in further progress. Such a result, however, only followed after my essays had appeared.

4. Objections to the Theory of the Germ-plasm

Important objections to this theory have been raised by several botanists; and at first sight the facts on which these are based may easily give rise to the impression that the theory cannot be carried out in the case of plants. If this were so, however, its correctness would be altogether doubtful, for the hereditary mechanism cannot be totally different in plants and animals. We must therefore make a closer examination into the facts as they concern plants, and I hope to be able to show that the fundamental ideas which I have assumed are applicable to plants as well as to animals, although they did not originate from the botanical point of view.

Certain misconceptions and inaccurate representations must first be put on one side. Many botanists deny the *existence* of the germ-plasm entirely.

Vines † considers the assumption of a special 'reproductive substance' unnecessary, as the capacity for reproduction is a fundamental property of protoplasm. A cutting gives rise to a complete plant, just as a broken crystal becomes complete when immersed in the mother-liquor, for it produces the missing parts, viz., roots. It is not necessary to assume the existence of a special 'reproductive substance' in either case.

I need not especially emphasise the fact that this stimulus which results in the completion of a part is not by any means a universal phenomenon, and that, for instance, some parts of plants cannot be reproduced from cuttings. I shall simply

---

* *Cf.* the account given in Geddes and Thomson's 'Evolution of Sex,' pp. 93 and 94.

† *Cf.* 'Nature,' Oct. 24, 1889, and 'Lectures on the Physiology of Plants,' Cambridge, 1886.

confine myself to calling attention to the fact that even if a universal reproductive power existed in protoplasm, it certainly would not explain matters. *For this power is just what has to be explained.*

If we know, for instance, that Infusoria are able to replace great losses of substance, — so that when the oral region is cut off, it, together with all the cilia and other minute structures, can be formed anew, — a proof is thereby obtained that these unicellular organisms actually possess the universal reproductive power which Vines wishes to ascribe to vegetable protoplasm. But does this help us in the slightest degree to understand the fact, or to explain why the ultimate particles of the cell-body become rearranged and transformed after a loss of substance has occurred, just as is necessary for the reappearance of the species? Do we thereby gain the faintest idea as to how and why the residual particles of the cell-body are compelled to give up their previous form and connection, and to reconstruct exactly that part which is required in order to render the whole complete? The assumption of a 'reproductive power' simply amounts to the statement of the fact that regeneration occurs; and this, it seems to me, is equivalent to saying that the reproductive power is a fundamental property of vegetable protoplasm.

In the case of the unicellular Infusorian we can, however, hardly venture at present to attempt an explanation of this problem, as we know very little of the vital units of which the cell-body is composed, and of the forces situated within them. But the case is different with regard to those organisms which consist of many physiologically differentiated cells: in these we are at any rate acquainted with the form and function of one arrangement of the vital units of which the whole aggregate is composed, and so we can attempt to deduce the functions of the whole body from those of the units, and conversely to refer the latter processes to a distribution of the forces amongst the units composing the whole. We need not therefore confine ourselves to the mere statement of the fact that a process occurs by means of which the whole is completed, but we may further inquire as to when this occurs, from what elements it proceeds, how the whole body arises at all, and how so complex a structure can be formed from the apparently simple substance of the germ.

In order to give a satisfactory answer to these questions, I have assumed the existence of a germ-plasm, but have not primarily regarded this as a 'special reproductive substance' which is very different from all other substances in the body; I have looked upon it, on the contrary, as *the substance which gives rise to all the other formative substances of the entire individual.* Every part of the body contains a *portion* of this substance, and the whole organism can only be formed anew when all the portions of this controlling substance (the idioplasm) are combined; that is to say, when germ-plasm is present. The assumption of germ-idioplasm or germ-plasm is, I consider, quite unavoidable, for we must at the present day take it as proved that the hereditary tendencies are connected with a substance. In my opinion, it is also an irrefutable fact that this germ-plasm undergoes regular changes from the ovum onwards: it must, indeed, undergo change from cell to cell, for we know that the individual cell is the seat of the forces which give rise collectively to the functions of the whole. The forces which are virtually contained in the germ-plasm can therefore only become apparent when its substance undergoes disintegration, and its component parts, the determinants, become rearranged. The difference in function seen in the various groups of cells in the body compels us to suppose that these contain a substance which acts in various ways. *The cells are therefore centres of force of different worth, and the substance (idioplasm) which controls them must be just as dissimilar as are the forces developed by them.*

The apparent similarity of many young plant-cells may account for the vagueness with which Vines, following Sachs's example, speaks of an '*embryonic substance*' from which reproduction is supposed to proceed in all cases, and which is assumed to be present in all 'young' cells. In my opinion the hereditary value of a cell can be estimated as little by its age as by its appearance. The mass of cells resulting from the segmentation of an animal egg certainly possess the character of youth, and in a certain stage of development these cells are all of the same age and all look alike. They have, however, entirely different hereditary values; and if we are accurately acquainted with the ontogeny of the animal in question, we can tell what hereditary tendencies lie hidden in each cell. The primary constituents of the entire endoderm, for instance, may be contained in one cell, and that of the ectoderm or mesoderm in another: or, again, in a later stage. only a rudiment of a particular part, organ, or portion of an organ belonging to the germinal layer in question, may be present in an individual cell. But if we inquire whether the entire body could arise from each of these cells, known facts give a very decided answer in the negative. Only one, or a few perfectly definite cells amongst them, which we speak of as germ-cells, can reproduce the whole animal under favourable circumstances. This is true of all the higher Metazoa: *the cells of the segmenting ovum are completely dissimilar as regards their hereditary value, although they are all 'young' and 'embryonic,' and are not infrequently quite similar in appearance.* It therefore seems to me to follow from this, as a logical necessity, that the hereditary substance of the egg-cell, which contains all the hereditary tendencies of the species, does not transmit them *in toto* to the segmentation-cells, but separates them into various combinations, and transmits these in groups to the cells. I have taken account of these facts in considering the regular distribution of the determinants of the germ-plasm and the conversion of the latter into the idioplasm of the cells in the different stages of ontogeny. All these cells contain 'embryonic substance,' but the determinants contained in one set differ from those in another, and therefore contain different hereditary tendencies. Hence it is comparatively meaningless to speak merely of an 'embryonic substance.'

De Vries regards some of my views in a very different way, and from an entirely different aspect. In an extremely able manner he brings forward a number of facts concerning heredity in plants, and finds that they usually do not fit in with my views. I have followed his deductions with great interest, and have gratefully made use of the facts which he has brought forward; but I nevertheless believe that the chasm which separates his views from mine can be bridged over.

In the first place, de Vries accuses me of having taken a one-sided view of the question by considering the processes as they occur in animals only: in these it may be possible, as I have assumed, to draw a sharp line of distinction between somatic- and germ-cells, but this cannot be done in the case of plants. In the latter, those series of cells which I have called germ-tracks may give rise to many other cells besides germ-cells, although this as a general rule is only exceptionally the case:

that is to say, it occurs in response to definite external influences. It would not, however, only take place in those parts of the plant which might be assumed to be specially adapted for this capacity, but might also occur in those in which adaptation is out of the question. We are therefore compelled to assume that *most*, if not all, of the *cells contain all the primary constituents of the species in a latent condition.*

I will first discuss the manner in which de Vries applies my hypothesis of the germ-tracks to the case of plants, and the conclusions at which he has arrived and has illustrated by describing a number of genealogical trees representing the various series of cells in plants.

De Vries draws a distinction between 'primary' and 'accessory' germ-tracks. The former correspond to the germ-tracks I have already assumed: that is to say, to those cell-series which normally lead from the fertilised egg-cell to the new germ-cells (ova, spermatozoa, pollen-grains). By 'accessory germ-tracks' are meant those cell-series which lead to germ-cells 'through adventitious buds.' These accessory germ-tracks are, according to de Vries, absent in the higher animals, but are of common occurrence amongst plants, and I am accused of not having taken them sufficiently into account. The 'accessory germ-tracks,' if I understand the term aright, are regular germ-tracks, which do not, however, always come into use. In many of the lower plants, such as mosses and fungi, 'almost all of the cells may develop into new individuals;' and in the higher plants, buds, from which entire plants possessing germ-cells may arise, can, under certain circumstances, be formed at any rate from certain kinds of tissue, which may consist of young (meristematic) cells or indeed even of full-grown cells.

Let us first consider the 'primary germ-tracks.' De Vries thinks that their behaviour is essentially different in the higher animals and in plants: in the former, the genealogical tree of the cells of the germ-track 'is straight, and only slightly branched at the apex,' while in the higher plants 'the branches are so numerous and subdivided from the base upwards that they frequently overtop the main stem; or, more accurately, the main stem is hardly recognisable.' No objection can certainly be raised to this statement, which we may illustrate by a blossoming apple tree, in which the blossoms which crown the top may be taken as corresponding to the germ-cells. But how is this difference to be proved, and on what does it depend? It is not based on the animal or vegetable nature of the organisms, for as de Vries himself incidentally acknowledges, we find a similar kind of branching of the primary germ-tracks in the Hydroid-polypes. It simply depends on the fact that a *higher individuality of the stock* exists in these animals, just as in the case of the higher plants. In both cases we have to deal not with a single person and the formation of its germ-cells, but with a number of persons which arose from the primary one by budding, each of which has a body of its own, and gives rise to its own germ-cells. The germ-track is concealed within the first person of the stock produced from an egg, and gives off a lateral branch as soon as this first polype develops a bud. Shortly afterwards, the polype gives rise to a second bud, into which a lateral germ-track likewise extends; and when these two buds have developed into complete polypes, they again give rise to buds, into which germ-tracks are once more given off, and so on. The copious branching of the germ-track is thus accounted for, and it is quite immaterial whether the separate persons of the stock are more or less independent and perfect, and to what extent they may be regarded as 'individuals.' In those cases in which a periodic segmentation of the body into serially homologous segments or metameres takes place, — each of which has almost a similar origin and is able to produce germ-cells, — the type of the genealogical tree of the germ-track, as described above, results.

If, however, we inquire as to the conclusions which can be drawn from the course taken by the germ-tracks in animals and plants, we receive a reply from de Vries which is very significant of the way in which this problem is at present regarded by many botanists: — *the whole question which I have raised with regard to the continuity of the germ-plasm is an idle one.* In his opinion, 'the whole question as to whether somatoplasm can become transformed into germ-plasm has no basis in fact.' 'A germ-track,' says de Vries, 'never arises from a somatic-track,' and 'a continuity of the germ-cells exists, not merely in the very rarest cases, but universally and without exception, although it often takes place through a long series by means of the germ-tracks.'

With the exception of the last one, these sentences merely repeat my own views, and the apparent contradiction of the latter is simply due to the fact that de Vries adopts expressions which

I have used in another sense. In stating that germ-cells arise from somatic cells in innumerable cases, I referred to the somatic cells which are situated along the germ-track, the existence of the latter being assumed for this special purpose. De Vries, however, disputes the somatic character of these cells, because he considers that they also contain 'germ-substance.' I should attach slight importance to a mere name, if a very definite idea did not depend on this name, the abandonment of which would lead to confusion. It appears to me to be dangerous to introduce a third category of cells — viz., those of the germ-track — between the somatic and the germ-cells. In the first place, it is unpractical to do so, for the appearance of a cell does not reveal to us whether it is situated in the germ-track or not; and secondly, it would lead to a total confusion of the ideas of somatic and germ-cells; for, as has been shown in the previous chapters, there are a number of cells in plant- and animal-stocks which are undeniably somatic, and which must therefore contain germ-plasm. Since we regard the 'blastogenic' idioplasm of plants and Hydroids as a modification of germ-plasm, we must also look upon a very considerable number of the cells which constitute these organisms as cells of the germ-track, and we should therefore arrive at the absurd conclusion that a soma (body) is not present at all in these cases. The soma nevertheless *is* present, and a contrast also exists between it and the germ-cells in plants as much as in animals.

De Vries contradicts himself when he states that a universal 'continuity of the germ-*cells*' exists through the germ-track; for in other passages he emphasises the fact that germ-cells do not as a rule arise directly from one another (p. 84), and that a distinction must be made between germ-cells and cells of the germ-track. The somatic character of the cells of a fern-prothallus, for instance, cannot be denied, for they function as somatic cells, and at first are all similar in appearance, so far as we are able to observe. But nevertheless some of them are situated on the germ-track, and give rise to male and female germ-cells.

If de Vries puts aside the whole question of the continuity of the germ-plasm because he is able to prove that germ-cells always arise from cells of the germ-track, it is evident that he must be labouring under a similar delusion to that which induced Sachs to claim precedence as regards the theory of the continuity of the germ-plasm. Both these observers consider it self-evident that each apical cell contains the germ-substance of the ovum, because in plants all growth takes place from the growing point and originates in the apical cells, which are derived directly from the egg-cell.

This, however, is at any rate only self-evident in the case of the first apical cell of the main shoot, and is certainly not so in that of the lateral shoots, which are, indeed, only derived indirectly from the former. All the cells of a plant are undoubtedly descended in a direct line from the ovum; but this fact does not imply that they must all give rise to apical cells or must all contain germ-plasm, nor does it in any way explain the fact that only relatively few of them can become germ-cells and the rest cannot. These were the very facts which the hypothesis of the continuity of the germ-plasm was intended to make comprehensible, to some slight extent at any rate. The origin of a cell from the ovum gives no clue to its nature; and, as de Vries himself says, the entire description in detail of the cell-series leading from the ovum to the first apical cell, although very interesting in itself, gives us no information as to the origin of the germ-substance present in certain parts of the plant-body. I do not understand therefore how de Vries can look upon the fact of the existence of these cell-series as constituting in itself an important explanation of the problem, without attempting to explain it further. It seems to me that the cell-series can only be of any explanatory value when they are regarded as germ-tracks in the sense in which I use the term, — that is to say, as those series through which the germ-substance is transmitted from the egg-cell to the remotest parts of the plant.

I must say that it seemed to me to be a somewhat crude idea to suppose that the same kind of idioplasm is contained in all the cells of the germ-track, including the apical cell, and that it is equivalent to 'germ-substance.' Why do not the apical cells in the sterile shoots of the horse-tail give rise to germ-cells, while those of the fertile shoots do so? This must be due to a difference in the idioplasm of the apical cells of these shoots. And although structures bearing germ-cells may become developed from the apical cell of a fertile shoot of the plant, all the cells of the latter do not nevertheless give rise to germ-cells: only *certain* cell-series lead from the apical cell to the new germ-cell, and these are the cells of the germ-track which contain germ-plasm! The process of the formation of a shoot from an apical cell is analogous to that of the production of a single polype by budding from a polype-stock. But both these processes are essentially the same as that of the development from the ovum in a higher animal. In all three cases the formation of the new animal originates in one cell. The latter must therefore possess an idioplasm which contains all the primary constituents of the organism; and, moreover, if the organism is to be 'fertile,'—in the sense in which this term is used by botanists,—the original cell from which it is derived must contain the primary constituents of *all* the structures characteristic of the species in its idioplasm: that is to say, it must contain germ-plasm. If we trace the development of such a shoot or organism, we shall find that it follows a precisely similar course to that which we have already described in the case of embryogeny; and that at each cell-division the primary constituents break up into smaller and smaller groups, until at last each cell only contains *one* such element. And yet all these very different kinds of cells are descended in a direct line from the original cells. How, then, can we account for the fact that one or several of them contain all the primary constituents of the species in a latent condition, in addition to one specific primary constituent of a particular kind of somatic cell, as must be the case in those which give rise to germ-cells? It would, indeed, be a very simple matter if a continuous series of cells which contain 'germ-substance' *only*, led from the original cell to the new germ-cells. But as simple a case as this only occurs in the Diptera: in all other instances *the intermediate cells which constitute the germ-track can be proved to contain perfectly definite somatic elements in addition to the germ-plasm; and this is the case in plants as well as in animals.*

To make this clear, it is only necessary to glance at the genealogical tree representing the ontogeny of *Rhabditis nigrovenosa* (fig. 16). How does it come to pass, for example, after the division of the primary endoderm cell into the first endoderm and first mesoderm cell, that the latter is nevertheless capable of producing cells subsequently which contain 'germ-substance,' *i.e.*, germ-cells? At its origin this cell gave up the primary constituents of the endoderm to the sister-cell; by what means do these primary constituents—even those of the ectoderm which were previously given up—reach the germ-cells which eventually arise from this cell? My answer to these questions has already been given, and is as follows: — in addition to their active meso-derm-idioplasm, the cells which will give rise to germ-cells carry along with them a certain amount of germ-plasm in an unalterable condition. De Vries, and those botanists who agree with him, consider my answer superfluous. Every one, of course, is at liberty to reject the solution of a problem, but in that case he must not claim to have explained it.

I now come to the consideration of de Vries's 'accessory germ-tracks.' As has already been stated, this term is used to describe those series of cells which give rise to germ-cells through the agency of 'adventitious buds.' According to Sachs,* adventitious buds correspond to those growing points which are not derived from those already present, or 'from embryonic tissue already present,' but which 'originate at places where the tissue has already passed over into the permanent condition, — in fully-developed roots, in the interfoliar parts of shoot-axes, and more particularly in foliage leaves, the tissues of which are already completely differentiated and developed.'

In my former essays I have endeavoured to account for these 'adventitious' buds, — such as are formed, for instance, in a Begonia leaf when it is placed on damp soil, — by supposing them to be adaptations of particular species of plants to this peculiar method of reproduction: I assumed that certain series of cells which in these species take part in forming the leaves contain unalterable and inactive germ-plasm in addition to their own active idioplasm.

In opposition to this interpretation much may be, and in fact has already been, said, and the principal objections must now be considered.

It has, in the first instance, been urged that the capacity possessed by leaves, roots, and so on, for producing adventitious buds, cannot be regarded as an adaptation, because so many cases are known in which this process only occurs exceptionally, and is of no advantage to the plant. There can be no doubt, however, that the power possessed by *Begonia*, *Bryophyllum*, *Cardamine pratensis*, and *Nasturtium officinale*, of giving rise to buds in those parts in which they are not formed in most plants, is due to an arrangement peculiar to these plants. Neither in *Begonia* nor in *Bryophyllum* can the buds and young plants arise from

---

* 'Lectures on the Physiology of Plants,' p. 477.

all parts of the leaf; they are only formed in *perfectly definite regions*, *e.g.*, on the margins of the leaves in *Bryophyllum* and in the angles between the points of origin of the large veins in *Begonia*. All the cells of the leaf do not, therefore, as de Vries supposes, possess this capacity, which is limited to perfectly definite though numerous cells. These therefore correspond to somatic cells, quite as much as do those which produce the several cells in the prothallus of a fern, which contain unalterable germ-plasm in addition to the active somatic idioplasm, the former only becoming active by the influence of particular external influences.

These conditions may be fulfilled in thousands of other leaves without resulting in the production of young plants. There are indeed a whole series of observations which apparently prove that 'every small fragment of the members of a plant contains the elements from which the whole complex body can be built up, when this fragment is isolated under suitable external conditions.' Phenomena of this kind are exhibited by cuttings and adventitious buds which arise on a twig the apex of which has been cut off. In the higher plants, the development of roots on a cutting, or the formation of adventitious buds, does not take place in all parts of the plant, but only in those which contain 'a number of cambium cells.' These cells alone therefore contain accessory idioplasm, which, according to the nature of the stimulus acting on them, renders them capable of growing in a manner which is very different from the normal. There can be very little doubt that the whole of the cambium layer of these plants is endowed with the capacity for reproduction. The only question is, whether this is a result of special adaptation, or only the outcome of the normal constitution of each plant-cell.

I should still, however, be inclined to consider it as a special adaptation, and will endeavour to state my reasons for this view; not being a specialist in botany, however, I am unable to deal with the various groups of the vegetable kingdom in such detail as I could wish.

The question to be decided is, whether each cell was provided with all the specific primary constituents in a latent condition at the time when the multicellular plant arose from the unicellular form; or whether, owing to the diversity of the differentiation of the idioplasm, a sharp distinction first arose between the somatic cells and the germ-cells, and the idioplasm of the somatic cells was only subsequently provided with germ-plasm in a latent condition in those cases in which this arrangement was a useful one. I hold the latter view, and de Vries the former one. It is important for the theory of the germ-plasm to decide between these two opinions; for it would be incompatible with this theory for germ-substance to be present as the idioplasm of the somatic cells at the phyletic origin of the soma. According to my conception of the germ-plasm, the phyletic origin of the somatic cells depends on the determinants contained in the germ-plasm being separated into groups. It would entirely contradict this assumption if those somatic cells which were phyletically the first to be formed, had contained all the other characters of the species in a latent condition in addition to their manifest specific characters. De Vries thinks that the marked distinction which actually exists between the somatic and germ-cells of the higher animals has led me to assume the universal existence of this contrast, which is not nearly of such a decided nature in the case of plants in which gradual transitions from somatic-cells to germ-cells can be proved to exist.

This, however, I believe is not the case: transitions between somatic and germ-cells never occur, and de Vries's opinion simply rests on the fact that he confuses germ-cells with the cells of the germ-track. That the latter must be regarded as somatic cells has already been shown.

In my opinion, germ-cells were sharply distinguished from somatic cells on their first appearance in phylogeny, and this distinction has since persisted. In no species, whether animal or vegetable, can there be any doubt as to the cells which are to be looked upon as germ-cells; and as regards the somatic cells, such a doubt can only arise when cells in the germ-track are regarded as germ-cells.

I know of no more convincing proof of my view than that which is furnished by the *Volvocinæ*. These organisms consist of communities of cells which may or may not exhibit a division of labour, and in which a contrast between the somatic and germ-cells may or may not exist. In *Pandorina* all the cells of the colony are similar to one another, and each performs all the vital functions. In *Volvox* the cells are differentiated: some of them have the function of maintaining the individual, and others that of preserving the species: that is to say, they are differentiated into somatic cells and germ-cells. The heteroplastid genus *Volvox* must have arisen phyletically from a homoplastid form; but we can hardly imagine that there can be

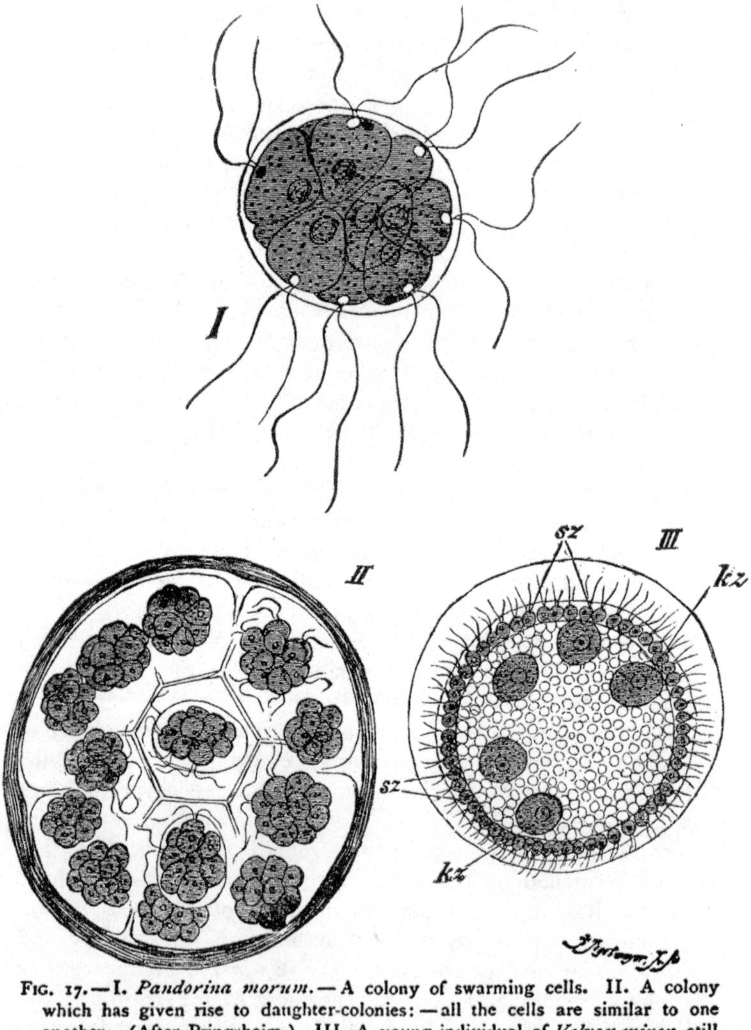

FIG. 17.—I. *Pandorina morum.*—A colony of swarming cells. II. A colony which has given rise to daughter-colonies:—all the cells are similar to one another. (After Pringsheim.) III. A young individual of *Volvox minor*, still enclosed within the parent (after Stein): the cells are differentiated into somatic- and germ-cells.

many intermediate stages between these two, for at the present day the two kinds of cells in *Volvox* hardly differ as much as do the somatic- and germ-cells in the case of the higher organisms. The somatic cells have nevertheless entirely lost the capacity of reproducing the entire organism.

Transitions between these two kinds of cells could naturally only arise by the germ-cells first becoming only slightly differentiated from the somatic cells, and could not have been produced, as de Vries thinks, owing to all the cells containing germ-substance in a more or less latent condition from the first. There is no germ-substance in the somatic cells of *Volvox*, which, figuratively speaking, have only just become differentiated from the germ-cells. If the latter are artificially removed from a colony, the somatic cells continue to exist for a long time, but they do not give rise either to new germ-cells or to a new colony.

And why should they do so? Of what advantage would this be to the species, since millions of individuals, each of which again produces daughter-individuals, exist in the same pond? The ordinary process of multiplication is so vigorous that special means for ensuring the existence of the species are unnecessary.

Such means have, however, been adopted in very many, if not by far the greater number, of the more highly organised plants.

The power possessed by fungi and mosses of reproducing a new individual from any bit of the plant under favourable conditions, has been supposed to contradict my view. But I do not see what prevents us from regarding this power as an adaptation for ensuring the existence of a species surrounded by dangers of all kinds. When the top of a toadstool is knocked off, a new one is formed (Brefeld); and this arrangement is obviously of great use in the preservation of the species. An entire liverwort can be regrown from the smallest fragments of the plant (Vöchting). Why therefore should the assumption be improbable that this power has been acquired in order to insure the persistence of a species the existence of which is threatened by every sudden drought? My knowledge of plant life is not sufficient for me to be able to support this statement in detail, but other facts will, I think, to some extent confirm my opinion from the opposite point of view.

Why is this power of regeneration not possessed by adult ferns and horse-tails? If a frond of a fern is cut off, it is not reproduced from the stalk, and even the individual pinnæ cannot be formed anew. In answer to this it might be urged that the somatic cells of ferns have become too highly differentiated: but this is contradicted by the fact that many, although by no means all, ferns can produce bulbils on their fronds. I must leave botanists to decide why this occurs; but were I asked whether the power of producing entire plants from somatic cells would not have been of advantage to the other ferns, and would therefore be expected to be possessed by them, I should be inclined to reply that all ferns are able to replace lost fronds by forming new ones, — not by the regeneration of the injured leaf, but by budding from the stem. This suffices to restore the plant when it has been injured.

We must now consider the Phanerogams in this connection. In these plants, again, 'accessory germ-tracks' are usually absent in the leaves: that is to say, the cells of the leaves are not capable of producing buds or even of restoring a piece which has been cut out. The axes of the shoots, on the other hand, possess this power in a high degree, and it depends on the presence of cambium cells, all of which are apparently capable of giving rise to new growing points, which produce new shoots with leaves and blossoms, and consequently also germ-cells. *The power of regenerating the leaves is, as a rule, useless; for the formation of new persons of the plant stock can take place to an unlimited extent by means of the cambial layer*; and this mode of compensation for losses sustained is more effectual than the restoration of defects in the leaves would be. The power of growing adventitious buds is probably unnecessary in the case of most leaves, on account of the enormous number and certain dispersion of the seeds produced by the plant. Amongst animals the same is true as regards polype-stocks. Cells are distributed throughout the stock which have the appearance and functions of ordinary somatic cells, but which can give rise to new persons under certain circumstances, such as, for instance, when the stock has become injured. In a living stock of *Tubularia mesembryanthemum*, which I once brought from Marseilles to Freiburg, the crowns of the polype died one after the other within a week, probably on account of want of nourishment; but within a few days afterwards all the stalks had given rise to new crowns; and though these were very small, they would undoubtedly have grown to the full size had it been possible to supply them with food. The capacity for regeneration is apparently provided for in this case; it is at any rate stated by Loeb * in a recent paper that a shedding and new formation of the crowns occurs periodically. The same writer also showed by experiment that, under favourable circumstances, crowns may bud out at any point of the stem, either at the distal end or at the base. On the other hand, he never succeeded in getting the root-like organs of attachment to be produced at the apical end.

No one will be surprised that such a growth did not occur who agrees with me in looking upon all these processes of regeneration and budding as resulting from adaptation. Under natural conditions the apex of a stem could hardly be situated in such a position as to render the formation of roots necessary, for it never comes to lie upside down in the earth, and consequently none of the cells in the apex contain 'rhizogenic idioplasm' ('Wurzel-Idioplasma'). On the other hand, however, it is easy to understand why the power of budding exists in such a marked degree in polypes, if one considers how liable the soft body is to be attacked by crabs, worms, gasteropods, Pycnogonids, and other small enemies. If these polype-stocks did not possess the power of continually producing new crowns,—*i.e.* new individuals,—when the old ones have been bitten off, the whole colony would soon perish owing to the absence of 'nutritive persons.' The fact that regeneration is possible to such an enormous extent results, at any rate *in part*, from the aggregation of persons to form the higher stage of individuality of the stock. For such a combination of individuals procures the advantage of permanent nutrition as long as *all* the individuals of the stock have not fallen victims to their enemies, and thus it is favourable to the production of new buds.

Amongst the *Polyzoa* the case is very similar. In many of these animals the normal form of gemmation takes place with great regularity, and the region at which the next bud will arise can be predicted beforehand: on this fact depends, as in the case of the Hydroid polypes, the characteristic form of the stock in the different species, which is sometimes branched like a foliage tree, and sometimes like a fir-tree or a feather. In these animals, therefore, definite cells must be provided with 'blastogenic' idioplasm in advance, and merely the stimulus due to ordi-

* Jacques Loeb, 'Untersuchungen zur physiologischen Morphologie der Thiere,' I. Uber Hetermorphose,' Wurzburg, 1891.

nary nutrition is required in order to incite them to form buds. The series of cells which lead directly to the cells of these buds must be looked upon as the main germ-tracks, using the term in de Vries's sense. According to Seeliger's researches, however, budding takes place from other regions as well as the ordinary ones in certain Polyzoa, *e.g.*, *Pedicellina*. If the crown is lost in this animal, so that only a stump of the stalk remains, new crowns are produced on the end of the stalk; and in this case, therefore, *budding originates in the flat epithelial cells characteristic of the ectoderm*, which did not previously appear to be capable of proliferating at all. This is another instance of the presence of accessory germ-tracks. The cells of the epidermis are provided with blastogenic germ-plasm, although they do not as a rule take part in the formation of buds, but only give rise to them in response to unusual stimuli. These cells are just as much exposed to destruction as are those of the Hydroid polypes, and we need therefore not be surprised that arrangements for budding should have been made in the stalk, even were we not aware of the fact that in *Pedicellina*, under normal conditions, the crowns drop off periodically from the stalk, and are replaced by others which bud out afresh. This process certainly occurs at the upper end of the stalk, but it is quite comprehensible that it would be advantageous for the lower end of the stalk also to be provided with blastogenic idioplasm.

The arrangement which exists universally in the higher plants for the production of adventitious buds is, in my opinion, to be explained in a similar manner. In this group of organisms the cambium layer in particular is provided with the means of replacing lost leaves and entire shoots. This obviously affords an important protection against numerous enemies — such as insects more especially — the number of which is often incalculable. It is therefore not surprising that such an arrangement — viz., the addition of unalterable germ-plasm to the cambium cells — is here met with.

5. Galls

De Vries has also brought forward the question of the formation of galls as furnishing an additional argument against my views. In his opinion the production of galls proves that a vegetable cell, even when it exhibits a specific histological differentiation, contains the primary constituents of every other kind of cell in a latent condition, and these are ready to become active as soon as a suitable stimulus is brought to bear upon them. This proof he considers to be incompatible with the assumption of the existence and continuity of a germ-plasm.

The development of galls is undoubtedly a highly interesting problem, which, in my opinion, has not yet been fully explained, in spite of the numerous and excellent researches on the subject which have been made within the last ten years. Amongst these, the contributions of Adler * and Beyerinck † in particular have materially helped to throw light on the problem.

The most important point in the consideration of this question is the fact that galls are not by any means exclusively composed of those kinds of cells which occur in the organs of the plants upon which they arise, but may also contain cells of other kinds. 'Cells which are usually only developed in the bark of a plant may also frequently be found in the galls produced by those *Cynipidæ* and *Diptera* which infest leaves.' It is therefore certain that the power of producing forms of cells which do not usually occur in the leaf, for instance, 'is not confined to those organs in which they are developed normally,' but is present also in certain cells of the leaf, and even indeed, de Vries thinks, 'in all other parts of the plant.'

This is not surprising if we look upon the formation of the gall as due to an adaptation of the plant to its parasites, such as we may assume to have occurred with regard to the peculiar arrangements exhibited by certain tropical plants for the protection of ants, which in their turn again protect the plant. Reciprocal adaptation has taken place in this case; the animal has become adapted to the plant, and the plant to the animal, because a joint existence is advantageous to both of them. In the case of the galls of the *Cynipidæ* and *Tenthredinidæ*, the advantage which might result to the plant from the presence of the parasite is not apparent, and we may therefore be inclined

* Adler, 'Beitrage zur Naturgeschichte der Cynipiden,' Deutsche entomolog. Zeitschr. xxi., 1877, p. 209; and 'Uber den Generationswechsel der Eichengallenwespen,' Zeitschr. f. wiss. Zool., Bd. xxxv., 1880, p. 151.

† M. N. Beyerinck, 'Beobachtungen über die ersten Entwicklungsphasen einiger Cynipidengallen,' Akademie d. Wiss. zu Amsterdam, 1882; 'Die Galle von Cecidomyia poæ,' Bot. Zeitung, 1885; 'Uber das Cecidium von Nematus capreæ,' Bot. Zeitung, 1888, No. 1.

to explain their formation as due to a reaction of the plant in response to the stimulus exerted by the animal. If, as was formerly supposed, the gall resulted from the action of a poison which is inserted into the tissues of the plant by the female during oviposition, this explanation would be totally inadequate; for it is not conceivable that the infusion of a poison, which happens only *once*, could with such regularity produce a gall which grows slowly, and only gradually attains its definitive and often complex structure. Moreover, several kinds of galls, differing very much from one another, may be produced from the *same* substratum, such as an oak-leaf, for instance. We know, however, from the researches of Adler and Beyerinck, that the formation of the gall is not due to the sting of the parent animal, but to the activity of the larva after it has been hatched. We must therefore suppose that this peculiar specific proliferation of the tissues of the plant is due, in the first instance, to the stimulus produced by the bodies of the larvæ when they begin to move about and to feed, the specific secretion of their salivary glands then also having an effect. The diversity of the galls arising from the same substratum must therefore be due to differences in these factors; and the conspicuous adaptations of the galls, which serve to protect, support, and nourish the parasite, must depend on adaptations of the latter as regards its mode of feeding and movement, and the chemical composition of its salivary secretion. We cannot help accepting this interpretation of the facts since no other is forthcoming; and we must therefore suppose that natural selection has operated so long on these factors, and has gradually effected such an improvement, that the kind of gall which provided the best protection and nourishment for the species was ultimately produced by the larva.

Beyerinck has, in fact, proved that cells and tissues often occur in galls which very closely resemble those in different parts of the plant, but which do not exist in the substratum (*e.g.* a leaf) on which the gall is produced. De Vries infers from this fact that the primary constituents of such tissues must have been contained in the cells of the leaf, although they could not previously be recognised as such. This inference does not seem to me to follow of necessity, for the stimulus produced by the parasite might conceivably have modified the idioplasm of the cells of the leaf so as to result in the formation of cells differing from those ordinarily present in the leaf. It will be shown in the chapter on Variation that changes of this kind do occur, and that somatic idioplasm *may* at times, owing to known or unknown causes, become so modified as to produce a deviation from the inherited form of the cells of the series. The sudden appearance of such peculiarities as those exhibited by the moss-rose may be taken as an instance. It is very possible, therefore, that owing to the specific stimulus produced by the larva, and more especially by its secretion, the idioplasm of certain layers of cells in the gall becomes modified and causes the cells to assume another character, such as that of woody fibres.

This view receives decided support from the circumstance that *the gall is by no means only composed of those kinds of cells which occur in other parts of the plant.* A similar statement to this is, indeed, made by de Vries, who, however, makes an exception in the case of 'the peculiar layer of sclerenchyma in some Cynipid galls, which afterwards becomes modified into thin-walled, nutritive tissue.' I cannot look upon this as being 'only an apparent exception' to the rule, for it seems to me to be a very valuable proof that no such rule exists, and that the above instance is to be explained as an apparent reversion to inherited forms of cells, such as were already contained in a latent condition in the cells of the leaf. I should rather be inclined to regard these 'exceptions' as a proof that *definite new formations occur in galls, and that these are due to modifications of the cells from which they arise* in response to the stimulus produced by the larva. It can hardly be a matter of surprise that a marked resemblance exists between these cells and those occurring in other parts of the plant, for the changes produced by the larva take place in an idioplasm consisting of determinants of the species in question; these changes would not therefore at first result in combinations of biophors (determinants) very different from those which ordinarily occur in the plant. *The new combination of the biophors in different ways results, I believe, from the action of the larva, and thus modifications of the determinants are produced.*

The galls of *Cecidomyia poæ*, which de Vries mentions when contesting my views, are probably to be accounted for in the same manner. In response to the stimulus produced by the larva, these stalk-galls become covered with a thick felt of rootlike outgrowths, which doubtless serve as a protection: these, when they gain access to the soil, become branched like ordinary roots. The assumption that under certain circumstances the idioplasm of certain somatic cells becomes modified in response to the stimulus produced by the parasite, so as to give rise to a structure similar to that of another tissue or even organ of the same plant, seems to me by no means to prove that the primary constituent of this tissue must previously have been contained in these cells. In animal tissues transformations of this kind are certainly not known to occur. Pathological anatomists are now of the opinion that only those kinds of cells occur in tumours which actually belong to the sort of tissue from which the tumour arises. This is not surprising, for tissues of animals are far more highly differentiated than those of plants, and corresponding elements in the idioplasm must also differ in a corresponding degree; and consequently, in spite of the displacements and re-arrangements which may be produced by stimuli, they can never form precisely the same combinations as those which occur in the various other tissues of the body.

I shall not discuss the case of the gall of *Nematus*, as Beyerinck's observations with regard to it are not yet complete. If it should be shown that a complete willow can be produced from the leaf-gall of the plant, as de Vries considers probable, it will then certainly have been proved that the cells in the leaf contain germ-plasm, just as in the case of the leaves of *Begonia*. At present, however, it is only known that the gall can give rise to roots, and although normal roots are always capable of forming adventitious buds, it cannot be said at present whether these abnormal roots are able to do so. In the willow, in any case, the primary constituents of roots are distributed throughout the stem in the form of invisible determinants, contained within visible cells, and this accounts for the fact that the production of new individuals by means of cuttings takes place exceptionally easily in this plant. This may perhaps be accounted for by a wider distribution than usual of the merely 'unalterable' group of determinants for roots taking place in the plant, in connection with the wide distribution of the corresponding primary constituents. But I do not by any means imagine that in all these cases in which the cells of a plant possess inactive germ-plasm, its presence is actually useful *at the present day*. If the distribution of unalterable germ-plasm once took place in an organism to such a considerable extent as has occurred in most plants, it would be a matter of slight importance in the economy of the plant whether the cells of those organs which at the present day are no longer in a condition to make use of this substance were provided with a minimum of germ-plasm or not. Such a provision might have been of advantage to the ancestors of the species; and if this were not the case, we know so little of the processes by means of which the various qualities of the idioplasm become separated mechanically in nuclear division, that we cannot altogether reject the assumption of an occasional chance admixture of germ-plasm to somatic idioplasm, especially in the case of the higher cormophytes, which must in any case possess a number of cells containing germ-plasm throughout the entire plant. Time will show whether we require this assumption.

The difference between my view and that of de Vries does not consist in the fact that I am compelled to deny the admixture of germ-plasm in the case of a large number of cells in the body on principle, but in my assumption that each somatic cell contains a definite somatic idioplasm consisting of a limited number of definite determinants, to which any other 'unalterable' accessory idioplasm may be added if required. De Vries, on the other hand, considers that the whole of the primary constituents of the species are contained in the idioplasm of every, or nearly every, cell of the organism. But he does not explain how it is that each cell nevertheless possesses a specific histological character. A new assumption, which would not be easy to formulate, would therefore be required to explain why only a certain very small portion of the total amount of idioplasm — which is similar in all parts of the plant — becomes active in each cell. This theory explains the differentiation of the body as being due to the disintegration of the determinants accumulated in the germ-plasm, and requires a special assumption, — viz., that of the addition of accessory idioplasm when necessary, — in order to account for the formation of germ-cells, and the processes of gemmation and regeneration. The reconstruction of entire plants or of parts from any point can be easily accounted for by de Vries's hypothesis, just as it can by Darwin's theory of pangenesis, for the pangenes or gemmules are present wherever they are wanted. But de Vries is unable, on the basis of his hypothesis, to offer even an attempt at an explanation of the *diversity* of the cells in kind and of the *differentiation* of the body.

These two assumptions appear to me to be of equal value in explaining the fact that in many of the lower plants each cell, under certain circumstances, can apparently reproduce an entire individual. The differences between the somatic cells are here only slight ones, and are so few in number, that we might be inclined to consider them as due to reactions of the same idioplasm to different kinds of influences exerted by the environment. Such is the case, for instance, in liverworts. But this assumption ceases to be tenable as soon as the soma can become variously differentiated, and any explanation must in the first place account for this differentiation: that is to say, the diversity which always exists amongst these cells and groups of cells arising from the ovum must be referred to some definite principle. De Vries's principle is of no use at all in this case, for it only accounts for the fact that entire plants may, under certain circumstances, arise from individual cells, and does not even touch the main point. In fact, no one could even look upon it as giving a partial solution of the problem, if differentiation is supposed to be due to that part alone of the germ-plasm always becoming active, which is required for the production of the cell or organ under consideration. But the higher we ascend in the organic world, the more limited does the power of producing the whole from separate cells become, and the more do the numerous and varied differentiations of the soma claim our attention and require an explanation in the first instance.

The presence of idioplasm in all parts containing *all* the primary constituents does not help us in this respect; and even in attempting to explain the formation of germ-cells, it is of very little use to assume that they arise from cells which, like the rest, contain all the primary constituents of the species. How is it that these cells, and these alone, in the entire soma of the animal, give rise to germ-cells? In the lower plants the fact of the differentiation of the soma is liable to be overlooked or underrated, but this cannot possibly be the case as regards the higher animals.

CHAPTER VII

Summary of Part II

We have now seen that the idioplasm of the fully-formed individual animal- or plant-cell may exhibit a considerable amount of difference as regards its degree of complexity; and before going further, it may be as well to state clearly in what this difference consists.

We suppose that the process in the idioplasm which brings about the ontogeny of a multicellular organism is due to the thousands of determinants, which constitute the germ-plasm of the fertilised ovum, becoming systematically separated into groups, and distributed among the successors of the egg-cell. This separation into smaller and smaller groups of determinants continues to take place, until each cell contains determinants of one sort only, and these then either control a single cell, or, in case the hereditary character ('determinate') is constituted by a group of cells with a common origin, the control is exerted over this whole group.

All the determinants are not active at the same time; every cell, in fact, which appears in the entire course of ontogeny is controlled by one determinant. This is effected by the disintegration of the determinant into its constituent biophors, which migrate into the cell-body. Even in the earlier stages of ontogeny, in which the idioplasm of a cell consists of a still larger number of different kinds of determinants, the cell is also controlled by only one of them in this manner. The rest of the determinants have in any case an important function with regard to the course taken by ontogeny, for each of them has its own rate of increase, and thus an alteration is produced in the proportion of the various determinants originally present in the germ-plasm, and consequently its definite architecture also undergoes alteration, the subsequent disintegration being controlled by the determinants which have thus been rearranged. These determinants are therefore only *inactive* with respect to the cell in which they are situated, and not as regards development as a whole.

Various circumstances may, however, produce complications in this simple course of development of the idioplasm.

In the first place, it is necessary that the organism should be able to replace losses of substance. In order that this may be possible, the final cells of ontogeny, at any rate, — *i.e.*, those of the various tissues, — must be rendered capable of producing others similar to themselves. The possession of this power necessitates that each cell shall be capable of unlimited multiplication, so that the *single* determinant which controls it can likewise grow and multiply. This would result in the cells being able to produce others of a like character by cell-division.

We find, however, that regeneration is not limited to this very simple form of restoration: more complex tissues can also be reproduced; and even entire organs, such as limbs, and still larger parts of the body, such as the head and tail, may, in certain groups of animals, undergo restoration when they have been lost. These facts are to be explained in terms of the idioplasm by supposing that in these cases also the determinants for the groups of cells which are to be capable of regeneration have undergone an increase, and have been supplied to certain cells in the course of ontogeny in the form of *inactive accessory idioplasm*. The equipment of such cells with determinants for regeneration is due to adaptation, and is only connected with the degree of organisation of the animal in so far as the difficulty of providing a large number of cells with accurately graduated determinants for regeneration increased as the number of cells from which regeneration had to proceed became larger, and as the number and degree of differentiation of the organs to be restored also increased. An increase, both as regards cells and organs, takes place in correspondence with the complexity of structure, and the 'regenerative power,' therefore, as a rule, gradually undergoes a proportionate decrease.

Cells which possess the regenerative power are therefore those which contain, in addition to their own active determinants, a larger or smaller group of inactive determinants: these latter belong to those cells and cell-series which are capable of taking part in the reconstruction of that part of the body which has been lost, and which is situated distally to them when the determinants become active. The occurrence of regeneration in a high degree is only rendered possible by the cells in a definite transverse plane of the body being regularly equipped with various groups of suitable supplementary determinants which are capable of acting together as a whole.

This form of regeneration leads directly to the process of *reproduction by fission*, which simply consists in the employment of a marked power of regeneration for the purpose of increasing the number of individuals.

While the possession of the regenerative power in a low, as well as in a high degree, is due to the equipment of cells with certain larger or smaller groups of determinants in the form of 'unalterable' accessory idioplasm, the *development of new persons by gemmation* depends either on the fact that a *single* cell contains all the determinants of the species in an inactive and unalterable condition, in the form of accessory idioplasm; or else is due to two or three cells in the different layers of the body containing large groups of determinants as accessory idioplasm, which *together* constitute all the determinants possessed by the species—*i.e.*, germ-plasm.

In cases where budding originates in a single cell, as in the Hydroid-polypes, the blastogenic idioplasm concerned in the process must be regarded as a modification of the germ-plasm, which consists of all the determinants of the species, though these have a different arrangement to that which obtains in the germ-plasm proper of the fertilised egg-cell. The fact that a bud may originate in two or three cells does not show that these cells contain exactly those groups of determinants which correspond to those of the two or three germinal layers of the Metazoon in question. In fact, the combination of the determinants differs more or less in all known cases, and is adapted to the circumstances under which budding occurs. This proves that in the embryogeny of the species, divisions of the accessory idioplasm occur quite independently of the ordinary divisions in the mass of determinants, and these result in certain cells being provided with a definitely constituted accessory idioplasm.

The blastogenic germ-plasm must be contained in the germ-plasm of the sexual cells in the form of special ids, for buds can vary independently of the persons which produce them. On the other hand, primary idioplasm must also be supplied to the bud during its development, and this may be effected by means either of special cells which contain the idioplasm in an unalterable condition, or else of certain ids of the primary idioplasm being added in an unalterable condition to the blastogenic germ-plasm when it becomes separated from the primary idioplasm.

As, apart from plants, gemmation only occurs in comparatively low forms of animals, viz., in the Cœlenterata, Polyzoa, and Tunicata, we may infer that the addition of this blastogenic idioplasm, consisting of accurately graduated groups of determinants, eventually reaches a limit, owing to the increasing complexity of the structure of the animal and to the surprising extent to which the number of determinants increases.

According to our view, the cells of the Metazoa and Metaphyta may not only be provided with the above-mentioned *accessory* idioplasm, but may in addition contain *primary germ-plasm*, and these cells are to be found along the '*germ-tracks*,'—that is to say, they are situated in the direct line of development which leads from the ovum to the germ-cells which are eventually formed from it. As each somatic cell is only controlled by one of the large number of determinants belonging to the germ-plasm, and as determinants cannot be produced spontaneously, those cells which are to give rise to germ-cells must contain unalterable germ-plasm in addition to the active determinants which control them; and the former can only be derived from the cell to which the whole organism owes its origin, for this alone contains the whole of the determinants organically united to form germ-plasm. A series of cells containing germ-plasm in the form of unalterable accessory idioplasm, must therefore be traceable from the egg-cell to that region of the body which sooner or later gives rise to germ-cells; that is, there must be a continuity of the germ-plasm.

The number of germ-tracks in the lower plants and animals is a very large one; under normal circumstances, germ-cells are not only found in very many parts, especially in the case of animal- and plant-stocks, but new persons may in exceptional cases be formed in many regions by budding, especially when injuries to the stock have occurred; and these persons can again produce germ-cells. In the case of the Hydroid-polypes and Polyzoa, a large number of cells of the stock must be provided with germ-plasm, although it is impossible to say whether these are the same as those which contain blastogenic idioplasm, or whether the latter is situated in other adjacent cells which also take part in the formation of the bud. In any case blastogenic idioplasm and the germ-plasm of the ovum are not identical, even if, as in plants and Hydroid-polypes, the former contains the whole of the determinants of the species. The determinants must at any rate have another arrangement: not infrequently, indeed, the blastogenic idioplasm must consist of entirely different kinds of determinants, and in the case of the alternation of generations of the Medusæ of far more numerous ones.

The cells of the germ-tracks are somatic cells: that is to say, each of them is controlled by a special determinant, and contains germ-plasm in an inactive as well as in an unalterable condition. In the latter state, it only again becomes capable of disintegration when the cells in which it is situated give rise to germ-cells, and begin to undergo development into embryos. This germ-plasm, like the unalterable blastogenic idioplasm, may be contained in young cells with only a slight amount of histological differentiation, as well as in cells with a sharply defined histological character.

We thus see that in many cases the cells of the adult organism contain an accessory idioplasm in addition to the determinants which control their special character, structure, and physiological activity for the moment: the former may become active in the ordinary course of development, — as occurs in the normal formation of germ-cells and in multiplication by fission and gemmation; or its activity may be due to abnormal causes only, — such as those resulting from injuries and mutilations, — from which the processes of regeneration or gemmation in the first place originated.

PART III

THE PHENOMENA OF HEREDITY RESULTING FROM SEXUAL REPRODUCTION

INTRODUCTORY REMARKS ON THE NATURE OF SEXUAL REPRODUCTION.

THE phenomena of heredity have so far been considered in connection with a purely asexual form of reproduction only: the complications of the germ-plasm arising from the intermingling of the hereditary parts of *two* parents have been left aside, and the composition of the germ-plasm has been assumed to be of such a nature as would result if *monogonic* reproduction were the only form in which the process existed. The advantage of this method of procedure is seen in the fact that it has only been necessary for us to bear the essential part of the processes in mind when analysing the fundamental phenomena of heredity, and this essential part has therefore not been lost sight of in the confusing and ever changing intermixture of individual variations which result from amphigonic reproduction. The course we have followed is justified by the fact that fundamental processes — such as ontogeny, regeneration, and multiplication by fission and gemmation — cannot owe their origin to amphigonic reproduction, but would take place even if this form of multiplication did not exist at all.

Bearing this fact in mind when considering the complications arising from sexual reproduction with which we have to deal in analysing the phenomena of heredity and their material substratum, it will now be profitable to consider the facts concerning this form of reproduction, and to see how they can be explained. I will now therefore give a short account of the processes in question as far as is necessary in order to render comprehensible the complications in the phenomena of heredity resulting from them.

Until far into the present century 'sexual reproduction' was considered to be the *essential* and *primary* form of the process; and although it gradually became more and more evident that several kinds of 'asexual reproduction' may also occur, these nevertheless only take place in the lower forms of animals and plants. As the details of the phenomena of reproduction were for a long time known almost exclusively with regard to the higher animals, and as in them sexual reproduction alone occurs, the special peculiarities of the latter were naturally considered to be necessary and indispensable in the process. *Fertilisation* was looked upon as an essential part of this process, and it alone was supposed to render life from one generation to another possible at all; in short, fertilisation was regarded as a 'process of rejuvenescence,' and sexual reproduction was considered to form the foundation from which all forms of reproduction have arisen. The existence of different forms of asexual reproduction was explained as an 'after-effect' of the process of fertilisation or rejuvenescence occurring in sexual reproduction.

This view appeared to receive support from the fact that sexual reproduction is of universal occurrence, from the lowest to the highest forms of animals and plants; and also that asexual reproduction never takes place in the higher organic forms, and even in the lower ones it only alternates with the sexual form of the process.

The present state of our knowledge of the process of fertilisation, however, justifies us in considering these earlier views to be totally erroneous. In no case does fertilisation correspond to a rejuvenescence or renewal of life, nor is its occurrence necessary in order that life may endure: it is merely *an arrangement which renders possible the intermingling of two different hereditary tendencies*. We shall deal later on with the question as to why such a mingling has been introduced and so extensively adopted by Nature; at present it is only necessary to prove that this is the case. Fertilisation consists in the union of two hereditary substances, *i.e.*, of the germ-plasms of two individuals: all the complicated and varied phenomena of differentiation — beginning with that of the two different kinds of reproductive cells, usually known as male and female, up to that of the individuals themselves into males and females, and including the innumerable other resulting adaptations and phenomena — take place solely for the purpose of rendering possible the union of the primary constituents of two individuals.

This process of the fusion of two germ-plasms, which constitutes the essential part of fertilisation and is as a rule connected with the fusion of two cell-bodies, I have designated as *amphimixis*. It is not always connected with reproduction, for these two processes take place independently of one another in all unicellular organisms. In the Infusoria, for instance, two individuals in the course of their life-history come into contact with one another, and then either fuse completely into one, or else undergo a partial or temporary fusion; in both of which cases half the hereditary substance is transferred from one individual into the other, and thus amphimixis is brought about. The latter process is only invariably connected with reproduction in the case of multicellular forms: this is necessitated by the fact that the union of two different germ-plasms cannot take place by the fusion of entire individuals, as the germ-plasm is enclosed in separate cells, a male and a female, the fusion of which takes place in a similar manner to that which occurs in the process of conjugation in unicellular organisms. This act of amphimixis must then be followed by the multiplication of the fertilised egg-cell, accompanied by the differentiation of its successors, — or, in other words, *by the ontogeny of a new individual*; for did this not result, the process of amphimixis would be useless. *Amphimixis is therefore always connected with reproduction in all multicellular forms*, and these two processes together constitute '*sexual reproduction*' or '*amphigony*' (Haeckel).

The process of amphimixis, as it occurs in amphigonic reproduction, is briefly as follows. The two kinds of germ-cells mutually attract one another, and then fuse together, the smaller male element always entering the larger female one. The nuclei of the two cells then approach each other, and so come to be situated close together, each being accompanied by its 'centrosome,' — *i e.*, that remarkable body, enclosed in a clear sphere, which, as already stated, constitutes the *apparatus for division*. The germ-plasm in both nuclei is at first distributed in the form of fine threads, as is represented in the case of the female nucleus in Fig. 18, I.; it subsequently, however, becomes contracted, so as to give rise to *nuclear rods* or *idants* (Fig. 18, II.). Edouard van Beneden was the first to prove that the number of these idants is the same in both of the conjugating-cells, and this dis-

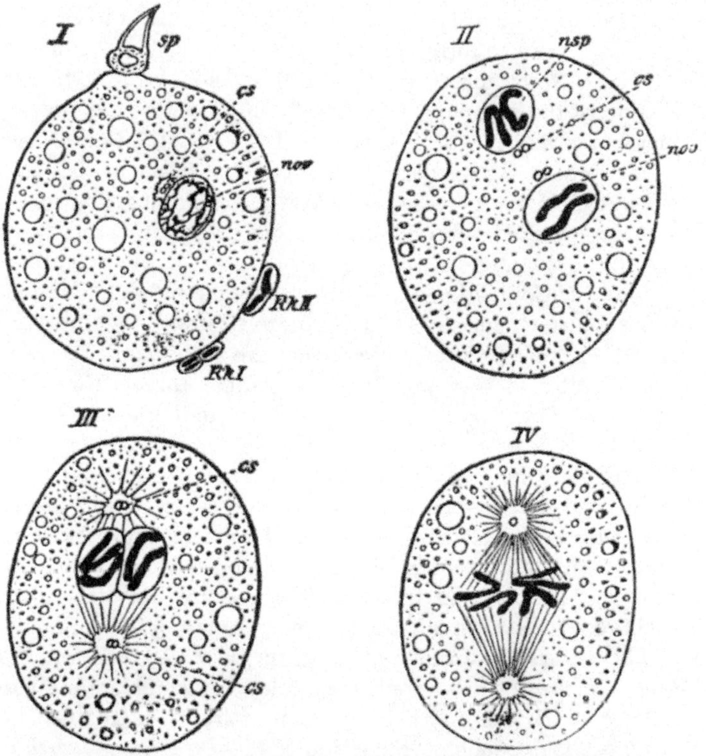

Fig. 18.—Diagram of the fertilisation of the egg in *Ascaris megalocephala.*—(Compounded from the figures and descriptions of Boveri and others.)

I.—The sperm-cell (*sp*) is about to enter the ovum, which contains a nucleus (*nov*) and centrosome (*cs*). *Rk I* and *Rk II*—the two primary polar-bodies, the first of which has divided into two; each contains two idants.

II.—The sperm-nucleus (*nsp*) has passed into the egg, near the nucleus of which it is situated. Each of these nuclei contains two idants, and also a centrosome, which has divided into two.

III.—The two nuclei are now close together: the centrosomes, with their 'spheres of attraction,' are connected together in pairs, and are situated at the poles of the spindle, which is already visible.

IV.—The nuclear membrane has disappeared, and the first embryonic nuclear division is now taking place.

covery—which has since been confirmed in the case of a large number of species of animals, and has been proved quite recently by Guignard to apply to plants also — is of decided importance in connection with the conception that idants constitute the hereditary substance. As the two nuclei are approaching one another, their centrosomes become doubled, and the corresponding pairs unite to form the two poles of a nuclear spindle (Fig. 18, III.), which direct the first cell-division leading to the formation of the embryo; this usually only occurs after the nuclear membrane has completely disappeared (Fig. 18, IV.).

The process of fertilisation therefore consists in the union of the nuclei of the two sexual cells within the maternal germ-cell, and also of the bodies of the cells, together with their apparatus for division. One half of the germ-plasm of the 'combination-nucleus' ('Copulationskern') thus formed by the union of the sexual nuclei consists of idants derived from the mother, and the other half of those derived from the father, and the resulting combination of two hereditary substances directs the ontogeny and controls the building-up of the new individual. The entire number of idants nevertheless always remains the same in all the cells of the body: thus, for instance, if eight paternal and eight maternal idants were brought together in the process of amphimixis, there would be sixteen idants in every * cell in the body of the individual arising from the fertilised ovum; and if, again, as represented in Fig. 18, there are only two idants in each germ-cell, each somatic cell will contain four idants.

The nature of sexual reproduction depends therefore on the intermingling of two hereditary tendencies which are individually different from one another; or, to pass from the abstract to the concrete, it depends on the union of *two hereditary substances* in the first rudiment of the individual. We must next investigate the manner in which this combination of hereditary substances affects the composition of the germ-plasm.

---

* A recent observation renders it doubtful whether 'every' cell contains the same number of idants; but this need not here be taken into consideration, as its importance cannot at present be estimated.

CHAPTER VIII

MODIFICATIONS OF THE GERM-PLASM CAUSED BY AMPHIMIXIS

1. The Necessity of a Halving of the Germ-Plasm

By the process of amphimixis the hereditary substances of two individuals become united into one substance in the offspring. If the process is repeated in every generation, a doubling of these individually different hereditary substances must take place each time, and the mass of germ-plasm and the number of idants must likewise be doubled. As a matter of fact this cannot and does not occur, for in every species the number of idants remains the same throughout all generations. The unlimited increase of the germ-plasm must therefore be prevented in some way or other.

The mass of germ-plasm might possibly remain constant if its growth stopped in the young germ-cells when only half the normal quantity had been formed. It is quite conceivable that a continual increase in mass might in this way be prevented, if, contrary to the theory of the germ-plasm here propounded, we were to imagine that the idioplasm merely consists of ultimate vital particles — 'pangenes,' 'primary constituents,' or whatever else we choose to call them — which are not combined into units of a higher order.

If, however, we assume the existence of a germ-plasm in the sense in which I use the word, — *i.e.*, an idioplasm in which the ultimate bearers of vitality (biophors) are combined to form units of a higher order, the determinants and ids, having a definite structure and size, — it is evident that the amount of germ-plasm would not remain constant, or at most it would only remain so for a few generations, as long, that is, as each kind of germ-plasm is represented by several ids. As soon as this stage was reached, a decrease in growth could no longer prevent a doubling of the mass; this could, in fact, only be prevented by the *removal of half of the number of ids present in the cell.*

This actually occurs before the germ-cells unite in the process of 'reducing division' of the nuclear matter of the germ-cells. This fact may probably be taken as indicating the correctness at any rate of the fundamental idea on which the theory of the germ-plasm is based, viz., that the hereditary substance is composed of ids. These parts of this substance, the existence of which I formerly concluded from purely theoretical considerations, and which I have called 'ancestral germ-plasms,' must exist in reality. I venture to make this assertion with all the more assurance, owing to the fact that at the time when I postulated the 'reducing division' merely on theoretical grounds, the existence of such a process could not be gleaned from recorded observations even in the case of the female germ-cells of animals, in which it can be observed comparatively easily, quite apart from that of the male cells of animals, or of the germ-cells of both sexes in plants.

We now know that this reduction of the number of ids by one half is of general occurrence, and is effected by means of the nuclear divisions which accompany cell-division. The divisions which result in the formation of the polar bodies perform the function of the 'reducing divisions' as regards the ovum, and the final divisions of the sperm mother-cells have this function in the case of the spermatozoa. In both cases the reducing division does not consist in the idants becoming split longitudinally, and in their resulting halves being distributed equally amongst the two daughter-nuclei as in ordinary nuclear division, but in one half of the entire number of rods passing into one daughter-nucleus, and the other half into the other. The process is somewhat more complicated than would appear from this statement, and it will be discussed more fully later on; but the final result is the same.

The following considerations may perhaps help to explain why the constant doubling of the germ-plasm could only be prevented by this method of removing entire nuclear rods, and will at the same time indicate what are the primary causes of the changes in the structure of the germ-plasm caused by amphimixis.

As already remarked, the nuclear rods must, before the introduction of the process of amphimixis into the organic world, have consisted of a number of *identical* ids, each corresponding exactly to the individuality of the organism in question. These ids must have been united into idants, which were all equal in value, their number, as well as that of the ids, remaining the same in subsequent generations. When sexual reproduction first arose, the same number of idants from both parents became enclosed in one nucleus, the total number of idants and mass of germ-plasm of which were thereby doubled. This may have been of no disadvantage if it occurred once only, but as the process was repeated, an arrangement for preventing the germ-plasm from increasing to an unlimited extent became necessary each time amphimixis took place.

Were the germ-plasm an unorganised, or even a perfectly homogeneous substance with no internal differentiation,—*i.e.*, were it not composed of units of different orders,—its doubling every time amphimixis occurred might have been prevented simply by a limitation of its growth in each germ-cell, so that the latter would contain only half the mass of germ-plasm formerly present. But as soon as the germ-plasm came to consist of a definite number of units, a diminution of the latter could not result from a mere limitation as regards growth, for their number would nevertheless remain the same. This result could only be attained by the appearance of a process by means of which the *number of units was reduced to half*, and we have seen that such a process occurs in the form of the remarkable 'reducing divisions' already described.

It is not difficult to ascertain what changes must result in the composition of the germ-plasm by the combination of this process with continued amphimixis.

Let us suppose that before the introduction of the latter process the germ-plasm of a species consisted of sixteen idants. When amphimixis, accompanied by the 'reducing division,' occurred for the first time, eight paternal idants A would unite with eight maternal idants B in the fertilised egg-cell to form the segmentation nucleus. In consequence of the reducing division, each of the germ-cells of the next generation would contain a combination of the idants A and B,—*e.g.*, 4 A + 4 B. These would again unite in the next amphimixis with eight idants—*e.g.*, 4 C + 4 D—in the germ-cell of another individual with different hereditary tendencies; and the ontogeny of the third sexual generation would therefore be controlled by a germ-plasm composed of the idants 4 A + 4 B + 4 C + 4 D. Let us assume, for the sake of simplicity, that the reduction always affected every kind of idant to the same extent; the germ-plasm of the fourth generation would then consist of the idants 2 A + 2 B + 2 C + 2 D + 2 E + 2 F + 2 G + 2 H, and that of the fifth, of a number of individually different idants, — provided, of course, that interbreeding had not occurred. The germ-plasm of this fifth generation would therefore consist of the idants A — Q.

This naturally does not imply that the process would really take place in such an even and systematic manner: it must, on the contrary, be a very irregular one. But although it may not in five generations have resulted in the germ-plasm being composed of a number of different ids, this result must certainly follow in the course of a greater number of generations.

The modification of the germ-plasm will not, however, then have reached its limit. If my view of the composition of idants out of ids is a correct one. and the id is really a unit which contains all the primary constituents of the species, — that is to say, if it contains all the determinants required for the construction of a single individual, — it follows that the *composition of the individual idants must gradually have become changed, so that each idant, instead of being made up of similar ids, comes to be constituted by dissimilar and individually different ids.*

The idants are not, in my opinion, perfectly invariable quantities; certain phenomena of heredity have led me to conclude that they are in any case only *relatively* constant, and that their composition becomes modified from time to time, so that the ids which previously belonged to the idant A may later take part in the composition of the idant B or C. Our present knowledge of the processes of the division of the nuclear substance does not enable us to say how frequently and regularly this occurs; but even if it only takes place at irregular intervals, during long periods of time, it must nevertheless have resulted in a very varied composition of the idants in the course of the enormous number of generations which have ensued since the introduction of the process of amphimixis into the organic world. As new idants are always added to those already present in one of the parents each time amphimixis occurs, a continual interpolation of new ids can take place in the idants; and as this process is repeated an indefinite number of times, a single idant must ultimately — if we neglect the repetition of similar ids which results from interbreeding — come to consist of a number of *individually* different ids.

The process of mingling the ids would proceed most rapidly if the paternal and maternal ids regularly combined each time amphimixis took place, so as to bring together the half of the different kinds of ids in the idants of both parents. If, for in-

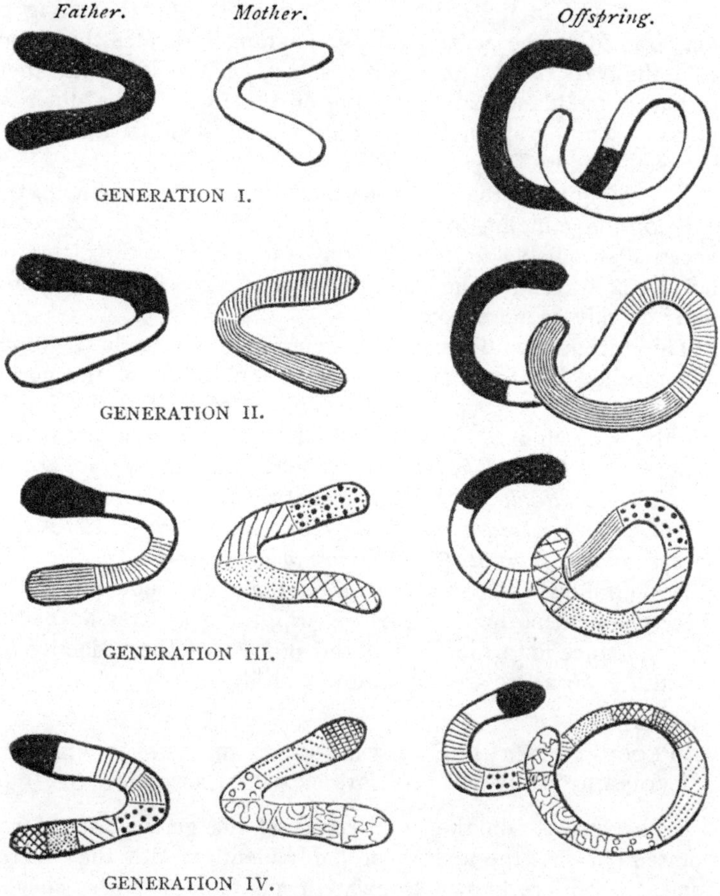

Fig. 19.—Diagram illustrating the composition of the idants out of individually different ids. (From Weismann's 'Essays,' Vol. I., p. 369.)

stance, each idant in an individual consisted of sixteen ids which were all similar to one another on the first appearance of sexual reproduction, the first occurrence of amphimixis would result in idants consisting of eight paternal and eight maternal ids, which are represented respectively by the black and white parts in the accompanying diagram (Fig. 19). (The boundaries between the single ids are only indicated in generation IV. in the figure.) In the second generation four groups, each consisting of four similar ids, would be combined; in the third, eight groups of two ids; and in the fourth, sixteen groups, each consisting of only one id. The accompanying diagram illustrates this process: the two parental idants are shown on the left, and their fusion to one idant in the offspring on the right. The different kinds of shading and dotting indicate the individual differences between the ids.

The mingling of the ids in the individual idants, just as in the case of the mingling of the idants themselves, will not have occurred so quickly and regularly as is indicated in the diagram; but the final result is the same, whether the process takes place more quickly or more slowly.

The introduction of sexual reproduction will thus have gradually resulted in a greater degree of complication of the germ-plasm, so that it is no longer composed of *similar* ids, but is mainly made up of ids which are individually different from one another. *All those phenomena of heredity which are spoken of as the intermingling of the characters of ancestors, such as degeneration or atavism of all kinds and degrees, depend, I believe, on this complicated structure of the germ-plasm.*

In the following chapter an attempt will be made to explain these phenomena theoretically. It will, however, first be necessary to glance for a moment at the process of the reduction of the ids, as far as we are acquainted with it.

2. Proof that the Essential Part in the Process of 'Reducing Division' consists in the Extrusion of Ids

In the chapter on the architecture of the germ-plasm, it was pointed out that the ids are probably identical with the 'microsomes' which are known to exist in many cases in the nuclear rods, and not with the *entire* rods, or idants. This conjecture was based on the fact that the rod-like chromosomes, the structure of which we are best acquainted with, consist of a series of granules, or microsomes, which are separate and independent structures. The composition of these rods evidently excludes the possibility of considering each of them to be equivalent to a *single id*. For an id is a vital unit, with a definite structure, and cannot be composed of a row of loosely-connected spherical bodies, each containing only a portion of its determinants. Moreover, the fact that the number of idants is on the whole a small one, speaks against their being regarded as ids: the phenomena of reversion alone, it seems to me, require the assumption of a larger number of ids.

The chromosomes are not, it is true, in all cases rod-like, and may have a more spheroidal form; the existence of microsomes has, moreover, not been definitely proved in all cases. We might therefore be inclined to look upon the chromosomes as structures which are not always and absolutely equivalent, and to regard some of them as single ids, and others as rows of ids. This conception receives support from the fact that a considerable variation as regards the number of chromosomes is seen in nearly allied species, in which we might expect the processes of heredity to occur in almost the same way. Thus, for instance, the usual number of nuclear rods in *Ascaris lumbricoides* is twelve, and in *Ascaris megalocephala* two or four; in other worms belonging to the same order the normal number of rods may be eight, twelve, or sixteen. I should not, however, consider these differences sufficiently great to warrant the assumption that these rods have a different value in different cases; and this view receives support from the observations of Boveri and Oscar Hertwig, which prove that in the same species (*Ascaris megalocephala*) two varieties occur, in one of which two, and in the other four, nuclear rods are present in the cells. In this case, then, the one variety likewise possesses twice as many microsomes as the other; and although it is not always easy to determine the number of microsomes in the case of other Nematodes, we may infer their existence from the form of the idants. For these reasons I am inclined *to regard the microsomes as corresponding individually to ids, and the nuclear rods as representing groups of ids*; for this reason I have called them *idants*.

The number of idants, and even that of the ids contained in each of them, is a definite one for each individual species, but it varies considerably in different species. Each id of any particular germ-plasm could direct the entire ontogeny if it were present in sufficient numbers; that is to say, every id contains all the determinants required for one individual: but, as has already been remarked, the ids contained in the idants of a species which multiplies sexually do not contain precisely identical determinants, but these differ more or less from one another, at any rate to such an extent that they correspond to the individual differences existing in the species at the present day. It results from the mechanism for nuclear division that *all the different kinds of ids pass into all the cells throughout ontogeny, and therefore the character of every individual cell occurring in ontogeny must be determined by an aggregate of ids; so that all, or at any rate the greater portion of the ids of which the idants are made up, determine the constitution of the cell in question, this determination resulting from the forces within the cell.* These preliminary remarks will serve as a general basis for the following considerations on the effects of sexual reproduction.

We can now consider the process of the 'reducing divisions' somewhat more closely. We require to know *what influence the reducing division exerts on the composition of the germ-plasm, and of what kind are the ids which are consequently respectively removed from, and retained in, the germ-plasm.*

Direct observation of the process is not *alone* sufficient to explain it; for not only do the ids and idants appear alike to our eyes, but we cannot even determine whether the idants of the young germ-cells of a new individual are the same as those of the fertilised egg-cell from which this organism arose; that is to say, whether an idant is a *permanent* structure, and whether a *particular* idant remains the same from one generation to another.

We know that during the process of amphimixis the paternal and maternal idants are situated close together, and are enclosed within a common nuclear membrane. There is often a small, though distinct space between the two groups of rods; and did this remain distinct during the whole period of ontogeny until new germ-cells were formed and underwent reducing divisions, we might be able to determine directly whether the paternal and maternal groups became separated, or whether half the number of the paternal rods remained in connection with the maternal ones, or also whether different combinations of rods are removed by the reduction.

The matter is, however, not so simple as this: the idants of the fertilised ovum only retain this form during the first division of the egg-cell, and then become broken up into a number of minute granules, which are distributed throughout the nuclear substance, and only recombine to form nuclear rods when the second division begins to take place. This process of disintegration and subsequent recombination of the idants is repeated every time a cell is formed by division during ontogeny, and thus it is impossible to decide whether a certain idant of any particular cell is derived from the father or from the mother. And further, we cannot even ascertain with any degree of certainty by mere observation, whether the idants of the subsequent cells are the same as those of the fertilised egg-cell, — that is to say, whether they contain the same kinds of ids in the same order. It is very possible that the ids may become entirely separated from each other whenever the idants undergo disintegration, and then become arranged in some other order subsequently. The number and nature of the ids contained in the entire idioplasm would then certainly remain the same as before, but the individual idants would differ, because the combination of ids would be different. It would then be immaterial whether the idants on the right were separated from those on the left in the reducing division, or whether the halving of the number of idants were effected in some other way; all the idants would consist of new combinations of ids already present, and their combination would necessarily differ completely from that of the idants of the fertilised egg-cell, which is almost always separated by a number of cell-generations from the new germ-cells, in each of which a rearrangement of the ids must have taken place. The removal of *entire* idants in the reducing division would obviously therefore be unnecessary, for the mere qualitative division of the whole of the idioplasm into two halves would be sufficient for the purpose.

As, however, the reducing division actually consists in the removal of half the number of idants, and as, moreover, this division is, as we shall see, a double one, I conclude that the *disintegration* of the idants after every nuclear division is *only an apparent one*, and that the separate ids of the idant, on the contrary, remain connected together by fine threads of the cementing substance, or 'linin'; and at the approach of nuclear division, they become rearranged in the same order as before.

That this is the case may be concluded from certain phenomena of heredity; a child, for instance, not unfrequently takes after one parent, *e.g.*, the father only, or at any rate to such an extent that the resemblance to the mother is unnoticeable. We must therefore suppose that the fertilised ovum from which the child arose contained a very similar combination of ids and idants to that which controlled the ontogeny of the father. It must therefore be possible, and cannot be altogether a matter of chance, that the germ-cell of the father contains these paternal or maternal idants, — or, in other words, almost precisely the same ids as those which directed the development of the father or mother, arranged in almost the same order. This is only conceivable, it seems to me, if the combination of ids into idants usually, at any rate, persists even during the disintegration of the latter in the nucleus.

Many recent observations support this conclusion, inasmuch as they show that fine threads of 'linin' connect the individual microsomata (ids), even when the idant has apparently undergone disintegration. In fact, Dr. Otto vom Rath * has just shown that such connecting threads even extend between the idants. It is therefore probably not too bold an hypothesis to assume the existence of such an arrangement for connecting the ids together.

I am therefore of the opinion that the idants only apparently undergo disintegration into granules during the 'resting-stage' of the nucleus, and I agree with van Beneden and Boveri in considering the idants to be *essentially permanent* structures. I do not, however, as already mentioned, wish this statement to be taken too literally: it must not be supposed that the structure of an idant must always remain the same throughout all generations, or that the reconstruction of an idant after its disintegration must *in all cases result in the ids being rearranged in the same order*. I imagine, on the contrary, that deviations from the original serial arrangement frequently occur in the ids. The fact of the constant change of individuality and non-recurrence of the same individual which can actually be observed in the human race in the course of generations, indicates, in my opinion, that an occasional change of the ids within the idants can take place in the course of generations, although this does not occur every time the idants are reconstructed.

If this is the case, and essentially the same idants persist during ontogeny from the fertilised ovum to the germ-cells of

---

* 'Zur Kentniss des Spermatogenese von Grylloptalpa vulgaris.' — Arch. f. Mikr. Anat., Bd. 40, p. 120.

the new organism,* we may conclude from certain phenomena of heredity that the reduction of the number of ids to one half does not result in the separation of groups of ids which are *always the same, and are definitely determined* beforehand, but in the removal of *different groups on different occasions. The germ-cells of one and the same organism must consequently contain very different combinations of ids, and consequently also of primary constituents, than those which were present in the parents of this organism.* The reduction affects the paternal and maternal idants in a precisely similar and equal way; it takes place in such a manner that any combinations may result from the halving of the number of idants. Let us take, for instance, four idants a + b and c + d; not only may the paternal group a + b and the maternal group c + d, as well as combinations of a + b and c + d, be present in the fully-formed germ-cell, but also the combinations a + c and b + d, or a + d and b + c,— that is to say, combinations each of which consist of one paternal and one maternal element.

A moderate amount of difference between the germ-cells of an organism as regards their contained primary hereditary constituents will thus result. In the case of the four idants taken above as an example, only six combinations would be possible, and consequently there could only be six kinds of germ-cells differing from one another in respect of their primary constituents. The number of possible combinations, however, increases very considerably with the increase in the number of idants; for example, 70 combinations are possible with eight idants, 12,870 combinations with sixteen idants.

---

* Appearances certainly seem to contradict this assumption, and I am fully aware of the fact that Oscar Hertwig, and more recently Guignard, have stated their opinion to the contrary. In many conditions of the nucleus it is, in fact, impossible to recognise the idants, and they certainly do not exist *as such*, — that is, in the form of compact rods. But it is quite conceivable that the connection of the ids in an idant may nevertheless persist, and that the individual ids are connected together by fine threads of 'linin.' An observation made by my assistant, Dr. Hacker, supports this view. He noticed that the microsomes of the rod-like idants of the growing egg in Copepods become separated from one another, but always remain connected by a delicate thread of linin, which in this instance can be stained: the linear arrangement of the microsomes certainly persists in this case. (*Cf.* Hacker, 'Die Eibildung bei Cyclops und Canthocamptus,' Zool. Jahrbucher, Abth. f. Anat. und Ontog., Bd. v., p. 237.)

In *Ascaris megalocephala* the number of idants is only two or four; but as far as we know, a greater number is present in the case of all other animals, and also in that of plants: thus there may be eight, sixteen, thirty-two, and even a hundred or more.* A simple and single reduction, such as we have hitherto assumed, will therefore in general secure a very considerable amount of variety as regards the combinations of primary constituents caused by the reducing division. Nature seems, however, to have aimed at a far greater degree of variety, at any rate in the case of animals, in which a *double instead of a single reduction of the number of idants to one half* always occurs; and this, as I have recently attempted to show, must have the effect of increasing the number of possible combinations of idants very considerably.†

The facts as they concern the Metazoa may be briefly summarised as follows. In all those species which have been investigated for this purpose, the germ-cells are formed by the mother-cell undergoing two consecutive divisions, each of which results in a halving of the number of idants, one half passing into the one daughter-cell, and the other half into the other. In the second division this would lead to the presence of only a quarter of the original number of idants, if the number in the mother-cell were not *doubled by each idant becoming split into two before the first division takes place*. Thus there is first a doubling, and then a halving, of the number of idants. It is a matter of secondary consideration in the question of heredity that in the formation of the female germ-cell or ovum three of the cells produced by the division of the mother-cell give rise to the evanescent 'polar-bodies,' one cell alone becoming an ovum capable of development, while all four of the male germ-cells become functional. The chief point which now concerns us is the proccess of doubling, and the two subsequent halvings of the number of idants: this is known to occur in all classes of the Metazoa from the lowest to the highest forms, and, as far as we know, is only wanting in those eggs which are adapted for parthenogenesis. Even in these cases the doubling also occurs, but it is followed by only a *single halving* of the number of idants, in

* Dr. vom Rath informs me that in the crayfish (*Astacus fluviatilis*) the number of idants reaches 108–125.

† *Cf.* 'Amphimixis,' Jena, 18 1 Essay xii. in the English Translation, Vol. ii., p. 105).

correspondence with the absence of amphimixis. For the full number of idants only appears a second time in an ovum adapted for fertilisation, by the union of the nucleus of the sperm-cell with that of the ovum.

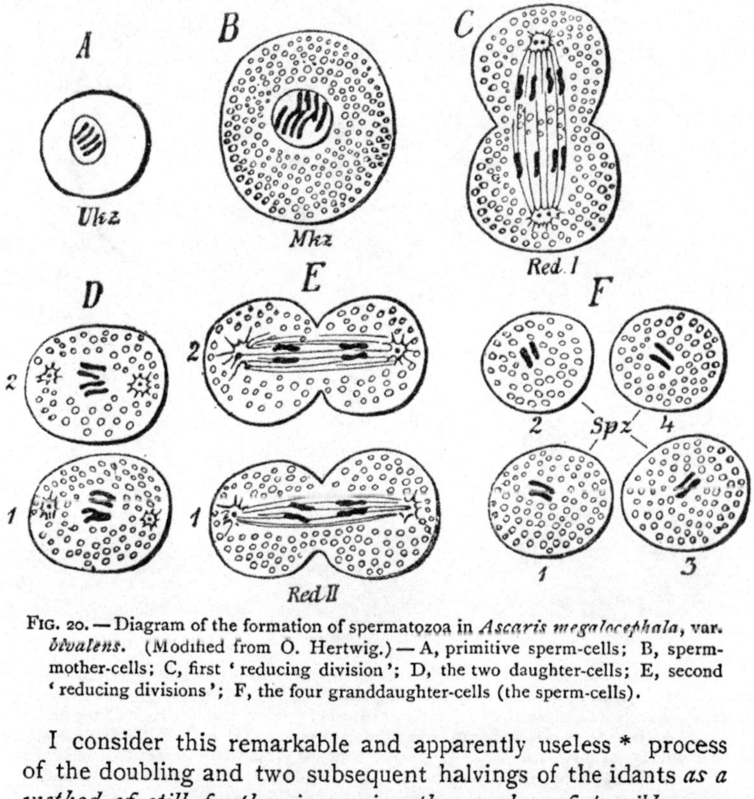

FIG. 20.—Diagram of the formation of spermatozoa in *Ascaris megalocephala*, var. *bivalens*. (Modified from O. Hertwig.)—A, primitive sperm-cells; B, sperm-mother-cells; C, first 'reducing division'; D, the two daughter-cells; E, second 'reducing divisions'; F, the four granddaughter-cells (the sperm-cells).

I consider this remarkable and apparently useless * process of the doubling and two subsequent halvings of the idants *as a method of still further increasing the number of possible combinations of idants in the germ-cell of one and the same individual*, and have given reasons for this opinion in the above-named

* It might be supposed from the observations of Ruckert on the ovum of the dog-fish, which were described in Chapters I. and II., that this doubling is simply concerned with a doubling as regards mass, and consequently with the *activity* of the idants: their activity must be very considerable in this case, for the egg of the dog-fish is very large, and requires a considerable amount of multiplication of the 'oogenetic' determinants. But a doubling of the idants occurs also in all other animal eggs, even in the

essay. As already stated, a single halving of four idants can only result in six combinations. But if, as actually occurs, each

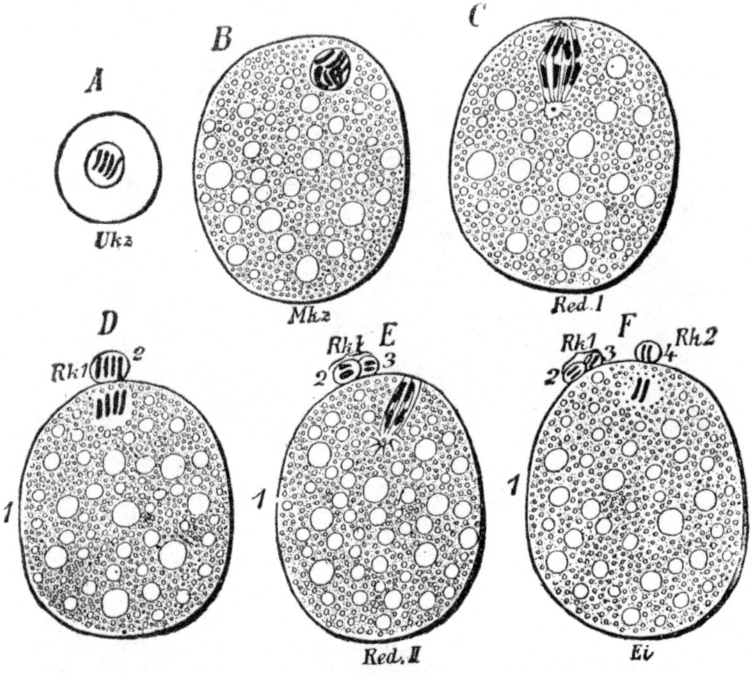

FIG. 21. — Formation of ova in *Ascaris megalocephala*, var. *bivalens*. — A, primitive germ-cell; B, fully-developed egg-cell, the number of the idants in which have increased from four to eight; C, first 'reducing division'; D, the egg with the first polar-body, immediately succeeding the stage represented in C; E, the first polar-body has divided into two daughter-cells (2 and 3), the four idants which remain in the egg giving rise to the second 'reducing spindle'; F, stage immediately succeeding the second 'reducing division' — 1, the ripe egg-cell; 2, 3, and 4, the three polar-cells; each of the four cells only containing two idants.

---

smallest — in which a very small amount of yolk is contained — as well as in the sperm-mother-cells, which never attain to such a size or structural differentiation as do the ova. The process cannot be concerned with an increase of the germ-plasm contained in the idants, for in the formation of the ova three-quarters of the mass of germ-plasm passes into the polar bodies and is again lost. The explanation of the process here given seems therefore to be the only possible one.

idant were doubled before the division, ten combinations would be possible. This means that one individual of any species possessing four idants in each of its cells can produce ten kinds of ova and spermatozoa differing from one another as regards individual hereditary tendencies. Two new idants are added to such an ovum when one of them is fertilised by the spermatozoon of another individual; and since each parent produces ten different kinds of germ-cells, as many offspring differing in character from one another may arise from these two parents as there are possible combinations of the ten kinds of spermatozoa of the father with the ten kinds of egg-cells of the mother, *i.e.*, $10 \times 10 = 100$.

With eight idants, 70 combinations are possible without, and 266 with, doubling; and following this up, twelve idants will thus give 924, or 8,074 combinations; sixteen, 12,870, or 258,570; twenty idants, 184,756, or 8,533,660; and with thirty-two idants, about five hundred times as many combinations would be obtained with doubling as without it.

Since the same number of idants from each of the conjugating cells come together in the process of fertilisation, and each of the parental germ-cells only contains one of the many possible combinations of idants, the number of variations in the germ-plasm which it is possible for two parents to produce must be an enormous one. It can be calculated by multiplying the number of possible combinations in the two conjugating cells together: thus in the case of twelve idants only, it would amount to 8,074 × 8,074. Unfortunately we are unacquainted with the number of idants in the human subject, in which we are best able to recognise individual differences in most minute detail. We may, however, suppose that this number is more than four. If, for instance, it were as high as twelve, we need not wonder that two children born consecutively are never identical, as must be the case if they had originated from the same combination of ids of the germ-plasm. Approximately identical children only occur in the case of twins, and we have every reason to believe that these originate from one sperm-cell and one ovum.

We cannot as yet judge with certainty as to how far the entire idants pass unchanged as regards their constituent ids from the germ-cells of one generation into those of the next. The phenomena of the reduction in the germ-cells which have recently been made known to us in the case of various Arthropods by the researches of Henking, vom Rath, and Häcker, indicate that even the idants *may* become changed during the process. If we suppose that in the mother germ-cell, when it is preparing for the first reducing division, the ids become arranged in their original order so as to form a long thread which doubles back on itself, and thus gives rise to a ring, the latter would become transversely divided in certain places. If the transverse divisions could take place at different points, it would be possible either for the old idants to be accurately restored, or for the new ones to differ from them to a greater or lesser extent.

This assumption is not, however, essential for a theory of amphigonic heredity, and we may here disregard it, although it will doubtless be found to apply to some extent, as was indicated above with regard to such a slow and slight change of the idants due to the disarrangement in the combinations of ids contained in them. It must be left to future researches to follow out this process in detail, and to show whether the differences in the combination of ids is merely due to the halving and rearrangement of the idants, or whether regular, or at any rate frequent, changes in the composition of the idants out of ids also occur. For the present we must be content with knowing that the *germ-cells of an individual contain very many different combinations of idants, and that a frequent repetition of amphimixis never indeed results in the germ-cells of the same parents containing the same combinations.* It therefore follows that the combination ot parental and ancestral characters continually varies, and this variation is characteristic of amphigonic heredity.

This statement also holds good for plants, in which we know that a reduction to half the number of idants takes place in the germ-cells. According to the researches of Guignard,* the

---

* L. Guignard, Compt. rend. May 11, 1891, and 'Nouv. études sur la fécondation,' Ann. scienc. nat. Bot. Vol. xiv., 1891, p. 163.

Particulars regarding Guignard's valuable researches cannot be entered into here. They have not only proved that in plants the mature germ-cells likewise contain only half as many ids as do the somatic cells, and that the normal number is again produced by the union of the nuclei of the male and female cells, but have also shown that the centrosome passes on from one generation to the next. In spite of the fact that these observations are obviously perfectly accurate, I cannot help doubting whether the reduction in the number of idants actually occurs *without a nuclear division*, as Guignard states is the case. I have arrived at this conclusion not somatic cells, as well as the mother-cells of both kinds of germ-cells in *Lilium martagon*, contain twenty-four idants, while the mature germ-cells contain twelve only. We do not as yet know whether this reduction is effected by a single reducing division, or by two such divisions preceded by a doubling, as in the case of animals. For evident theoretical reasons I consider it extremely unlikely that the reduction occurs in the mother-cell while it is preparing for division, as Guignard thinks is the case. It is very possible, however, that only one reducing division takes place in this instance.

The details of these processes in the lower plants are quite unknown, probably owing to the minute size of the idants, which till now has rendered the difficulties of such investigations insurmountable. It has, however, at any rate been ascertained that in many marine algæ (*Fucoideæ*) the development of the egg-cells is accompanied by the formation of 'polar bodies,' which certainly correspond to stunted and phyletically degenerated ova: this was first shown to be the case by Bütschli and Giard, and the fact has long been recognised by other zoologists besides myself. In the genus *Fucus* these polar bodies do not occur, and eight eggs are formed from the primary ovum,— if I may venture to apply this term to the original cell of the so-called oogonium or ovary; in an allied species of wrack, *Ascophyllum nodosum*, only four eggs are formed from the primary ovum, but four polar bodies are also produced; in *Pelvetia canaliculata* the primary ovum gives rise to two eggs and six polar bodies;

---

merely from a comparison of the analogous process in animals, but also because I cannot help thinking that it is possible, and even probable, that in this respect these otherwise admirable observations are not quite complete. In the formation of the male germ-cells, a reducing division may perhaps take place between the 'cellules mères primordiales' and the 'cellules mères définitives;' and as regards the female germ-cell, it will occur in the division which gives rise to the 'cellule mère du sac embryonnaire.' In both cases even the most acute observer might fail to notice the reducing division if his attention were not specially directed to this point. Why should an arrangement for a 'reducing division' have been made in the case of animals if the reduction could take place without nuclear division, and could produce the same result? Of all the other numerous observations which have been made on the process of karyokinesis, not a single one supports the view that the single (?) chromatin band of the 'skein' stage can become disintegrated into half as many idants as were previously present in the nucleus.

and in *Himanthalia lorea* only a single egg and seven polar bodies are formed.*

We have here, however, no information as regards the reducing divisions, which, as I pointed out long ago, need not by any means be connected with the degeneration of several germ-cells. We can only state that the three successive divisions of the primary ovum, which occur in all the above-mentioned cases, affords more than sufficient opportunity for one or even two reducing divisions, and that it is extremely probable that one, at least, actually occurs.

We may therefore assume that in plants very varied combinations of the germ-plasm derived from each parent usually take place in the germ-cells of the offspring, and that perfectly '*identical*' germ-cells can very rarely occur either in plants or in animals.

---

* *Cf.* Oltman's, 'Beitrage zur Kentniss der Fucaceen,' Cassel, 1889.

CHAPTER IX

ONTOGENY RESULTING FROM THE UNION OF THE GERM-PLASM OF TWO PARENTS

1. The Nature of the Offspring Determined by the Process of Fertilisation

The first question which presents itself in the consideration of 'amphigonic heredity' is concerned with the relative share taken by the germ-plasm of each parent in the control of ontogeny.—whether the paternal and maternal ids always co-operate simultaneously, and the forces contained within them together form a single resultant, or whether one group only is active while the other remains passive. This question cannot at present be decided from the results of observations on the nuclear substances themselves; the phenomena of heredity, together with what we know concerning the composition of the idioplasm resulting from amphimixis, can alone help to elucidate this problem. These phenomena must therefore be analysed as accurately and minutely as possible.

We must base our analysis on the fact which we have already proved, that the germ-cells of an individual differ from one another as regards the hereditary substance they contain, and that the proportion of paternal and maternal ids in a germ-cell varies between wide limits, the degree of variability being greatly increased by the union of the germ-cells of two individuals in the process of amphimixis. This fact is sufficient to account for the difference existing in the human race between children of the *same* parents. The fundamental law of amphigonic heredity enunciated by Victor Hensen follows directly from this fact:—'*the individual is determined at the time of fertilisation*;' or, in other words, *the individuality of an organism results from the fact that the germ-plasm is composed of the paternal and maternal ids which are brought together in the egg-cell.*

This law is not self-evident, for we might have believed, *a priori*, that the development and mingling of parental characters in the offspring is due entirely, or at any rate to a great extent, to external influences of nutrition, &c., to which the germ is subject after fertilisation. The existence of 'identical' human twins, however, proves the contrary. Some twins do not resemble one another more closely than do children of the same parents which are born consecutively; and, in fact, this is apparently true of the greater number of twins, in which the dissimilarity may even be very considerable. We have every reason to suppose that such 'dissimilar' twins are usually derived from *two* ova, which must of course have been fertilised by *two* different spermatozoa. On the other hand, in the case of those twins which I speak of as 'identical,' the resemblance, although not perfect, is much closer than has ever been observed in children born successively. There is every reason to suppose that such identical twins are derived from a single ovum and spermatozoon. If this is actually the case, it furnishes a proof of the above statement that heredity is potentially decided at the time of fertilisation, or, expressed in terms of the idioplasm, that *the nature of the combination of the parental ids which takes place during fertilisation predetermines the whole subsequent ontogeny.* The slight differences which exist between identical twins would therefore probably indicate to what extent the course of development may be affected by external influences. These differences are generally so slight that it is difficult to observe them at all, unless they are specially sought for; as a rule such twins can only be individually recognised by the parents or brothers and sisters, and cannot be distinguished from one another by strangers.

These slight differences might, however, be due to an imperfect predetermination of the influence which is exerted at every ontogenetic stage by the idioplasm of each of the parents. We can hardly decide between these views from the consideration of identical twins only. Mr. Otto Ammon, of Karlsruhe, has kindly furnished me with two photographs of identical twins taken in consecutive years at the ages of seventeen and eighteen, as well as with exact measurements of all parts of their bodies. In spite of a striking resemblance, not only in face but in all parts of their bodies, certain differences are nevertheless recognisable between them. For instance, the height of the one marked No. 507 on Mr. Ammon's list, measured, when lying down, 172 cm., and that of No. 508 only 170 cm.; and again, although the length of the hand and of the left arm is the same in both, —

the latter measuring 74 cm., — the right arm of No. 507 is only 71 cm. long, while that of No. 508 reaches 74 cm. The relative lengths of the upper arm and fore-arm are also different, that of the left upper arm of No. 507 being 27 cm., while in No. 508 it is 27.5 cm.; and consequently the length of the fore-arm in No. 507 is also 27 cm., while in No. 508 it is only 26 cm. Even if we possessed the measurements of the parents at the same age, we should probably be unable to draw any definite conclusions as to whether these slight differences in size are due to a corresponding difference in the combination of the germ-plasm, such as might arise from a slightly inexact division of the nucleus of the fertilised ovum in the process of doubling or at a later stage, or whether they simply owe their origin to slight general or local differences of nutrition taking effect during ontogeny.

Other facts are, however, known, which prove that although the nature of the combination of the parental idioplasms during ontogeny is in general, as a matter of fact, determined at the time of fertilisation, it is nevertheless liable to slight individual fluctuations. Instances of this kind are furnished by the *hybrids of certain species of plants*, many parts of which exhibit a considerable degree of variability, and fluctuate between the specific characters of the two parents. The blossoms of the hybrid plants obtained by crossing *Digitalis lutea* and *D. purpurea*, for instance, 'vary in colour; in some instances they are pale, with a slight pink tinge, which latter, again, may be entirely absent; and in others they have a more or less bright purple colour.'* These observations appear to me to be particularly important, owing to the fact that we may assume with certainty in this case, in which two distinct and sharply defined species were crossed, that both parents possessed the specific characters in the same degree of purity and strength, and that consequently the relative proportion of the parental idioplasms does not remain quite constant during ontogeny, owing either to slight irregularities in the nuclear division, or — and this is less probable — to inequalities in nutrition and in the growth of the idants derived from the two parents. Owing to the kindness of Professor Hildebrandt of Freiburg i. Br., I have had an opportunity, in the case of hybrids of two species of *Oxalis*, of observing how extremely detailed the process of predetermination is. The flowers of

* Focke, 'Die Pflanzen-Mischlinge,' Berlin, 1881, p. 316.

one of the parent-species were large, and of a pale lilac colour, while those of the other were smaller, and their colour was red, with a dark crimson ground. The flowers of the different hybrids were by no means quite similar, but three principal forms could be distinguished according to the combination of colours in the flowers, which I shall not describe in detail: *the flowers of the same hybrid, however, resembled each other in their most minute details.* One plant, for instance, had violet petals of a rather pinker tint than those of one of the parent-species, and all the petals were strongly tinged with red on one and the same lateral margin. As far as I could observe, *all* the flowers were similarly coloured on this stock. On another stock, all the sepals had brown rims, and on a third there was a narrow dark orange-coloured band in the centre of each flower. *In these cases, therefore, the combination of the colours of the parents which appeared in the petals of the hybrids must have been decided at the time of fertilisation.* It will be shown later on how this combination may vary somewhat in different plants.

Even although the slight differences in identical human twins, which can be proved to exist, are certainly due in part to minute differences in the idioplasm itself, some of them must nevertheless with equal certainty be attributed to the effect of various external influences. My photographs of the above-mentioned identical twins show that No. 507 has particularly white hands, while those of No. 508 are much browner. No one would attribute this fact to dissimilarity in the respective germ-plasms, or to an alteration in the proportion of paternal and maternal idioplasm which occurred during ontogeny: it must be due to the fact that the hands of No 508 had been more exposed to the sun than those of No. 507; and, as it happens, the former of the two had been more employed in the open air than the latter before the photograph was taken. Several differences in the proportional sizes of parts of the body may possibly have been brought about in a similar way.

2. The Share taken by the Ancestors in the Composition of the Germ-plasm

If, then, it is certain that the characters of the developing offspring are essentially decided by the mingling of parental idioplasms which takes place in the process of fertilisation, we must next try to ascertain whether *the entire parental idioplasm*, *with all its constituent determinants, or only a portion of it*, is passed into the germ-cell which will give rise to a new individual; and also *what proportion of it is constituted by the germ-plasm of the grandparents, great-grandparents, and more remote ancestors*. The fact that the reducing division, which takes place both in the male and female germ-cells before fertilisation, removes half of the idants from each, leads us to conclude that *only half the normal number of ids can be contained in each germ-cell;* and this could only be the case if two of each kind of parental idant were present in the germ-plasm, and if the reduction resulted in each individual germ-cell containing a similar group of idants. But this cannot be so, for the germ-plasm must consist of a number of entirely different idants, unless, in consequence of interbreeding, two of the same kind are present in certain of the groups. *The whole of the idants of both parents evidently cannot possibly be contained in any one germ-cell*, because the total number would be twice as great as that which actually occurs in a ripe germ-cell. If in Man, for example, there were thirty-two idants in the fertilised ovum, sixteen of them would be derived from each parent. Of this latter number, sixteen at most could be derived from *one* grandparent, and this could only occur if no idants at all from the other grandparent had passed into the germ-cell in question. It is evidently more than inaccurate to fix the limit of the hereditary power—as is done by animal-breeders—of a parent at $\frac{1}{2}$, of a grandparent at $\frac{1}{4}$, of a great-grandparent at $\frac{1}{8}$, and so on.* These numbers do not even represent the maximum or minimum share in heredity which may be taken by the respective ancestor in the constitution of the fertilised egg. The

* Galton has also emphasised this fact in the concluding chapter of his book on 'Natural Inheritance' (p. 187 et seq.). According to his view, the 'personal heritage' of each parent = $\frac{1}{4}$, and the heritage of 'latent elements' of the parent likewise $\frac{1}{4}$, the two together thus making up $\frac{1}{2}$. Naturally I cannot agree with this calculation, for in my opinion the latency of the characters of a parent does not result from the 'primary constituents' of these qualities, but from the struggle between the primary constituents of both parents; and I do not in the least suppose that the primary constituents which practically give rise to the individual become separated from those which form the latent germ for the germ-cells of the next generation. But I fully agree with Galton that all the 'characters' of the ancestor—the grandparent, for instance—are never present in every germ-cell from which a grandchild may arise.

parent is certainly always represented by one-half, but the share varies even in the case of the grandparent; in the instance just given it would vary between 0 and 16. For the reducing division may, for instance, cause the sixteen paternal idants resulting from the reduction of the thirty-two originally present in the sperm-cell, to contain idants derived from the grandfather only, and none from the grandmother; or, again, there might be fifteen from the grandfather, and one from the grandmother, or fourteen *pp* and two *mm*, or thirteen *pp* and three *mm*, and so on.* This would be so, at least, if any kind of combination of the idants may result from the reducing division. It may perhaps not be the case absolutely, but the capriciousness with which reversion to a grandparent may occur nevertheless indicates that a considerable latitude exists with regard to this combination.

In passing back to the third, fourth, and fifth generations, we cannot in the least determine, *a priori*, to what extent an individual ancestor of the animal in question is still represented in the germ-plasm of a germ-cell; we can only state the maximum which *might* be possible in the most favourable case. In the above-mentioned instance, an ancestor of the third generation *might* still be represented by sixteen idants, for the sixteen idants which this ancestor furnished for the purposes of amphimixis in the second generation might, all in fact, possibly have passed into *one* germ-cell in the process of reducing division in this generation, and the same, again, might have occurred in the first generation. Such a case can only be of rare occurrence, but it apparently accounts for the instances of reversion in Man to ancestors more remote than grandparents, which, though rare, certainly occur occasionally. The more remote the generation, the greater are the chances against the entire half of the total number of idants remaining together through several generations in individual germ-cells, and the probability of such an occurrence will very soon be reduced to zero.

We may suppose that, as a general rule. the number of ancestral idants contained in a fertilised egg-cell becomes less in pro-

* I shall now denote the paternal ids or idants by the letter *p*, the maternal ones by *m*, and those derived from the grandfather by *pp* or *pm*, and from the grandmother by *mp* or *mm* &c. The first letter in each case signifies the parent, the second grandparent, the third the great-grandparent, and so on.

portion as the ascendancy of the ancestor concerned decreases. Any more exact calculation of the share taken by a certain remote ancestor in the composition of the germ-plasm of its descendant would be erroneous. Hitherto the customary method of making such a calculation has been to assume that the following shares are taken by the various ancestors in the predisposition of the offspring:—parents, $2 \times \frac{1}{2}$; grandparents, $4 \times \frac{1}{4}$, and so on; that of the sixth generation of ancestors being $32 \times \frac{1}{32}$. Thus, in the last-mentioned generation, *one* idant out of the thirty-two assumed to be present in Man would, according to the theory of the germ-plasm, still remain. This does not by any means imply that each of the thirty-two ancestors of the sixth generation is still represented by one idant in the germ-plasm of the descendant; it is quite as probable that thirty or even twenty of these ancestors take part in its constitution, and the number may possibly, though improbably, be still less than this. In treating of the phenomena of reversion, I shall have occasion to refer to this subject again.

It is at any rate certain that *in no case can more than the half of the idants of one parent be present in the germ-plasm of the fertilised egg-cell.*

This statement is, however, apparently contradicted by certain facts.

Plant-hybrids frequently keep to the mean between the two ancestral species; that is to say, they contain *all* the characters of these two species in equal proportions. Thus all the primary constituents of each parent would be contained in the fertilised egg-cell, although, according to our theory, only half of the parental idants are concerned in its constitution. This contradiction is easily accounted for, if it be borne in mind that we are here concerned with the mingling of the characters of two *species*, and not of those of two *individuals* of the same species. The *characters of the species must be contained in the majority of ids in each idant*, if not in every id, *and half the idants may in this case produce the same effect as would result if all the idants were present:* that is to say, they contain *every* specific character. In cross-breeding, specific characters are opposed to specific characters, and in comparison with the greater differences between these, the lesser individual ones disappear.

The reverse is true in the case of reproduction in Man, especially within one and the same race. The specific characters are probably contained in all the ids of the father as well as of the mother, and the differences between the parents refer to individual characters only. Our theoretical conception of the idants as a collection of ids seems incompatible with the above-mentioned statement that the child can only closely resemble *one* parent, for only half of the idants of this parent take part in the construction of the child. We shall, however, be able to explain this apparent contradiction later on.

The facts of the case may be stated in general terms as follows. Half the number of parental idants always reach the germ-cells of the offspring, but this half may consist of all possible combinations of the parental idants: that is to say, either of idants derived from the grandfather or grandmother only, or of a combination derived from both, in which one or the other may predominate. Nothing will be gained by taking the ancestors of the third or fourth generation into consideration until we come to consider the phenomena of reversion.

3. The Struggle of the Ids in Ontogeny

a. Plant-Hybrids

The structure of the offspring results from the struggle of all the ids contained in the germ-plasm.

That this statement must be in general correct is to some extent indicated by the fact that all parts of hybrid plants, produced by crossing two species or varieties, usually possess the characters of both parents. The details concerning hybrids are of far greater value for theoretical purposes than are those relating to the normal offspring of any particular species, as we know for certain that the characters which compete with one another or combine, so as to result in the production of a hybrid, must be contained *in every idant* of one or other of the parents; for these characters are those of the *species*.

The difference as regards the idioplasm between *individual* and *specific* characters, seems to me to be due to the determinants of the latter being present in an overwhelming majority in all the ids of every idant of the germ-plasm, while the determinants controlling the structure of individual characteristics are only contained in a portion of the idants of which the germ-plasm consists: *at most* they can only be present in all the idants of one of the parents,— that is, in half the entire number of idants. The extent to which the determinants of any individual character are represented — whether they are contained in all the ids, or only in a small portion of them — could only be ascertained from the phenomena of heredity if we knew the cause of the predominance of any particular 'successful' character. This, however, can only be inferred from crosses between species, the great constancy of the specific characters in which leads us to presuppose that their determinants predominate in all the idants of the parental germ-cell.

In plant-hybrids, paternal and maternal idants come together in the process of amphimixis, and each group of them may be assumed to consist of *similar* idants. The effect this arrangement will produce on the phenomena of heredity must now be considered, and conclusions drawn from these considerations.

From a very large number of observations on hybrid-plants, we find that, in the first place, parental characters may be variously intermingled. From a comparison of all the cases observed up to the year 1881, Focke * concludes that these combinations of characters may be divided into three principal groups, viz.: — (1) *a mean between both parents is maintained in all parts of the plant*; (2) *the characteristics of the father or mother predominate*; and (3) *certain parts of the hybrid exhibit the maternal, and others the paternal, characters.*

The first-mentioned case is by far the most frequent: we may take as an example the hybrid obtained by Köllreutter from two species of the tobacco plant, — *Nicotiana rustica* ♀ and *N. paniculata* ♂. Köllreutter † himself stated that this plant was exactly intermediate between its parents, while it was considered by Gärtner to bear a slightly greater resemblance to *N. paniculata*, and by Focke to *N. rustica*. Köllreutter's opinion is probably therefore a fairly accurate one, for in any case the point can only be estimated, and cannot be decided with mathematical precision. According to Focke, the corolla-tube of *N. rustica* is 14 mm. long, that of *N. paniculata* 26 mm., and that of the hybrid 19 mm. The exact mathematical mean between these is 20 mm.; and hence in this character the hybrid approaches *N. paniculata*. As regards the diameter of the widest part of the

* 'Die Pflanzen-Mischlingen,' Berlin, 1881.

† Joseph Gottlieb Kollreutter, 'Vorlaufige Nachricht von einiger Geschlecht der Pflanzen betreffenden Versuchen u. Beobachtungen' 1761

corolla-tube, on the other hand, the hybrid inclines more to *N. rustica*, while in the narrowest part it exhibits the exact mathematical mean. This case is instructive, for we cannot recognise the true *physiological* mean, because the length, as well as the diameter, of the corolla-tube is determined by the same cells. Any estimation of the mean between the colours must be still less precise, for the different shades depend on entirely different morphological constituents. If the yellow and red of two different species were blended in the flower of a hybrid, the intensity of both these colours might conceivably be as great as in the parent plant, and yet one of them might predominate because it happened to cover the other. For the yellow is due to special pigment-granules, while the red occurs only in the cell-sap, which might possibly be nearly hidden by a superficial layer of chromatophores.

In any inquiry with regard to the factors which control the struggle of the parental characters, it must, above all, be borne in mind that *the cells are always the determining agents.* The determinants of the father and mother come together in the cell, and in the cell only; and all characters, whether relating to a large part of the organism, or merely to a single cell, can only be determined by processes taking place within the substance of one or of many cells.

This does not mean that the *visible* differentiation of every individual cell always constitutes a 'character' of the organism. The nature of the histological differentiation of the cells — that is to say, whether muscle- or nerve-substance, or chlorophyll granules, for instance, are produced in the cell-body or not — only comes into consideration in connection with the *number* of cells in the *definitive* cell-aggregate constituting the organism. Very many characteristic qualities cannot be due to this fact, but must chiefly depend on the number and arrangement of the cells in an organ, which again must be due to qualities of the embryonic cell which are invisible to us — principally those which relate to its *method of division*, and its *vigour and rate of multiplication*.

We must suppose that these factors are wholly determined by the idioplasm of the cells quite as much as is the visible differentiation of the latter. The division of a cell certainly originates in its apparatus for division, and primarily in the 'sphere of attraction' and its contained centrosome; but we should have to give up the whole idea of the controlling influence of the nuclear substance were we to suppose that this apparatus actually *directs* the process of division. The entire process of development of the animal from the ovum depends so essentially on the rate of division of the cells, that the nuclear substance could no longer be considered to correspond to the hereditary substance if it merely caused the visible differentiation of the cell. But in the Introduction I have already stated the reasons which prove beyond a doubt that the nuclear matter actually contains the primary hereditary constituents, and it follows from this fact that there can be no question of a self-determination of the apparatus for division. We must, in fact, suppose that the ultimate structure of the cell-body, which is invisible to our eyes, controls its entire growth as well as its method and rate of division, this structure itself being controlled by the nuclear substance or idioplasm. Ultimately, therefore, everything depends on the determinant of the cell; and in the case of sexual reproduction, the co-operation of the paternal and maternal determinants determines the character of the cell, whether this character is visible or invisible. As every cell in the entire ontogeny is, according to our view, controlled by *one kind* of determinant only, irrespective of the fact whether it also contains other determinants in a latent condition or not, the co-operation of maternal and paternal determinants always determines the character of the cell, and controls the development of the individual as far as the influence of this cell extends. In spite of its greater histological differentiation, a slighter influence is therefore obviously exerted by one of the final cells in ontogeny, — that is, one of the tissue-cells, — than by one of the first four segmentation-cells, or by the primary cell of the entire germinal layer, or, again, by any cell from which many and various kinds of cells may subsequently arise. On the other hand, we must not forget that each embryonic cell only determines its *own* method of division, and not necessarily that of its daughter-cells, and that consequently in these a new counter-balance or co-operation of the paternal and maternal determinants again takes place.

The determinants which control the daughter-cells are, however, derived from the *latent* ids of the mother-cell, and it therefore essentially depends on the methods of division and on the architecture of these ids as to which determinants are to control the daughter-cells. *That portion of the ids of the mother-cell, therefore, which is for the time latent, exerts a definite influence on the determination of the subsequent development*; on it alone, indeed, depends the number and order of succession of the determinants which will subsequently become active, and all those characters which result from the number and grouping of the cells, as well as from the histological nature of the individual cell, are also determined in the first instance by their ids.

We can thus explain why cross-breeding is only successful in the case of nearly-related species, and not in that of members of entirely different families. If it were possible, for instance, for the ovum of a sea-urchin to be fertilised by the spermatozoon of a worm, such as *Rhabditis nigrovenosa*, the disintegration of the ids of germ-plasm derived from the father and mother respectively would take place in an entirely different manner even in the first stage of segmentation; the maternal ids might become disintegrated into the groups of determinants for the right and left halves of the body, while the paternal ids became separated into the groups for the external and internal germinal layers. But dissimilar groups of determinants of this kind could not co-operate and give rise to intermediate structures; and even if the ontogeny advanced a few more stages, embryonic structures could never be produced which would work together harmoniously.

The term *homologous* determinants and ids may be applied to those which control homologous cells and groups of cells. It may therefore be stated that *structures which are intermediate between those of the two parents may arise whenever homologous determinants and ids come together*. If a small spot is present on a certain region of the wing in two allied species of butterflies, for instance, and this is represented in the germ-plasm by a *single* determinant, the homologous determinants of the two cells in which this spot was first contained will be brought together when these two species are crossed, and will eventually be able to share in the control of the cell in question. They need not, however, necessarily be exactly alike; the spot may be brown in species A, and red in species B. In this case the determinants would be homologous but not *homodynamous*, and might possibly combine to form a reddish-brown spot. The essential point in amphigonic heredity is, that *the idioplasm in each cell in the entire ontogeny contains ids which are individually different from one another, and does not consist of a number of identical ids; and that the cell may take on an intermediate character in consequence of their co-operation.* In normal reproduction the active ids of the idioplasm are all *homologous*,—that is to say, their function is to control the same part of the body,—but they differ among themselves, *i.e.*, are *heterodynamous*, or, in other words, they tend to impress a somewhat different character on the same part of the body. In the crosses between different species, the idioplasm of a cell in many stages will be composed of homologous as well as of *heterologous* ids, and, as has just been shown, it then remains to be decided whether a common control of the cell is possible at all, and if so, over how many generations of cells it can extend.

The scales of butterflies are cells belonging to the final stage of ontogeny, and their ids are made up of determinants of one kind only. The further back we pass towards the beginning of ontogeny, the more numerous are the determinants composing the ids; only one of them, however, breaks up into biophors and controls the cell. If the determinant which actually controls the cell is homologous in both parents, an intermediate form of cell may result; but as soon as the rest of the determinants in the id no longer correspond to one another, the further development gradually becomes checked, and will ultimately be brought to a stand-still. The processes which must be supposed to take place from the beginning of ontogeny onwards in the case of the supposed cross between a sea-urchin and a worm may very possibly be only *partial, i.e., they may refer merely to special parts*. Let us suppose that a certain species of insect, which is normal in other respects, possesses limbs on the dorsal side instead of wings, and that this is crossed with another insect possessing normal wings, development of the fertilised egg taking place up to the stage in which the wings arise. The idioplasm of the primary cell of the wings and 'dorsal limbs' respectively would then contain two perfectly *heterologous* ids, one derived from the father and the other from the mother, and none of the determinants in these two ids would be respectively homologous. In the first stage of the development of the wings or 'dorsal limbs,' as the case may be, antagonistic determinants would be opposed to one another in each cell-generation, and would prevent a common determination of the cell.

Such extreme cases do not actually occur, as very different

species do not interbreed; but the principles which may be deduced from these imaginary cases may be applied to the production of hybrids. *Homologous* determinants and ids co-operate, while *heterologous* ones do not; and the larger the number of heterologous determinants present in the homologous ids of the parents, the more do the hereditary tendencies of the latter tend to nullify one another.

We must now attempt to give a theoretical explanation of the *first of the three kinds of combination of parental characters in the offspring mentioned above*, viz., that in which '*a mean between both parents is maintained in all parts of the plant*.'

If we assume that the two parental species are so closely allied that each determinant in the one corresponds to a homologous determinant in the other, a form exactly intermediate between the two species must result, supposing that the number of ids in either parent is the same, and that a similar controlling force acts on the homologous determinants.

If the idioplasm of one parent is represented by a larger number of ids, its controlling force must obviously be greater; and as regards the controlling force of the individual determinants, we may state as follows:—the control of the cells is in our opinion effected by the disintegration of the determinant into biophors, which. like the pangenes of de Vries, migrate into the cell-body, multiply at its expense, and give rise to definite cell-structures. This multiplication must take place with a certain amount of energy, the degree of which varies in the different kinds of biophors. Thus, whenever such controlling biophors, possessing the energy of growth in different degrees, migrate into the same cell-body, *a struggle of the parts* (Roux) must ensue, in which the stronger part will be successful, and the weaker will be more or less, or even completely, suppressed.

The formation of structures which are strictly intermediate between those of the two parents, implies that the homologous determinants possess a similar controlling force. All the determinants of any two species are, however, never homologous: this follows from the fact that the *number* of cells in homologous parts is often very different. The characters of the species do not by any means depend only on the histological nature of the individual cells, but, as already stated, are due in almost a greater degree to the number and arrangement of the cells, to the repetition and position of certain organs, and so on Thus the flower of *Nicotiana paniculata* is decidedly longer than that of *N. rustica*, and the former species is more extensively branched and possesses a larger number of glands than the latter. The above assumption that the individual determinants correspond to one another in the two species cannot therefore be quite an accurate one; the germ-plasm of *N. paniculata* must, on the contrary, contain a larger number of determinants than that of *N. rustica*, and the process of disintegration of the two species must differ in many ways.

If an equal number of ids from each of these species occurs in the hybrid, the two kinds of ids will only be able to co-operate in ontogeny as long as their determinants still correspond to one another. As soon as a point is reached in which the ids of *N. rustica* begin to decline in number, and their last determinants have become disintegrated, the ids of *N. paniculata* will alone be able to produce series of cells; but as only half the normal number of these ids are present, the structures arising from them cannot be as complete as they are in the pure ancestral form: and, apart from this, the ids of *N. rustica* may possibly not have disappeared entirely, but the succession of the determinants, which are, properly speaking, the final ones, may persist beyond the normal period, and may in this way interfere with the development of the pure characters of *N. paniculata*. This would, at any rate, render the fact comprehensible in principle that intermediate forms may also arise in cases in which the struggle of the parental ids does not extend into the final cells of ontogeny, and the characters of the species come into contact at an earlier stage, as in the case of the greater or less degree of ramification in plants. The following considerations will make this more evident.

In the *second kind of combination*, either the paternal or the maternal characters predominate in all parts of the hybrid, so that the latter bears a closer resemblance to one parent than to the other: in this case, therefore, *the transmission is apparently monogonic*.

Several cases of this kind have been recorded, in some of which the paternal, and in others the maternal characters predominate. Instances of both kinds occur in the genus *Nicotiana*. 'The hybrid *Nicotiana paniculata* ♀ × *vincæflora* ♂ bears so close a resemblance to the last-named species that the character of *N. paniculata* can hardly be recognised at all.'* In this case, therefore, the paternal plant predominates, while the cross between *N. suaveolens* ♀ and *N. langsdorffii* ♂ 'bears little resemblance' to the latter species. Here, then, the maternal characters predominate, and the hybrid plants are 'extremely similar' to *N. suaveolens*, and are only to be distinguished from it by a partial separation of the stamens from the corolla-tube, a slight difference in the colour and size of the flowers, the violet or bluish coloration of the anthers, and by complete sterility.'

On p. 474 of his book, Focke gives further instances. 'In many cases the resemblance of the hybrid to one of the ancestral forms is so close, that it might easily be taken for a minor variety of this form.' Thus in the following instances the hybrid resembles much more closely the parent form mentioned after each:—*Dianthus armeria* × *deltoides*, *D. deltoides*; *Dianthus caryophyllus* × *chinensis*, *D. caryophyllus*; *Melandryum rubrum* × *noctiflorum*, *M. rubrum*; *Verbascum blattaria* × *nigrum*, *V. nigrum*; *Digitalis purpurea* × *lutea*, *D. lutea*.

These cases seem to me to be of special importance on account of the conclusions which can be drawn from them respecting perfectly similar cases of *individual* transmission in Man. This '*pseudo-monogonic* transmission,' as I shall call it, must be explained in terms of the idioplasm somewhat as follows.

The predominance of one parent—*e.g.*, the mother—might be due to the presence of a *larger number of idants and ids* of the species in question. If, for instance, *Digitalis lutea* possessed thirty-two idants and *D. purpurea sixteen only*, the idants in both cases consisting of the same number of ids, and if the controlling force of the ids were the same in both species, the ids of *D. lutea* would then predominate over the others in every cell during the entire ontogeny; that is to say, the character of *D. lutea* would be impressed more strongly than that of *D. purpurea* on the cell. The objection might be raised that the cell which is thus produced must possess an intermediate character, and cannot be a pure cell of *D. lutea*, even though it resembles the latter most closely. As yet, however, we are unable to determine to what extent intermediate forms of individual cells may occur in individual cases; and the expression 'resultant,' in connection with the unknown forces of the

* Focke, *loc. cit.* p. 289.

biophors, does not indicate anything beyond a mere abstract idea, which is in any case totally insufficient as an explanation of the phenomena. We must be satisfied with the statement that, *when the controlling forces of the two determinants acting together in the cell* are very unequal, the effect of the weaker will be extremely small under certain circumstances. *A 'struggle between the biophors'* takes place, in which we may suppose that the stronger assimilates, grows, and multiplies more quickly, and thus deprives the weaker of room and nourishment, prevents its multiplication, entirely destroys it, and even makes use of it as nourishment. Without a very considerable multiplication, the crowd of biophors which migrate from the nucleus into the cell-body cannot, indeed, exert any determining influence on the latter. It therefore seems to me to be conceivable that an apparent pseudo-monogonic transmission—*i.e.*, a complete suppression of the elements of one parent — may take place even when exactly the same number of ids are derived from both parents.

This is still more likely to be the case if the number of ids derived from one parent is greater than that from the other. We know that the number of idants may vary considerably, even in allied species, and it is therefore not improbable that 'pseudo-monogonic' heredity is sometimes due to this fact. In many plant-hybrids this assumption can be tested directly by ascertaining the number of idants present.

The *third* of the above-mentioned kinds of combination of parental characters seems to me to be theoretically almost the most important of all, for it is most intimately concerned with the ultimate processes which take place in the idioplasm. *In this case the parts of the hybrid plant most nearly resemble those either of the paternal or the maternal form.*

Such cases are apparently not often very pronounced, but fluctuations from the paternal to the maternal side occur to a slight extent in almost all those hybrids which are usually described as intermediate forms. The hybrid between *N. paniculata* ♂ and *N. rustica* ♀, which Kollreutter considered to be a pure intermediate form, bears, as mentioned above, a somewhat closer resemblance to the former species as regards the length, and to the latter as regards the diameter, of the corolla-tube. Such slight fluctuations to one side or the other from the pure intermediate form seem to be of frequent occurrence. Cases are, however, also known in which these fluctuations are so considerable as to become conspicuous. Sometimes, for instance, the hybrid resembles one species as regards its leaves, and the other with regard to its blossoms. Brandza * has recently examined the general, as well as the microscopic structure of certain hybrids, in order to ascertain the parental characteristics exhibited by them, and has succeeded in proving that a fluctuation of this kind exists. *Marrubium vaillantii*, a cross between *Leonurus cardiaca* and *Marrubium vulgare*, for instance, exhibited the winged leaf-stalk characteristic of *M. vulgare*, while the arrangement of the vascular bundles of the petiole resembled that existing in *Leonurus*. The upper surface of the petiole bore branched hairs like those of *Leonurus*, while on the lower surface the hairs resembled those of *Marrubium*.

I shall speak of this fluctuation in the resemblance of the *organs* to those of the parents as '*the shifting of the hereditary resultants in ontogeny*.'

We might imagine, *a priori*, that such a shifting cannot possibly occur. If, as has been proved, transmission is virtually completed at the time of fertilisation, and if therefore the relative proportion of the two parental idioplasms is also fixed *for all the subsequent ontogenetic stages* when they have once been combined, we might expect a *similar combination* of the characters of the parents to appear in all parts of the young plant, so that these would either be exactly intermediate between those of the two parents, or else that the paternal or maternal characters would everywhere predominate in a similar manner. The fact that this may not be, and in fact is usually not the case, may be accounted for in several ways.

In the first place, we must bear in mind the statement made above, that a combination of two characters in equal proportions need not always give rise to one which is apparently intermediate, and that it is impossible to give an exact definition of what is meant by such a character, because we only observe the final effect of forces acting within the cell, and not the actual processes by which this effect is produced. This, however, would not be a true shifting of the hereditary resultants, but only an *apparent* one.

Genuine cases of such a shifting, however, undoubtedly occur;

---

* Brandza, 'Compt. rend.' 1890, T. III, p. 317.

and I believe that they are principally due to the fact that the *number of homodynamous determinants in the idioplasm of a cell may vary in the course of ontogeny, and that, in fact, it must always do so.* In one stage or in one organ the paternal, and in another the maternal ids will contain the majority of homodynamous determinants, and the fluctuations in the predominance of the maternal or paternal characters, *which is definitely determined in advance*, must depend on this fact.

To make this clear, I must intrench somewhat upon the chapter on *Variation*.

b. Intercalary Remarks on Variation

Hitherto we have assumed that the germ-plasm of a species is composed of ids, each of which contains *all* the characters of the species. A brief consideration, however, shows that this cannot be the case. Not only is it conceivable, but it is even necessary to assume that the development of the characters of the species, as well as those of the individual, is only the expression of the forces to which these processes are due, the ids in which the forces are situated being by no means perfectly similar to one another. A great majority of the ids certainly contain all the determinants of the species, — that is, they are capable of giving rise to all the specific characters; but in a minority of the determinants, to which, indeed, the origin and development of this species was due, will not yet have begun to undergo a phyletic transformation. All the determinants in this minority need not necessarily be similar to one another; one id, for instance, may contain unmodified determinants of the ancestral species; while another, although still retaining some of the old determinants, may exhibit a greater resemblance to the pure ids of the existing species, and so on. Such a *gradual transformation of the ids* affecting the majority of their determinants must constitute, indeed, the process of the formation of the species; and it will be quite in accordance with the principle of variation to assume that modifications must take place in the invisible vital units of a lower order, viz., the ids and biophors, in exactly the same way as they occur in different degrees and directions in different individuals in the visible vital units, — the unicellular organisms and persons of stocks.

Thus the provisional hypothesis put forward above, that the germ-plasm of a species — as far as it concerns the specific char-

acters — consists of a number of identical ids, cannot be a strictly correct one: the germ-plasm must, on the contrary, contain a majority of completely modified ids provided with new determinants of the species, and of a minority of ids of the ancestral species which are only slightly or not at all modified. The number of the latter will diminish in the course of time in consequence of selection of the individuals, and the new specific characters will gradually lose their original variability. Owing to natural selection, the germ-plasm will gradually be relieved of those of its ids which have become only slightly or not at all modified in the new direction, for those individuals in which it still contains a large number of unmodified ids are as well adapted as the others for the conditions of existence. As these individuals are gradually eliminated in the struggle for existence, the number of unmodified ids in the subsequent generations must gradually become reduced; and *this process of selection in the germ-plasm will only reach a limit when the number of unmodified or incompletely modified ids has become so small that their influence on the development of the essential characters of the species is inappreciable.*

This process of transformation of the ids of the germ-plasm will, however, reach a limit, as do all processes of selection, when its continuance is no longer of any advantage. All adaptations in an animal remain stationary and are not further perfected, directly further improvement becomes useless; and in the same way the process of modification and elimination of the ids, which forms the basis of other adaptations, will cease as soon as the completely modified ids are present in such a majority that the others can only exert an inappreciable influence on the nature of the offspring. The useful and adaptable characters will always undergo complete development in the normal process of reproduction, — that is to say, in the intermingling of individuals of the species, — even though some unmodified ids may be present in the germ-plasm. Let us take as an example the well-known case of the butterfly *Kallima parallecta,* which resembles a leaf. The resemblance is very marked, although not a complete one. The form of a leaf, with its midrib and veins, — some regions appearing more or less faded, and some dry or wet, and even of the appearance of a dewdrop, — is indicated on the folded wings, but *only certain of the secondary veins on the right and left of the middle line are represented.*

This is quite sufficient to deceive the birds which pursue the butterfly, and a more accurate copy of the markings on a leaf would not increase the deception; for it only needs to be effective at a certain distance, and therefore has not increased in perfection, but has reached its limit at this stage. A precisely similar occurrence must take place as regards the modification of the ids of the germ-plasm if a new adaptation of the species is concerned. In such a case, again, only a *relative* and not an *absolute* perfection will be attained. The majority, but not *all*, of the ids may become modified by selection, while a minority must accompany these through long periods and generations in an unmodified or slightly modified condition.

I have had to make this digression on the transformation of species in its relation to the germ-plasm, in order that my explanation of '*the ontogenetic shifting of the hereditary resultant*' may be comprehensible. It follows from the process of the gradual transformation of the ids of germ-plasm and the transference of unmodified ids from one cell-generation to another, *that the germ-plasm of every species must consist of a combination of ids of a somewhat different nature*. For although specific characters must have appeared simultaneously, and many others must have arisen successively, in the course of many generations, all the characters of the species will not be represented by determinants in the same number of ids. The oldest character, in fact, will be contained in almost all the ids; those which are somewhat younger, in a considerable majority; and still younger ones, perhaps also in a slight majority; while those which have only just begun to be of use to the species will only be present in a minority of the ids of each individual.

This circumstance is evidently connected with the degree of variability of characters, which may, in fact, be a very different one as regards the different characters of a species. Characters on which selection is only beginning to act, can only be represented by a majority of ids in a minority of individuals, and the less variable characters are those which have been selected for a greater number of generations, and are therefore present in a large number of ids of numerous individuals; and again, those characters which have long become definitely fixed in all or almost all the individuals of a species, must be also represented in the majority of the ids in almost all these individuals. *Conversely, those characters which are beginning to become useless to the species must be contained only in a gradually increasing minority of the ids;* the number of the latter must gradually decrease, until it finally becomes so small that it can no longer give rise to the character in question in consequence of an overwhelming majority of the other ids.

The above conception of the germ-plasm enables us to understand why the force of heredity may vary *in the course of ontogeny* in the case of crosses between two species. and why the paternal tendency may dominate in one character and the maternal in another. For, quite apart from the possible degree in which the force of heredity may be present in the individual determinants, and assuming it to be equal in the two species, the *number* of ids which contain homodynamous determinants will nevertheless vary *according to the age of the character in question.* A greater number of ids containing homodynamous determinants will indicate a greater force of heredity. If the form of the flower in a species A was acquired long ago, while that of the leaf is a new acquisition, and if the reverse is true in the case of a species B, the hybrid which would be obtained by crossing these two species would resemble species A in the form of its flowers and species B in that of its leaves. A larger number of homodynamous determinants of species A, for the rudiments of the flowers, will be opposed to a smaller number of homodynamous determinants of the species B, which, on the other hand, will contain a larger number of homodynamous determinants for the rudiments of the leaves. The importance of the whole principle will be made still more evident in the following section.

c. The Struggle of Individual Characters

The question of *individual* characters in the two parents has not been taken into account in our previous investigations: in crosses between different species they may be looked upon as insignificant when compared with the specific characters. We must now consider those cases in which the two parents differ in respect of slight individual characters only, confining ourselves to the human race, the individual characters of which we can recognise most clearly.

What strikes us most forcibly in connection with the process of reproduction and transmission in man, when compared with that of the formation of hybrids in the vegetable kingdom, is *the dissimilarity between children born of the same parents.* In the case of plant-hybrids, a striking constancy is observable in the offspring of a cross, and this is true not only of the offspring of the same parents, but also of all the hybrids produced by crossing any individuals of the two species in question, if the latter also are constant.

The dissimilarity between the children of the same parents has been already mentioned, and was explained as resulting from the halving of the germ-plasm in the process of 'reducing division,' which takes place in a different manner each time, and, when a larger number of idants are present, gives rise to a surprising number of combinations. As the idants are very different with regard to the *individual* primary constituents they contain, new combinations of the latter are thus continually being formed without affecting the characters of the species.

Three principal kinds of combination have to be considered in any attempt to explain the blending of parental characters in the child, — that is to say, to refer it to processes which take place in the idioplasm: these are (1) *the characters of the child are intermediate between those of the parents*; (2) *the child exclusively or principally resembles one parent*; and (3) *the child resembles the father as regards some characters and the mother in respect of others.*

The first case, if it ever *strictly* occurs, must be attributed to the presence of a precisely similar controlling force in all the determinants.

The number of idants derived from each parent must be the same in this case, as the parents belong to the same species, and there will certainly also be very little difference as regards the number of ids; the number of determinants, moreover, will be the same, or almost the same, in the germ-plasm of the two parental germ-cells. Theoretically, an exactly intermediate form would therefore result if each determinant of the father and mother were homologous to one another, and if the homologous determinants were controlled by precisely similar forces, — *i.e.*, if they contained the same number of biophors, and the homologous biophors of either side possessed the power of assimilation and reproduction to the same extent. These conditions which we have presupposed will, however, hardly exist all together, but they may concur approximately with regard to a certain number of characters.

It can hardly be doubted that the *second kind of combination* occurs, and that in some cases the *offspring take after one or other of the parents only*, — not only in respect of those characteristics with regard to which the term 'resemblance' is generally applied, such as those which concern the form and expression of the face, but equally as regards stature, form of the body, proportions of the limbs, nature of the skin and hair, character, and temperament.

We are met with a two-fold difficulty in attempting to explain these facts: in the first place, how is it possible for all the primary constituents of one parent — *e.g.*, the father — to be present in one of the germ-cells of this parent, as the germ-plasm was halved by means of the reducing division before these germ-cells become ripe? — and secondly, how can it happen that the maternal germ-plasm exerts no influence on the formation of the child?

Let us first consider the former of these two difficulties: how is it possible for all the characters of the father to be contained in one paternal germ-cell in spite of the reducing division? If the latter process resulted in a quantitative halving of the germ-plasm, no further explanation would be necessary, for the quality of half the mass might be exactly similar to that of the whole. But a reduction of the units of the germ-plasm to half their number occurs in this process; the number of ids is reduced by one half, and the structure of the offspring results from the combination in the germ-plasm of the ids of both parents, as was shown in the case of plant-hybrids; it is difficult, therefore, to understand why half the number of ids can nevertheless give rise to all the characters of the parent. Strictly speaking, it is immaterial whether we concern ourselves with *all* the characters, or with only a *single* one; for many characters, in fact, depend on the co-operation of all the ids of the ontogenetic stages in question.

There is only one way out of this difficulty, — we must accept the assumption, which has been confirmed by fact, *that the controlling power of the ids of one of the parents may become nullified at every ontogenetic stage.* Observations on plant-hybrids are invaluable in this connection, for in them we know for certain that the ids of another species are present, even although they may not produce a perceptible effect We may conclude from the phenomena of 'pseudo-monogonic' heredity exhibited by these hybrids, that the ids of one parent may, as has already been explained, prevent those of the other from taking part in the control of the cell, group of cells, or part of the body.

We may therefore assume that if by means of the reducing division, all the idants *which controlled or dominated the ontogeny of the mother*, for instance, reach one of the germ-cells produced by this parent, this germ-cell will be capable under certain circumstances of reproducing the maternal 'type'* in the child. But in order that this may happen, it is necessary for this cell to unite with a sperm-cell, the germ-plasm of which possesses on the whole a weaker controlling power than its own, so that the germ-plasm of the father is dominated by that of the mother.

As in the case of crosses between species, the controlling power of the idioplasm will not always be dependent on the same cause.

We cannot enter more deeply into those cases in which a more marked force of heredity of individual characters occurs. Darwin mentions, for instance, that the 'white' colour, in flowers as well as in animals, is very commonly transmitted to the offspring when white individuals are crossed with dark-coloured ones, the majority of the descendants inheriting the white colour. We can in this case only assume that those biophors which dominate the cell, and give rise to the white colour, must be 'stronger' than those which cause the formation of pigment, and this 'strength' must, in fact, be attributed to the possession of a greater power of assimilation.

The case is different in many other instances in which the greater hereditary power is attributable to *quantitative* differences in the constitution of the paternal and maternal groups of idants.

The number of ids contained in each idant is certainly constant, or nearly so, in all individuals of the human spe-

---

* I shall make use of the term 'type' ('Bild'), to express the whole aggregate of essential characteristics which together constitute the individuality of a human being.

cies;* but a predominance of paternal or maternal ids may nevertheless exist, so that there may be *a greater number of homodynamous determinants* in one line of ancestors than in the other, as has already been shown to be the case in hybrids.

By the term *homologous* determinants, we understand those elements of the idioplasm which are capable of so controlling homologous parts of the body or determinates; *homodynamous* determinants, on the other hand, are those of the homologous determinants which have the special function of impressing *a like character* on any part of the body, and which, taking an example we have already made use of, serve to produce a *particular form and colour* in a certain region of a butterfly's wing. That parent in which the ids contain numerous homodynamous determinants, or at any rate a larger number than are contained in the ids of the other parent, must undoubtedly exert the greater controlling power. The power of homodynamous determinants is simply cumulative, whereas dissimilar or heterodynamous determinants may, in the most favourable cases, co-operate to form a single resultant, but may, under certain circumstances, counteract or even neutralise one another. The larger the number of homodynamous determinants which the entire aggregate of ids of a parent contains at any ontogenetic stage, the greater will be the likelihood that these will predominate in the struggle of the parts which takes place in the cell, and will therefore stamp the latter more or less distinctly.

The colour of the eyes may be taken as an example. Let us take a case in which those of the mother are blue, and those of the father brown. The number of ids in the idioplasm of the pigment cells of the iris will be the same in both parents, but in the case of the father nine-tenths of them, let us say, are composed of 'brown'† determinants, and only one-tenth of deter-

* It is conceivable that individual fluctuations in the number of ids may occur, although the number of idants always, indeed, remains the same—if one can judge by its constancy in many animals and plants. Unfortunately we do not yet know the number of idants in the case of man; I have, at least, been unable to obtain any information on this point.

† 'Brown' and 'blue' determinants are spoken of in this and subsequent passages simply for the sake of brevity. I am not ignorant of the fact that the blue colour of the iris is not due to blue pigment. The above terms merely indicate that the determinant produces a structure in the iris which causes it to appear blue or brown, as the case may be, quite apart from the histological details on which this depends.

minants of other colours; two-thirds of the ids of the mother on the other hand consist of blue, and one-third of brown determinants. The iris of the child will therefore most probably be brown, for in its formation nine-tenths of the determinants of the father will co-operate with one-third of the determinants of the mother, which are homodynamous with these. The predominance of the brown determinants would also, however, still be assured if the maternal ids contain no determinants at all of this colour, but only red or green ones,—supposing that red or green pigment cells occurred in the human iris. For in this case nine-tenths of the paternal determinants would be opposed to various small groups of heterodynamous determinants of the mother. The latter might possibly modify the brown colour which would be produced by groups of paternal determinants *alone*, for they also control part of the cell-body; but it is quite as conceivable that they might be completely overcome by the paternal determinants, and thus excluded from the control of the cell. We cannot, as already stated, at present judge as to the result of the struggle of the parts in individual cases, but there is no doubt that, provided the controlling power of the determinants is similar, the *number* of the latter is of the first importance.

Since, therefore, the nature of the combination in the germ-plasm is different in every individual in consequence of sexual reproduction, the number of homodynamous determinants of any particular characteristic must also be different in every case. It will be shown in the chapter on variation that processes of selection may even bring about an increase or decrease in the number of the homodynamous ids of individual characters, although these are never of sufficient biological importance to give rise to specific characters.

The same competition of forces must take place in the case of individual as in that of specific characters. A child may inherit the colour of the eyes from its father, and the shape of the mouth from its mother, just as in a plant-hybrid the form of the leaf may resemble that in the paternal, and the flower that in the maternal plant. In both instances the character is inherited from that parent in which the group of idants contains a *preponderating majority of homodynamous determinants* of this character. When this is the case, this majority preponderates over the scattered minority derived from the other parent.

In neither instance, however, does the preponderance of the father, as regards a *single* character, necessitate that *all the other* characteristics will be controlled to the same degree by this parent. The germ-plasm consists of an equal number of paternal and maternal idants and ids, which remains constant throughout ontogeny. We suppose that every id of the germ-plasm contains all the determinants of the species, *e.g.*, the determinant a for the character A, the determinant b for the character B, and so on. But all ids of the germ-plasm need not contain a number of homodynamous determinants only; for if id i, for instance, contains the determinant a^1, id ii may perhaps contain the determinant a^2 for the homologous character, and id iii the determinant a^3, id iv the determinant a^4, and so on. There is no reason why id i, in which the determinant a^1 represents the character A, should not contain the determinant b^2 instead of b for the character homologous to B. If we denote the determinants of corresponding characters — *i.e.*, those which may become substituted for one another — by the same letters, a certain id of the germ-plasm — *e.g.*, id i — might contain the determinants a^1, b^2, c^4, d^3, e^1, f^5, g^6, h^8, i^5, &c. If in the germ-plasm of the father id i, id ii, id iii, id iv, and so on to the last, which I will call id xx, all contain the determinant a^1 for the character A, and none of the variants a^2, a^3, &c., are present in them, this determinant a^1 will be more powerful than any other variant of a derived from the mother, which may exist in the idioplasm of the cell in question, provided that the total number of each of these variants is less than twenty. Hence the character A^1 will be impressed on that particular cell, or group of cells, and not A^2 or A^3. The case may be entirely different as regards the character B; the determinant b^3 or b^4, for instance, may be the dominant one in most of the ids and idants: and in this case, one of the other variants of B, such as B^3 or B^4, will be produced.

If we may compare the groups of paternal and maternal idants to two centres of force each of which attempts to obtain control over the cell, each of these two forces will be determined by the individual forces of the idants in which they are contained, while the force in each idant will be controlled by the individual forces of the ids of which it is composed. If, for instance, there are two idants derived from each parent, each consisting of ten ids, the following considerations would enable us to ascertain which variant of the character must actually be produced. The paternal idants P^1 and P^2 might each consist of ten ids, the same determinant a^5 being contained in six ids in P^1, and in eight ids in P^2. These two idants would then combine in the attempt to give rise to the character A^5, with a controlling force of $6 + 8 = 14$ ids. Each of the maternal idants M^1 and M^2 might also contain ten ids, M^1 being composed of two determinants a^1, four a^3, three a^7, and one a^{10}; this idant will therefore tend to produce the character A^3 with a power of four ids only. If now, the other maternal idant M^2, with all its contained ids, tended to give rise to the character A^6, with a power, that is, of ten ids, the group of paternal idants would nevertheless predominate over that derived from the mother, as fourteen paternal are opposed to ten maternal homodynamous determinants. In this case both parental groups might possibly control the character of the child together, but the paternal group would be the stronger of the two. If, however, eighteen homodynamous ids of the father were opposed to four homodynamous ids of the mother, the influence of the latter would be entirely suppressed as regards the character A. We must at any rate conclude from the facts of the case, that the characters of one parent may be strictly inherited without any apparent intermingling of the corresponding characters of the other parent. As already mentioned, this very point in the theory seems to me to be the most reliable one, and known facts concerning plant-hybrids compel us to accept this assumption. The controlling force of the groups of paternal or maternal idants may be entirely different in respect of individual characters and groups of characters, according to the number of homodynamous determinants by which these are respectively represented. Moreover, this depends not only on the fact as to whether the individual character is derived from the father or the mother, or is a mixed one, but also on the entire number of homodynamous determinants present from each parent.

Although I have all along spoken of idants, I do not wish to imply that every idant behaves as an *independent whole.* Neither do I suppose that the resultant of the forces of the whole aggregate of paternal, is opposed to that of the maternal ids. It is quite conceivable that the same homodynamous determinants occur in the ids of both parents, and that their forces are cumulative, just as they would be if present in *one* or *the other* parent only. Homodynamous ids *must*, indeed, frequently occur in both parents in cases of close interbreeding, as well as in species which exist in comparatively small numbers in small isolated districts.

It is obvious that this struggle of the parental ids takes place at every stage in ontogeny, and that its result is different according to the state in which the forces exist at this stage. This accounts for the frequent changes as regards the resemblance to one or other of the parents, and for the combination of parental characters which occurs in different parts of the body.

The facts, so far as they are known to me, apparently agree very satisfactorily with the above explanation. I have attempted to collect fresh evidence bearing upon this question, which till now has not been closely kept in view. This, however, has unfortunately not been so easy a matter as might have been expected, but I will mention some of the facts relating to this point.

In some cases a child resembles its father in most respects, and takes after its mother as regards a few minor parts. This would be inexplicable if the whole part were not controlled by the resultant of a determinant other than that for the individual portions of this part. The single determinant and its immediate successors, which control the primary cell of the whole part, determine, in the first place, the rate of the cell-division and the primary form of the entire organ; but in each subsequent stage, one of the succeeding determinants takes on a controlling function, and as its influence is always the resultant of the homologous determinants of all the ids of the cell, the successors of this cell may at any time differ from one another with regard to their resemblance to those of the father or mother.

A person of my acquaintance resembles his father very closely in respect of that portion of the skin which is derived from the external germinal layer. In both the father and son the epidermis is thick and inclined to be horny. The nails of the hands are much thickened, and the skin on the soles of the feet is especially hard. In this case, therefore, the force-resultant of the paternal determinants must continually have been stronger than that of the maternal ones from the primary ectoderm onwards, through a long series of cells. On the other hand, the resultant of the maternal determinants seems to have predominated at the formation of the brain, which likewise arises from the external germinal layer, for the person in question resembles his mother as regards most of the mental qualities, such as intelligence, talents, and strength of will. This can only be accounted for on the supposition that the determinants of the subsequent descendants of the primary ectoderm-cell were unable to exert any influence *upon this cell itself*; in it they were still latent, and in this condition were merely passed on to succeeding cells. If the ids of subsequent generations of these primary ectoderm-cells which formed the rudiments of the brain, contained more maternal than paternal homodynamous determinants, a resemblance to the mother instead of to the father would arise at this point.

In the above case, the *entire* brain does not seem to have taken after that of the mother, for marked paternal traits also exist in it. According to our theory, it may well be conceived that even such a repeated alternation of hereditary tendencies as has occurred in this instance is predetermined in the germ, for the power of the paternal, and naturally also of the maternal ids, varies throughout ontogeny at every further stage in the division of the ids, and the relation between the controlling forces of the father and mother may be transposed. In general we might even expect that one or other group would predominate in most cases, and that the child is consequently composed of a combination of parental characters which varies in different parts. The parts or organs which resemble those of the same parent may also vary very considerably as regards size: this is possibly true of a single cell, as well as of a whole organ or an entire germinal layer, or even of the entire organism.

This theoretical deduction is in general supported by facts, for a child is rarely or never an exact repetition of either parent. It is by no means easy, indeed, to form a correct estimate with regard to the resemblance between parent and child, for in order to do so, an exact knowledge of both at the same ages would be necessary, and a detailed comparison is only possible between father and son, or mother and daughter. It would therefore be essential to compare photographs of the father and son, *at the same ages*; and this, as far as my knowledge of the observations which have been made on heredity extend, has never yet been done. It would, moreover, be necessary to photograph the whole body, and not merely the face.

As far as we can at present gather from the facts, even in those cases in which the resemblance is a close one, the child always differs from the parent either as regards individual characters, or, as is more frequently the case, in undefinable slightly characteristic details, such as the length of the limbs, colour of the hair or eyes, or the quantity of hair. These parts cannot be said to resemble those of the other parent, but they give the impression that the main direction in which heredity has tended has been slightly changed in an undefinable way. A daughter may resemble her mother, for instance, so closely that she is universally said to be an exact image of her mother; and yet a close comparison will show that the likeness is by no means an exact one, and although the child may not display a single paternal character, there are nevertheless a number of parts which respectively differ from those of the mother. In the case of identical twins, there can be no doubt that many of the minor differences existing between them are due to differences in the germ-plasm, and not to the diversity of external influences. The germ-plasm of both parents, that is to say, has taken part in determining the different parts of the child, although in the case of one parent this determination is slight and little marked, and has caused a slight deviation from the maternal characters rather than the development of specifically paternal ones.

If this view is correct, and the germ-plasm of *one* of the parents alone never determines the formation of the child, it becomes more obvious than ever that even when heredity tends to follow in the direction of one of the parents in the greatest possible degree, the mother and daughter can never resemble one another so closely as do identical twins. Owing to the slight influence of the germ-plasm of the father, the type of the daughter deviates somewhat from that which would have been produced from the maternal germ-plasm *alone;* and similarly, if the mother owes her nature to the predominance of the germ-plasm of one parent, a slight deviation must have occurred owing to the weaker influence of the other parent. But the whole of the germ-plasm of *both* grandparents cannot possibly have been contained in the maternal germ-cell from which the daughter arose, for the reducing division causes the removal of half the germ-plasm from the egg-cell before fertilisation takes place. Even if the idants which materially determine the type of the mother remained in the ovum from which the daughter was developed, the group of idants of the other grandparent, which would modify this type somewhat, must necessarily be absent; the types of the mother and daughter consequently cannot exactly correspond, for the double reason that the influence of one grandparent was wanting at the development of the daughter, while that of the father was present in addition.

The following examples may serve to show *in how many different ways the hereditary tendencies of the parents may*, in accordance with our theory, *be interchanged in the course of ontogeny*. In the bilaterally symmetrical human being, all those parts which are not situated in the median line are paired, and the corresponding organs generally behave nearly, if not quite, similarly as regards heredity. If one hand bears a decided resemblance to that of the mother, the same will, as a rule, be true of the other also; and if the left leg is intermediate between the character of both parents, the right one will also be so in exactly the same degree. Even such a subtle characteristic as the colour of the eyes usually corresponds in both organs; and even in those cases in which it is intermediate between that of the two parents, the colour only varies slightly in shade in the two eyes. One might be disposed to conclude from this fact that paired organs are represented by a single primary constituent in the germ-plasm. This, however, would be an erroneous conclusion; for, apart from the facts already mentioned which contradict such a view, there are *exceptions to the rule that paired organs are similar to one another*. One brown and one blue eye sometimes occur in dogs, especially in boar-hounds, and I know of a similar instance in the human subject: the father, a brewer in a small Suabian town, has blue eyes, and the mother brown ones, while a daughter of twelve years of age has one blue and one brown eye.

In addition to these cases, the frequent *inheritance of birth-marks* and other minor characters on one side of the body only, necessitates *the assumption of double determinants for the corresponding parts of each half of the body*. We must therefore imagine that each id of the germ-plasm of bilateral organisms is primarily bilateral, and that all the determinants, indeed, of the whole body are double, even of course including those for the organs which are apparently situated in the median plane, but which actually consist of corresponding halves. In a large number of animals the two first segmentation-cells of the egg, or blastomeres, correspond to the future right and left halves of the body. The first division of the nucleus of the fertilised ovum must therefore separate the determinants for the right and left halves of the body; and the occurrence of this process is rendered all the more probable from the fact that a longitudinal division of the idants actually occurs, and results in each of the spherical ids being halved.

The explanation of the striking correspondence of the homologous parts of *antimerous* or paired organs must be referred to the history of the transformation of species. The permanence of this resemblance during the continued crossing of individuals, is naturally due to this approximate similarity of the corresponding determinants for the right and left sides in both parents. It is evident, according to our conception of the structure of the idioplasm, that the resemblance of the corresponding parts of the child originates in the antimerous determinants of both parents, for the latter determine the comparative number of homodynamous ids on the paternal and maternal sides, together with the relative 'controlling power' of the idioplasm of the father and mother in the organ in question. And since this must be the same on both the right and left sides, the organ itself must display the same combination of paternal and maternal characters on either side: that is to say, its two sides must be alike.

I do not think that these facts can be understood by the aid of any other theoretical assumption with regard to the structure of the hereditary substance. The assumption of the existence of pangenes, for example, might certainly explain the circumstance that a combination of the paternal and maternal characteristics does indeed occur in the organ in question, — in the external ear, for instance, — but it cannot account for the fact that this combination is similar in the right and left ears.

These very facts seem to me to furnish a further welcome proof of the correctness of the view arrived at by other methods, that the hereditary combination in each part is predetermined from the germ onwards. The right and left ears could not possibly resemble each other, if the relative strength of the hereditary tendencies on both sides were not predetermined for all parts of the child by the nature of the paternal and maternal idants.

There are, however, exceptions to this rule. As already stated, the homologous parts of the two antimeres may differ from one another; and such a difference is even frequent in certain animals, though it only affects characters which are of minor biological importance. Little attention has hitherto been paid to the fact that many of our *domestic animals have lost the original symmetry of the marking of their coats.* Piebald cats, dogs, horses, cows, and guinea-pigs are not uncommon, and show that the symmetry of the markings may become completely lost by domestication. This must be owing to the fact that these originally symmetrical patches of colour are, in consequence of domestication, no longer of biological importance. If the determinants for these characters varied in different ways on the right and left sides, and the animal in question thus became spotted, no disadvantage would thereby result, and the animal would nevertheless be able to exist and produce offspring. If two individuals with different piebald markings then paired, the asymmetry in the coloration and markings would be increased; and as a matter of fact in many of our races of cattle no parts of any two animals are alike in this respect, and the same is true with regard to many dogs and guinea-pigs. We know how important these markings and coloured patterns may be for the preservation of individuals and species *in the natural condition*, and are therefore justified in attributing the retention of the symmetry to natural selection, and its loss to panmixia.

Certain facts with regard to the *metameres*, or successive parts of which the body of a segmented animal is composed, show that the maternal or paternal characters may preponderate in different segments. This naturally cannot be proved in the case of Man, as the metameric segmentation only affects the bones, muscles, and nerves, which are not externally visible. But I think I have observed that consecutive parts, even when they are homologous, may occasionally exhibit different hereditary types. A child may closely resemble its mother as regards the arms and hands, and nevertheless may take after its father in respect of the legs and feet. I have endeavoured to ascertain whether any definite rules are followed with regard to certain organs which are closely related to one another in ontogeny, according to which these organs would exhibit a similar combination of parental characters; but the only rule I could discover is that which relates to the symmetry of the two halves of the body. Apparently all possible combinations may actually occur. The form of the skull may resemble that of the father, and the face that of the mother; or the form of the entire head and face may be like the mother's, while the eyes may be similar to the father's in every detail. The son may, like his father, possess a dimple on the chin, although he takes much more closely after his mother as regards the shape of the face and nose. That the combination of parental characteristics may even extend into far greater details, is shown especially by the remarkable amalgamation of the mental qualities of the parents which often occurs. The intellect and practical talent may be inherited from the mother, and strength of will and unselfishness from the father; and all these qualities may be contained in one skull, the form of which essentially resembles that of *one* of the parents only. These combinations of mental characteristics of the parents cannot, however, always be definitely analysed, owing in the first instance to the fact that they are not always sharply contrasted in the parents, but more frequently are only different in degree. We may, however, at any rate consider it certain that *the brain rarely resembles that of one parent only in all its parts and as regards the most minute details of its structure*; it usually, on the contrary, exhibits a combination or alternation between that of the two parents, and this combination is of the most varied kind.

In connection with this statement it may be mentioned that no part of the human body is so important as the brain in the struggle for existence, and its importance extends even to the minutest details; its parts must therefore be subjected to incessant processes of selection. In other words, the number of homodynamous determinants in the various parts of the brain must be extremely different in the individual, and must vary extremely in different individuals.

These statements with regard to the struggle of the individual characters may perhaps be objected to on the ground that they contradict the suppositions with which we started. It may be contended that an alternation of paternal and maternal hereditary parts is rendered possible on the basis of my theory, because the transmission of a paternal character implies that the whole dominant group of idants of the father passed into the germ-plasm of the child at the reducing division, and the transmission of a maternal character necessitates the presence of the whole of the dominant group of the mother. It might be considered improbable that both of these groups should come together in one germ-cell, or that this can occur as often as must be the case in reality, considering the frequency with which a combination of the two parental characters exists.

In reply to this it may be stated that the meeting of two germ-cells in the process of fertilisation, one of which contains the dominant group of idants of the mother and the other that of the father, must take place from time to time, for every possible combination will occur at one time or another. It must, moreover, not be forgotten that it is extremely difficult to distinguish between the pure individual characters of a parent and those of one of the immediate ancestors of this parent; but the production of these characters in the latter does not necessitate the presence of the whole of the dominant group of parental idants, for a portion of them will undoubtedly suffice if the character in question is represented by a majority of homodynamous determinants. The characters which alternate are very often not the specifically paternal or maternal ones, but those which in general characterise the mother's or the father's family. These, however, must be represented in most of the ids of the dominant group of idants, and may therefore become apparent even when the reducing division only causes a certain number of idants, instead of the entire dominant parental group, to pass into the germ-cell of the child.

On the other hand, the frequency with which a child bears a closer resemblance either to its father or mother — and this in many families is actually the rule — can also be easily explained by our theory. In such cases, not only can the whole number of dominant idants produce the type of the parent in question with approximate accuracy, but even a majority of them will be sufficient to do so, provided that a large number of ids with homodynamous determinants are contained in them. Many germ-cells will, therefore, contain a sufficient number of the dominant idants of the parent, while others, although perhaps containing an equally large number of these idants, will consist of various combinations of ids enclosing relatively few homodynamous determinants. If, then, an idant of the former kind, derived from the father, should, in the process of amphimixis, meet with one of the latter kind, derived from the mother, the type of the father would predominate in the child, and *vice versa*. If two germ-cells of the first kind, derived from the father and mother respectively, came together, a mixture of the characters of the two parents would result; and if two of the latter kind came into connection, none of the special characteristics either of the father or mother, would be recognisable in the child, which would only possess such characters as are common to the two families.

4. The Force of Heredity

We have now seen that 'pseudo-monogonic' heredity is to be explained in terms of the idioplasm as follows:—the dominating group of idants from one parent—the mother, let us say—passes over completely into the germ-cell of the offspring, and there meets with a weaker group of idants from the father in the process of amphimixis. Although not all, but only a large number of the determinants in the maternal group of idants predominate, a marked resemblance between the mother and child will result.

The fact that such cases as this occur at all, was taken above as a proof that the combination of ids in the idants persists during ontogeny,—*i.e.*, from germ-cell to germ-cell,—and that the idants often, or even generally, remain unchanged in the reducing division. It naturally does not follow, however, that precisely that combination of idants which predominated in the ontogeny of the parent must remain unaltered in the germ-cell of the offspring: it *may* do so, and such cases will frequently occur amongst the thousands of ova or still more numerous sperm-cells which are produced by a single individual in the course of its life.

In some families it certainly appears as if the perfect type ('Habitus') of an ancestor had been transmitted with great constancy to the children through a great number of generations, and we must therefore assume that the dominant group of idants in the ancestor reappears very frequently in the germ-cells of the offspring. Thus the high forehead, widely-separated eyes, and small mouth of the imperial family of the Cæsars, the large and peculiarly hooked nose of the Bourbons, and the projecting lower lip of the Hapsburgs, can all be traced through several generations. It is, however, difficult to say whether such similarities are not accidental, or whether our recognition of them is merely due to an incomplete knowledge of the facts, only those descendants being taken into consideration in whom these family characters were prominent; but similar observations with regard to animals lead to the conclusion that they are not *all* due to chance.

The phenomenon which breeders describe as '*individual prepotency*' comes under this category. It seems that a marked tendency occasionally exists in certain individuals to transmit their special individual characters to the majority of the offspring. It has often been observed that individual horses, cattle, sheep, and other domestic animals possess this capacity in a high degree; and breeders pay enormous prices for such individuals, which must, it is true, excel not only as regards this supposed special power of transmission, but also in respect of certain particular and desirable characters. It is, however, believed that similar observations have also been made with regard to plants. Vilmorin,* one of the most eminent raisers of plants, at any rate distinguished in his experiments between individuals possessing the capacity of transmitting their own characters to the offspring in a greater and in a lesser degree. The former he called '*bons étalons*,' and made use of them alone for purposes of propagation. He could not, however, find out by a mere examination of the plant whether it belonged to this preferred group. This could only be ascertained by examining the offspring, which therefore served as the guides in the selection of the plants for purposes of propagation.

Darwin,† Prosper Lucas,‡ and Settegast§ give many instances of this kind, one of the best known of which is that of the 'otter sheep.' This race was descended from a ram which was remarkable for having short, crooked legs and a long body. It transmitted this peculiarity to many of its descendants, and so enabled the owner to breed a special race of sheep with crooked legs, the advantage of this peculiarity being that they were unable to leap over fences. Similarly, English thoroughbred

---

* Quoted from de Vries, *loc. cit.* p. 88.—L. Lévêque de Vilmorin 'Notices sur l'amélioration des plantes par le semis,' Nouvelle édition, 1886, p. 44.

† Darwin, 'Animals and Plants under Domestication,' Vol. II., p. 40 *et seq.*, London, 1888.

‡ Prosper Lucas, 'Traité philosophique et physiologique de l'hérédité naturelle dans les états de santé et de maladie du système nerveux,' Paris, 1850.

§ Settegast, 'Die Thierzucht,' Breslau, 1878, p. 197.

horses owe their superiority to three individuals,—viz., to the Turkish horse 'Byerley,' and the Arabs 'Darley' and 'Godolphin'; and the celebrated race of Orlow trotters can be traced back to the stallion 'Bars the First.'

If these animals really possessed a stronger 'force of heredity' in the sense indicated, it must not be confounded with the property of *faithful transmission* in a race. This property of 'breeding true' must be due to the presence of a large majority of homodynamous determinants in the germ-plasm, or, what amounts to the same thing, to the existence of similar, *i.e.*, of 'racial,' determinants for every character in most of the ids. The longer a pure race has been kept up, all the individuals which exhibit variations being carefully eliminated, the greater will be the number of ids containing 'racial' determinants, and the more rarely will variations appear in individuals.

At present, however, we are concerned with *individual*, and not with *racial* characters. These cannot possibly have been contained in a preponderating majority of the ids of the germ-plasm from which the individual arose, for the germ-plasm is composed of paternal and maternal ids. The transmission of the proper 'type' can in this case therefore only be due to the fact that the group of idants which preponderated in the development of the parent is once more present in the germ-cell. I should consequently prefer to account for the so-called *prepotency in transmission* by assuming that in some individuals the reducing division simply occurs in such a manner as to separate the paternal and maternal groups of idants, while ordinarily it results in combinations of idants of all kinds. It is impossible to say at present on what peculiarity of the idants themselves or of the apparatus for nuclear division, this must depend; but it can at any rate be stated that the dominant group of idants cannot possibly be contained in every germ-cell of such an individual, even in the most favourable case. On the contrary, it can only be present in half of them; for, according to our assumption, the reducing division always causes the dominant group of idants to pass into *one* of two germ-cells only, the subsidiary group passing into the other. This supposition is in accordance with the facts; for, so far as I know, it has never been observed that *all* the offspring resemble the parent which exhibited 'individual prepotency,' but this, on the contrary, was only the case as regards some of them. In fact, it is expressly stated as regards the 'otter sheep,' that the offspring of the first ram 'closely resembled either the mother sheep of the ordinary breed, or the ram'; and this statement is in correspondence with the theory. Those spermatozoa of the ram which contained the dominant group of idants preponderated over the group of idants of the egg-cell, and an 'otter sheep' thus resulted; while those which contained the subsidiary group of idants could only tend to produce an ordinary sheep of the ancestral breed.

A greater 'force of heredity' is also spoken of in the sense of the prepotency of one race over another. According to Darwin, the short-horn race of cattle seems to possess a particularly marked power of transmission in contrast to other races; and this power is more marked in the pouter pigeon than in the fantail, so that when these two races are crossed, the characters of the pouter preponderate in the offspring. This preponderance of one race over the other must be due to the same causes as those which produce a much greater resemblance to one of the parents in the case of plant-hybrids, which were discussed in another section of this chapter. In both cases the preponderance may be due to the presence of a larger number of idants, of ids, or possibly even of biophors only, in the individual determinants.

5. Summary of Chapter IX

It may be advisable before proceeding further to give a short summary of the results arrived at in the present chapter, and to test the soundness of the assumption on which they are based.

According to my view, the co-operation of the hereditary substances of the two parents in the fertilised egg depends on the presence in each parental germ-plasm of a large number of units, and not of a single one. These units or ids are, moreover, not all similar to one another in the case of each parent, and although in *normal* sexual reproduction they all contain homologous determinants, they exhibit slight *individual* differences. The differences between the ids of the two parents need not in any case be greater than those existing between the ids of the father or the mother alone; it may, indeed, happen that individual ids derived from both parents may be similar to one another, and this is more likely to be the case the oftener inter-breeding has taken place in previous generations.

Each id of the germ-plasm passes through all the ontogenetic stages; that is to say, the number of paternal and maternal ids remains constant throughout ontogeny, and every cell is therefore controlled by an equal number of ids from both parents; and, moreover, in normal reproduction between individuals of the same species, the same number of paternal and maternal ids take part in the process.

The fact that the structure of each cell, organ, or part of the body of the offspring is nevertheless not exactly intermediate between that of the corresponding parts of the parents, must be due to the following causes.

In the first place, an exactly intermediate structure need not necessarily arise even if all the active paternal and maternal determinants of a cell were exactly alike, or homodynamous; for even in such a case, the 'controlling forces' of the maternal determinants might preponderate over those of the paternal ones, owing to a superiority as regards the rate of assimilation and multiplication, or in some other respect. The cell-body would then become filled more rapidly by the biophors of the maternal determinants which pass into it from the nucleus, and the multiplication and distribution of the paternal determinants would thus be checked. Hence the controlling force itself may be different with regard to the homologous paternal and maternal determinants.

The differences are, however, still further increased, owing to the fact that the ids, and more especially the homologous determinants derived from either parent, are not always homodynamous, but are, in part at any rate, almost always heterodynamous. The controlling force of the homodynamous determinants must, however, necessarily be cumulative, and the inequality in the force of heredity of the father and mother at any particular stage in ontogeny, is essentially due to the fact that although the number of ids is the same, the number of homodynamous determinants—*i.e.*, those which have a cumulative effect—is different.

It was deduced from the consideration of the phylogeny of variation, that the determinants must have varied independently of each other in the different ids of a germ-plasm, so that the homologous determinants may be present in very different variants in their ids; and that, moreover, these variants of the determinants in the different onto-idic stages may be combined in a special way in each id. It therefore results that the number of homodynamous determinants may vary in the different stages of ontogeny, and consequently the paternal and maternal hereditary tendencies may preponderate according to the stage or organ under consideration.

The fact that the offspring may resemble one parent much more closely than the other, is not incompatible with the fact that only half the total number of ids of this parent are present in each of its germ-cells, for *all* the ids only co-operate when strictly intermediate structures are formed; but in those cases in which the ids of *one* parent are overpowered and rendered inactive by a preponderating majority of homodynamous ids of the other parent, *the control of the cell is effected by the successful or 'dominant' ids*, and the others exert no influence. If, therefore, the determinants of one parent, in very numerous stages of development, preponderate in this manner over those of the other, an instance of apparently *monogonic heredity* will result, and the offspring will bear a closer resemblance to this parent; and if, moreover, the 'reducing division' should happen to take place in one of the germ-cells of this offspring in such a way as to result in those ids which were 'dominant,' and controlled the development of this offspring remaining together in the germ-cell, they might possibly preponderate in the next generation over the ids introduced by another parent in the process of amphimixis.

The problem concerning the possibility of the offspring bearing a much closer resemblance to one parent than to the other in spite of the fact that the hereditary substance of both parents is contained in the fertilised ovum, has already been stated in my essay on 'Amphimixis,' and its solution is to be sought in the struggle of the ids which takes place in every cell in the entire course of ontogeny. This struggle, however, only occurs when the determinants become active, and presumably concerns the biophors which pass into the cell-body, the stronger ones annihilating those with a lesser power of assimilation. It does not concern the determinants which are still 'unalterable,' and are inactive as regards the control of the cell. Moreover, the struggle does not occur between the elements of the *'reserve germ-plasm,'* which brings about the formation of the germ-cells of the offspring; and we can therefore understand that the offspring does not by any means only produce germ-cells containing the group of ids which controlled, or was 'dominant' at, its ontogeny, but many other combinations of ids may be contained in its germ-cells.

In this connection I should like to call attention to an interesting essay which appeared when I had almost finished putting the final touches to my manuscript. It bears the pseudonym 'Josef Muller'* on the title-page, and contains in particular an attempt to solve the problem discussed above. The ingenious author, who is accurately acquainted with the subject he treats of, doubts my hypothesis of the ids, but endeavours to account for the very remarkable disappearance of the hereditary tendencies of *one* of the parents in 'pseudo-monogonic' heredity by supposing that the two homologous primary constituents ('Anlage') of the father and mother respectively take part in a struggle ('gamomachia'), which results in the destruction and complete consumption ('gamophagia') of one of them. In principle this explanation of the problem approaches very closely to the solution I have attempted to give, and though I consider the fundamental idea it contains to be correct, I do not think that we may suppose, as the author does, that this struggle occurs at the *beginning of ontogeny*. Basing this conclusion on a statement made by Oscar Hertwig, from which it is conceivable that the homologous 'primary constituents' of the parents unite in the process of fertilisation, he further concludes that the struggle takes place during this union, and leads to the destruction of one of them. Apart from the fact that the paternal and maternal idants remain separate during fertilisation, it seems to me that a large number of the phenomena of heredity contradict the idea of such a union and subsequent struggle. The reappearance of the 'destroyed' primary constituent in the germ-cells — and consequently in the next generation, the phenomena of reversion — which show that every primary constituent must be present in more than two variants in the germ-plasm, and, finally, sexual dismorphism, the occurrence of a large number of very different hermaphrodite structures in certain cases, and sexual reversion, — all tend to disprove such a hypothesis. Moreover, apart from my theory of the ids, I believe that this struggle of the homologous primary constituents

* Josef Muller, 'Uber Gamophagie, ein Versuch zum weiteren Aufbau der Theorie der Befruchtung u. Vererbung,' Stuttgart, 1892.

must occur in the individual cells, in which the decision both as regards the preponderance or suppression of certain of the primary constituents, and the *number* derived from each parent which are to become effective, takes place. For it does not appear to me to be essential that any one of them *must* be entirely suppressed, although this will probably occur in most cases.

If I am not mistaken in my interpretation of a statement made by de Vries, there is no doubt that the primary constituents from *both* parents may undergo development in *one and the same* cell. By crossing a red-flowered with a white-flowered species of bean, this observer obtained a hybrid with *pale red* blossoms, on which the red colouring matter could be recognised in solution in the vacuoles of the cells.* If *parts* only of the cells were coloured, while other parts were colourless, it proves that at least two different (heterodynamous) kinds of biophors, derived from both parents, may control the same cell. There is here, however, an extensive field for further investigation.

My explanation of the process of mingling of the parental characters is based on the assumption of hereditary units or *ids*, each of which contains the whole of the 'primary constituents' of the species, which are, however, modified in the individual. In this connection it may therefore be as well once more to summarise the reasons which lead to this assumption.

In the first place, such an assumption naturally follows from the view that the germ-plasm is made up of 'determining parts' or determinants, for the latter necessitate a definite architecture of the germ-plasm. There must therefore be at least one limited unit of the germ-plasm, to which nothing can be added and from which nothing can be removed without producing an alteration in its capacity for directing ontogeny. But since the process of amphimixis unites the paternal and maternal germ-plasms, *each of which contains all the primary constituents of the species*, each being which is produced sexually must contain *at least two ids* in its germ-plasm.

The phenomena of reversion, which will be treated of in greater detail in the following chapter, show that there must be

---

* *Cf.* de Vries, *l.c.*, pp. 177, 178. The two species referred to are *Phaseolus multiflorus* and *Phaseolus vulgaris nana*.

several, and indeed many, ids in the germ-plasm of each individual. We know that the personal characters of the grandparents, as well as those of the parents, may reappear in the offspring, and we may therefore conclude that hereditary units or ids derived from the grandparents must be present in the germ-plasm of this offspring, and that it must therefore be composed of more than two ids.

A similar conclusion is arrived at on other grounds. If the assumption of hereditary units in the form of ids is once made, it follows as a matter of course that their number must be doubled in each process of amphimixis; and it becomes evident that this number must have increased enormously, in arithmetical progression, if the 'reducing division' had not intervened and reduced it to the half before each occurrence of amphimixis. This 'reducing division' must have appeared at a certain stage in the phylogeny of amphimixis. If it arose in the germ-cells of the first animal which was produced sexually, — supposing that the germ-plasm of each of the parents previously consisted of only *one* id, — it would always have caused the removal of the id of *one* parent from each germ-cell of the offspring, and thus no grandchild could ever have inherited characters from *both* grandparents. According to our theory of the presence of a large number of ids, such a case would seldom occur, although it is apparently not impossible. A further consequence, however, would be seen in an unusually great uniformity in the structure of consecutive generations; for if only two ids were present, one of which was always removed in the next generation, the same individual ids would pass through a great number of generations, and the diversity of the individual, such as occurs to so great an extent in the human race, would be extremely limited. It is, however, just this extraordinary individual diversity which seems to me to be due to the multiplicity of the ids; it could not have been produced by only two ids taking part in the process of amphimixis.

Finally, as soon as we have recognised, on theoretical grounds, the existence of ids at all, the fact that a number of them exist in the cell is supported by direct observation. For whether they correspond to the 'chromosomes' of other writers, which I speak of as idants, or to the 'microsomes,' of which the chromosomes are composed, as I assume to be the case, a large number of ids may always be observed to be present in the cell.

CHAPTER X

THE PHENOMENA OF REVERSION IN THEIR RELATION TO AMPHIMIXIS

1. Reversion to Racial Characters in Plant-Hybrids

By the term reversion, is meant the appearance of characteristics which existed in the *more remote* ancestors, but were absent in the *immediate* ancestors—*i.e.*, the parents.

The facts relating to these phenomena are familiar enough, and I shall therefore only refer to as many of them as are necessary for the further development of my theory.

The simplest case of reversion occurs in hybrids. It occasionally happens that hybrids which have been fertilised by their own pollen, produce offspring some of which more or less resemble only *one* of the two ancestral species. In such cases, therefore, a simple reversion to a grandparent takes place. Instances of this kind certainly occur, though not in all hybrid-plants; nor are they often met with even in those species in which they do occur. On this point Darwin quotes two contradictory statements made by Wichura and Naudin respectively, the former of whom never observed instances of reversion in his specimens of willow-hybrids, while the latter insisted strongly on the frequent occurrence of reversion in the *Cucurbitaceæ*. Darwin thought that this contradiction is explained by Gärtner's statement that reversions seldom occur in hybrid-plants raised from wild species, but are of frequent occurrence in those produced from cultivated species. Opinions on this point have since undergone some modification, for Focke states that 'without the influence of the pollen of the parent-species, complete reversions to the ancestral form occur practically only in hybrids of closely allied races.' Instances of reversion of this kind do therefore at any rate occur.

Such cases can easily be explained on the basis of our theory. The germ-mother-cells of the hybrid contain a group of idants derived from the paternal, and another from the maternal ancestral species If therefore the 'reducing division' halves the germ-plasm of these mother-cells in such a manner that idants of the mother alone reach one ripe germ-cell, and those of the father alone are contained in another, it is possible that two such germ-cells may unite when fertilisation takes place between these hybrids. In such a case a plant completely resembling one of the ancestral forms would arise, for it would have been produced from a germ-plasm which contains idants of this species only. As however such cases do not often occur, we may conclude that the reducing division only rarely effects such a complete separation of the paternal and maternal groups of idants, and that, in fact, as a rule, both paternal and maternal idants are distributed to each of the four germ-cells produced by the mother-germ-cell. As this halving of the germ-plasm occurs, as we have seen, in a different manner in different instances, we may presuppose that it will also exhibit differences with regard to the proportion of paternal and maternal idants which come together in each germ-cell in consequence of the reducing division; and this supposition is most satisfactorily borne out by the facts, for it is well-known that the offspring of *hybrid-plants, produced by fertilisation with their own pollen, become very variable in the following generation.* It is evident, indeed, that they *must* vary greatly, according to whether each one has received a greater number of maternal or paternal ids, or an equal number of both, from the two germ-cells which combined in the process of fertilisation to produce this particular individual. Thus Focke describes the offspring of hybrid-plants of the first or second year as being 'as a rule unusually diverse and rich in forms,' and gives as examples the genera *Pisum*, *Phaseolus*, *Lactuca*, *Tragopagon*, and *Datura*, mentioning especially in this connection the hybrid of *Nicotiana alata* and *N. langsdorffii*. De Vries * also refers to these facts, and describes them very aptly in the following passage:—'The hybrids of the first generation have perfectly distinct characteristics in the case of every pair of species. If a hybrid is produced from two species which have already been crossed successfully by previous experimenters, we may be sure that the description given by them will as a rule apply exactly to the intermediate form in question. If the hybrid is fertile without the help of the parent-

* Hugo de Vries, 'Intracelluläre Pangenesis,' Jena, 1889, p. 25.

forms, and if in a few generations thousands of specimens of its offspring are raised, it will always be found that scarcely two of them are alike. Some revert to the paternal and some to the maternal form, while others, again, are intermediate between the two. The remainder present the most varied alternation of paternal and maternal characteristics, and show almost every degree of mutual intermingling.'

De Vries states this merely as a proof of what he calls the 'free miscibility of the characters,' without attaching importance to the fact that the hybrids of the first generation behave quite differently from those of the second, or attempting to account for this fact theoretically.

Professor Liebscher * has recently brought forward the following interesting instance, in which the details were very accurately investigated. He crossed two species of barley, *Hordeum steudelii* ♀ and *Hordeum trifurcatum* ♂, in the former of which the spikelets are arranged in two rows and are black, while in the latter they are arranged in four rows and are white. The hybrid is as nearly as possible intermediate between the two forms, '*all the ears*,' moreover, '*being strikingly uniform*,' as one would be led to expect theoretically. In all the hybrids the spikelets are arranged in two rows, and in the main spikelets the tips are black, while in the lateral ones they are white, and the 'Löffel'—which are peculiar to *Hordeum trifurcatum*—are black and white. The offspring produced from these hybrids *were exceedingly variable in the first, as well as in the second generation*.

Liebscher has attempted to account for this variability by assuming that a 'loosening ('Lockerung') of the structure of the germ-plasm,' as well as 're-combination of the individual characters,' is produced by the process of reproduction. The former results in 'a weakening of the power of faithful transmission in the generative products,' *i.e.*, 'an inclination to individual variation in the descendants.' This statement indicates that Liebscher certainly had some kind of idea of the actual process which occurs in the idioplasm, although it is not made clear in what this 'loosening' consists.

A definite meaning, however, underlies this expression as soon as it is recognised that the germ-plasm consists of a large

* Liebscher, 'Vererbung,' etc., Jen. Zeitschrift, Bd. 23, 1888.

number of ids. The 'loosening' depends on the repeated removal of half of the ids, which occurs every time the germ-cells are formed. Half the germ-plasm of the hybrid consists of maternal ids of the species A, and the other half of paternal ids of the species B; in the formation of the germ-cells, this perfectly uniform composition becomes extremely diversified, owing to the fact that the 'reducing division' halves the germ-plasm in different ways. If we suppose that the ids, or even idants, are arranged in a circle, the plane of division will sometimes cut the circle across one diameter. and sometimes across another, and the combination of the germ-plasm in the germ-cells containing ids of A and B will thus be very diversified. If the hybrid is then fertilised by its own pollen, so that amphimixis occurs between two of the differently constituted germ-cells, a still greater diversity in the idic combinations will result, and a high degree of variability in the offspring must inevitably ensue.

The offspring of hybrids are also very variable, even in those cases in which they are produced by *a recrossing with one of the ancestral species*. 'Fairly dissimilar offspring generally result from the fertilisation of a hybrid with the pollen of an ancestral form; intermediate forms between the hybrid and the ancestral species in question are as a rule the most numerous; while a smaller number of examples resemble either the original hybrid or the ancestral species.' * This statement agrees perfectly with our theory, for a consideration of the reducing division renders it evident that in recrossing with one of the ancestral species, a very dissimilar numerical proportion of idants of the two ancestral species must prevail in the fertilised egg-cell of the following generation. Such hybrids are generally spoken of as '$\frac{3}{4}$ hybrids,' because in them the force of transmission of one of the ancestral species is assumed to be one-quarter only, and that of the other three-quarters. This term is probably indispensable in practice, but is obviously totally inaccurate. 'Three-quarter hybrids' do not by any means contain all the hereditary substances of the two ancestral species in the proportion of three to one; the proportion is in fact a very variable one. A '$\frac{3}{4}$ hybrid' is produced, for instance. from two species of pink, *Dianthus chinensis* and *D. barbatus*, by fertilising the hybrid

* Focke, p. 485.

D. chinensis ♀ × *barbatus* ♂ by the pollen of *D. barbatus*. The formula of this hybrid would therefore be—*Dianthus* (*chinensis* × *barbatus*) ♀ × *barbatus* ♂, or in more general terms by (A + B) ♀ × A ♂.

For the sake of simplicity let us assume that the number of idants and force of transmission is similar in the case of both ancestral species. We will suppose that sixteen idants* are arranged in a circle, as represented in Fig. 22, I. In the production of the ½ hybrid, eight idants of A combine with

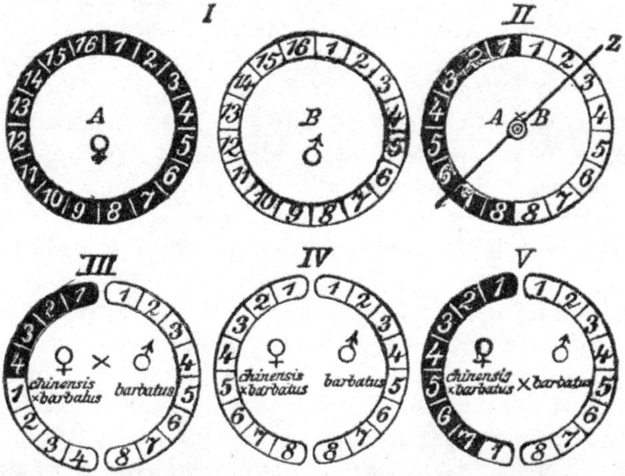

Fig. 22.—***Diagram of the composition of the germ-plasm in hybrids.***—**I.** The germ-plasm of the parental species, each composed of sixteen idants; A, *Dianthus chinensis;* B, *Dianthus barbatus.* II. Germ-plasm of the hybrid, composed of eight idants from A and eight from B: *z*, the plasm of section, which may be rotated. III., IV., V. Three of the possible combinations of the germ-plasm which might arise by crossing hybrid II. with the parental species B. III. A true three-quarter hybrid. IV. Contains idants of the parental species B only. V. Contains nine idants from B and seven from A.

eight idants of B; the mother germ-cells of this hybrid will therefore always contain 8A × 8B idants (Fig. 22, II.). The reducing division then occurs, and bisects the circle at some point. Fig. 22, II., will make it apparent that by rotating the

* We might just as well speak of 'ids' as 'idants' in this illustration. I have referred to the idants simply because they are visible units, and not merely hypothetical structures, and also because the number of idants may be assumed to be less than that of the ids, and is thus more easily controllable.

plane of section, the circle may be separated into nine different combinations of the black idants A of *D. chinensis* and the white ones B of *D. barbatus*, viz.: 8A; 7A + 1B; 6A + 2B; 5A + 3B; 4A + 4B; 3A + 5B; 2A + 6B; 1A + 7B; and 8B. Nine kinds of germ-cells, differing in quality, may therefore be formed, the egg-cells only being taken into consideration in this connection.

In the formation of the so-called $\frac{3}{4}$ hybrid, one of these egg-cells unites with a germ-cell of the pure ancestral species B. The following combinations of idants may then result:—8B × 8B, which would produce the pure ancestral species B; (7A+1B) ♀ × 8B ♂; (6A+2B) ♀ × 8B ♂; (5A+3B) ♀ × 8B ♂; (4A+4B) ♀ × 8B ♂; (3A+5B) ♀ × 8B ♂; (2A+6B) ♀ × 8B ♂; (1A+7B) ♀ × 8B ♂; and 8A ♀ × 8B ♂. Theoretically, therefore, all stages from the pure ancestral form B (Fig. 22, III.) to the form which is intermediate between the ancestral species may occur, but no individual can arise which inclines more strongly towards A than does the exactly intermediate form. Whether all these are actually produced, and in what relative frequency they occur, can only be decided by further researches carried out from this point of view. Those which have till now been made are insufficient, as the number of seedlings raised has always been too small. We may, however, infer from the facts we already possess, that the different combinations of the two kinds of idants *do not occur with equal frequency*, and that the intermediate combinations are the most usual. Were this not the case, 'the form which is intermediate between the $\frac{1}{2}$ hybrid and the ancestral species' could not be 'the most frequent.' This is also most probable theoretically, and becomes more so as the number of idants is greater. The cells of many Phanerogams possess far more than sixteen idants, and even if we also assume that the position of the line of section is entirely a matter of chance, the rarest case will always be that in which it accurately separates the idants of A from those of B, and it will happen much more frequently that it divides them in some other plane. This is equivalent to saying that germ-cells of $\frac{1}{2}$ hybrids rarely contain idants of A or of B *only*, and that in most cases there is a combination of the two.

If the so-called $\frac{3}{4}$ hybrids are again crossed with the ancestral species B, $\frac{7}{8}$ hybrids (the third hybridised generation) are obtained, and these 'as a rule bear a close resemblance to that species of which this is represented, although the individual specimens still display considerable differences in form.' * The result of these observations also agrees with our theory, for even in the second generation of hybrids, eight was the highest possible number of ids of the ancestral species A in the mature germ-cells of the hybrid, and in the process of fertilisation these met with the same number of idants of the ancestral species B, which was used for recrossing. The germ-mother-cells of the $\frac{3}{4}$ hybrid cannot contain more than $8A \times 8B$, and, as a rule, a smaller number of the idants of A will be present. The reducing division then again halves these sixteen idants in various ways; and in the case which is most favourable for the preservation of the hereditary substance of A, the halving of one of these germ-mother-cells, which was most favourable for A, might again lead to the formation of germ-cells with eight idants of A and eight of B. But in all other $\frac{3}{4}$ hybrids the germ-mother-cells already contain more idants of B than of A, as is shown in Fig. 22, III., and the reduction therefore results in the germ-cells containing either idants of B only, or, at any rate, some of B in addition to those of A. If these egg-cells are then fertilised by germ-cells of the pure ancestral form B, the largest conceivable number of idants of A which can be contained in the fertilised egg-cell will again be eight. Hence the form exactly intermediate between the two ancestors may possibly also appear in the third generation, but this will occur *much more rarely* than in the second generation. On the other hand, those individuals the idioplasm of which contains only a few idants of A will preponderate; and in others, even idants of B only may be present, for the idants of B were in the majority in most of the mother-germ-cells of the second generation: the halving due to the reduction must therefore have frequently resulted in the distribution of idants of B *only* to one of the germ-cells, and these then once more combined in the third generation with idants of the pure species B. Theoretically, therefore, the $\frac{7}{8}$ hybrids must in part revert to the ancestral form B. This conclusion is supported by facts, inasmuch as Kölreutter and Gärtner, who made a large number of experiments on recrossing, found 'that three to six generations (four to five as a rule) were required for the complete transformation of the

* Focke, p. 485.

hybrid into one of the ancestral forms.' As these investigators were only concerned with obtaining a complete return to the ancestral form, they always selected for purposes of propagation those plants which bore the closest resemblance to the desired parental type. They, nevertheless, still obtained individual plants which bore traces of the other ancestral species, up to the fifth generation, just as would be expected on theoretical grounds.

The agreement with our theory extends even further than this:—*individual cases of complete reversion to the ancestral form must theoretically occur in the second generation, and such cases have actually been observed.* Godron, indeed, found that '*Melandryum album* × *rubrum*, even when fertilised by its own pollen, reverted in the second generation to the ancestral species.'* This is easily explained on the basis of our theory. If the reducing division took place in certain of the mother-germ-cells of the hybrid in such a manner that each of the two mature germ-cells contained $\frac{n}{2}$ idants † of *M. album* or *M. rubrum only*, two possible combinations might occur in the fertilisation of these germ-cells:—either a germ-cell with idants of *M. rubrum* would meet with another containing idants of *M. album*, or else both the conjugating cells would contain idants of the same species. In the former case, the hybrid form would again be produced, while in the latter, *the pure ancestral form* would result. Such reversions seem to be of rare occurrence, and this may indicate that when they do occur, some unknown conditions favour the halving of the germ-plasm in the mother-germ-cells along a *definite* plane of section, viz., that which passes between the groups of paternal and maternal idants.

Although not proved by actual observation, it is apparent from our theory that *the hybrid may revert completely to the ancestral form.* This would be impossible if the reducing division did not occur, or if the germ-plasm consisted of a homogeneous mass which became completely mixed with that of the other species in the process of hybridisation. The germ-plasm of the one species would then, moreover, never be removed by means of the reducing division:—the reduction would be quantitative, and not a qualitative one. Reversion to the ancestral form could then in no case occur on fertilisation taking

* Focke, p. 485.

† n signifies the normal number of idants in the species in question.

place with the pollen of the hybrid; and even by continued recrossing with one of the ancestral species, the combination of the germ-plasms of the first generation would gradually become weakened, and perfectly pure germ-plasm of the one ancestral species could never be produced. But as soon as we assume that the germ-plasm consists of units which remain separate in the form of idants or ids, it is evident that the whole of the units of one species may be removed from the germ-plasm of the hybrid-offspring, either with or without the occurrence of recrossing; and even if recrossing does take place, reversion *must* occur sooner or later in individual descendants.

Our theory also enables us to make certain predictions, which, as far as my knowledge extends, have not up to the present been confirmed by facts. If sometimes, though rarely, all the idants of A pass into one germ-cell, and all those of B into the other, in the process of halving of the germ-plasm of the primary hybrid A × B a reversion to both the ancestral forms will occur when the second generation is fertilised with the pollen of the hybrid, as has just been shown: that is to say, such a reversion will take place when a pollen-grain containing idants of A only comes in contact with an egg-cell which also contains idants of A only, or when idants of B meet with others of similar origin. The same must also be possible in the hybrids of the third generation, even when all the complete reversions of the second generation are left out of consideration, simply for the reason that certain individuals of the second generation are produced from the combination of *n*A with *n*B idants, and are therefore exactly intermediate forms like all the hybrids of the first generation. The occurrence, even if only in individual cases, of such intermediate forms, renders it possible that germ-cells may again arise which contain idants of A or B only, so that ultimately reversion to one or other of the two ancestral forms becomes possible. These reversions will certainly be of rare occurrence; and as it is entirely a matter of chance that germ-cells which contain the most infrequent combinations of idioplasm should come together in the process of fertilisation, it would be necessary to examine a very large number of seedlings before such cases would be found.

Before passing on to the consideration of reversion to individual characters, it should be remarked that the same law of reversion which can be recognised in the case of plant-hybrids, holds good as regards the crossing of the different human races. According to this law, the intermingling of a racial character is uniform in the first generation, but subsequently becomes quite irregular when recrossing with one of the original races occurs. The skin of mulattoes, which are a cross between white races and negroes, is never quite white, its colour being as a rule an approximate mean between that of the two parents. When mulattoes and white races are recrossed, the skin does not regularly become less black, but the descendants of the third and fourth generations are sometimes white and sometimes fairly dark: this fact indicates the absence of uniformity in the process of 'reducing division.'

2. Reversion to Individual Characters in Man.

The essential difference between the process of reversion in plant-hybrids and in Man, consists in the fact that the former is concerned with the intermingling and subsequent separation of *specific* or racial characters, while in the case of the reproduction of human beings of the same race *individual* characters only are intermingled. As regards hybrid-plants, the idants of each parent might be assumed to be similar,—although this may not be strictly true in all instances, as will be seen later on,—but in the case of individual differences the idants of each parent cannot be regarded as similar. Each of these idants consists of a number of separate ids, which may differ in many respects. In all of them the determinants are as similar as the retention of the specific character renders necessary: that is to say, all the determinants of the same ontogenetic stage are homologous, though at the same time they are never all homodynamous, but differ in many respects owing to slight individual deviations. Hence different ids may contain different variations of any particular homologous determinant. In the following examples, each homologous determinant is indicated by a letter, and variants of a determinant are distinguished by dashes after each letter. Thus id i. of the germ-plasm might, for instance, contain the determinants a, b, c, d, e, - - - n; id ii., the determinants a, b′, c, d′, e, - - - n′; id iii., a′, b″, c′, d‴, e′, - - - n, and so on. The total effect of the idants is decided by the struggle of the ids; the laws which regulate this struggle cannot at present be determined more precisely, and until we know more about them, we may suppose that those variants which are present in the largest numbers always have the best prospect of controlling the cell entirely, or at any rate chiefly; their effect must be cumulative, and a small minority of homodynamous determinants will not be able to take effect against a large majority of some other variant. The control of the cell therefore results from this struggle of the determinants, which must naturally not be conceived to take place in such a manner that the group of paternal ids struggles with the maternal group, but so that all the active determinants which are contained in the idioplasm migrate into the cell body and there strive to obtain the control. If the parents of the organism in question are closely related, the same homodynamous determinants may very possibly be contained in the idants of both parents, and the forces of these will then combine just as would be the case if they had been contained in one idant of the father or mother. It will, however, happen more frequently that homodynamous determinants are present in the ids of *each* parent respectively, and a majority of homodynamous maternal determinants will then compete with a majority of paternal ones; they will then either control the cell together, or the prepotency of one of the parents will be so great as to suppress the influence of the other entirely.

In order to test to what extent these ideas of the co-operation of parental idioplasms may be applied to the phenomena of reversion in Man, we must again consider the simplest of these phenomena, viz., that of *reversion to a grandparent*.

It is well known that a child not infrequently resembles its father or mother in a high degree, and it is also assumed that a father may beget a child which does not resemble him, but it and its father's mother are as 'like as two peas.' This presupposes that the father himself bore no resemblance to his mother, for otherwise it would not be an instance of reversion of the child to the grandmother.

This case may be explained theoretically by assuming that the 'reducing division' of the respective germ-cells of the two generations accurately separated the paternal from the maternal idants, and that, as was shown above, the group of idants of one of the parents may possibly have had no influence on the formation of the child, the other group being the dominant one. The fertilised egg-cell from which the father was developed must therefore have contained the two groups of idants A and

C, A being derived from the grandmother, in the development of whom this group dominated. C was derived from the grandfather, and preponderated to such an extent over A that it alone determined the type of the son. Hence A remained latent in the idioplasm of the father, who therefore resembled the grandfather and not the grandmother. If now, that germ-cell of the father, from which the child of the third generation was developed, just happened in the reducing division to receive the group of idants of A only, which remained without influence on the development of the father, but which was dominant in the grandmother, a reveısion to the grandmother would then become possible. It would occur when this sperm-cell unites with an egg-cell in which the group of idants D has a much weaker controlling force than A. The child would then resemble neither the paternal nor the maternal type, but would be like his grandmother, for the group of idants A determined the type of the grandmother.

I nevertheless consider it doubtful whether reversion of this kind ever occurs so thoroughly and completely as is conceivable theoretically. The facts are unfortunately by no means so decisive as one could wish. No one, so far as I know, has hitherto attempted to ascertain whether *complete* reversion to a grandparent ever occurs, and from a theoretical point of view I should consider this to be improbable. Similarity to a grandparent may undoubtedly often be observed, but it does not therefore by any means follow that a correspondence exists between all, or at any rate the greater majority of individual characters, such as is seen in 'identical' twins. All the parts of the child and grandparent have never been carefully compared, not to mention the fact that such a comparison has never been made at corresponding ages. Moreover, to use the same example, only those cases could be utilised in questions of reversion in which the father bears no resemblance to the type of the grandmother. Cases of this kind are, however, not quite reliable, and are certainly not common.

I am therefore inclined to think that in all these cases we are only concerned with a partial and not a complete reversion to the grandparent,—that is, with the reappearance of a more or less extensive aggregate of characters of the grandparent; and this is certainly sufficient in many cases to produce what appears to be the type of the latter. More or less extensive aggregates of the characters of the grandparents undoubtedly reappear in the child, and these may be more easily explained theoretically, without the necessity of supposing that chance plays so great a part. For in order that perfect reversion may be possible, numerous most infrequent occurrences must take place together. To use the same example, in each individual of three consecutive generations, only one of the two groups of parental idants could determine the type of the child, and the reducing division, moreover, must affect one of the mother-germ-cells of each of four individuals—*i.e.*, the two grand-parents and the two parents—in such a manner as to separate the two groups of parental ids. All these possibilities may sometimes concur, but we can only imagine this to be so if it is proved that an exact resemblance exists between the child and grandparent.

A *partial* resemblance is theoretically far more likely to occur. Let us suppose, for instance, that the germ-plasm consists of sixteen idants, eight of which are derived from the mother and eight from the father, and that the reducing division causes six idants of one parent and two of the other to be present in one germ-cell: the former might possibly contain many of those determinants to which the type of this particular parent was due. This need not necessarily be the case, for all the sixteen idants may have had a farely equal share in the production of this type, and six out of the sixteen idants could not then possibly give rise to the same type even approximately.

It might, on the other hand, happen that the type of the parent was essentially determined by the group of idants derived from the father or mother of this parent only, and a resemblance of the child to its father could then only occur when the whole or a great part of that group which controlled the development of the father was also present in the germ-cell from which the child was developed. Even then, the production of the type of the father would not absolutely be ensured, for in the process of fertilisation by the other parent a group of idants might be added, in which the controlling force preponderates over those already present, either entirely, or at any rate as regards many or most of the determinants.

We may, however, conclude from all these considerations, which are unfortunately still very vague, that a large number of idants of the grandfather, for example, may be present in the germ-cell which will give rise to the father, and still may not result in the production of the type of the grandfather; and that these may then pass into a germ-cell of the next generation, and partially determine the type of the son, provided that they are opposed to a group of idants which has a weaker controlling force. At every ontogenetic stage, then, the struggle of the ids and idants decides which group of the latter is to control the cell. According to the hypothetical principle on which we have supposed this struggle to take place, the majority of homodynamous determinants would always represent the greatest controlling force, so that certain of the more comprehensive or special characters of the grandparent might very well reappear in the grandchild, even if only six or eight idants which controlled the development of the grandparent were present in the germ-plasm from which the grandchild arises.

We may even assume that, as regards Man, *ontogeny is hardly ever passed through without reversions occurring to one or other of the grandparents;* for determinants of one of the grandparents will almost invariably have been suppressed in the development of the parent by stronger ones derived from the germ-cell of the other grandparent, and will predominate in the formation of the grandchild, because they are here again opposed to other combinations of ids over which they may preponderate under certain circumstances. It may therefore happen that some of the individual characters of the grandparent may reappear in the grandchild, although there may be no general resemblance between the two.

The answer to the above question relating to the causes on which reversion to a grandparent depend, may be briefly stated as follows:—*such a reversion is due to the fact that the whole or part of the group of idants which determine the type of the grandparent were present in that germ-cell of the parent from which the grandchild was developed, and that it was there opposed to a weaker group derived from the other parent.*

We do not know what number of the controlling idants of the grandparent must be present in the germ-plasm of the grandchild in order that reversion to the grandparent may occur. *Complete* reversion can only take place if none of the determining idants are absent; but, as we have seen, such complete reversion has not by any means been proved to occur in Man. Still less is this the case in *the next following generation*, *viz.*, *as regards the great-grandparent.* Instances of a descendant of an ancient family reverting to a great-grandparent whose characteristics were not present in the intermediate generations, are certainly occasionally mentioned in novels, but it is only natural to doubt the accuracy of such cases, even when they claim to be true. A great-grandchild certainly often resembles its great-grandparent, but when this is so, *this resemblance had not disappeared entirely in the intermediate generations.* As has been shown above, it would certainly not be incredible, from a theoretical point of view, that the group of idants which controlled the development of the great-grandparent should remain intact in certain germ-cells of two generations, and should be suppressed by more powerful groups derived from germ-cells of other parents, once more to predominate in the third generation. If reversion of this kind could be proved to occur, it might be explained in this manner, and we should be justified in assuming that in many cases the idants of the two parents may again be separated into their original groups by the reducing division. The great variability of hybrids of the second generation proves that this only occurs very rarely, and in most instances not at all.

We have seen that the difference between reversions in hybrids and in human beings of the same race is simply due to every idant of one parent, in the case of hybrids, containing specific characters, so that all the idants of the one parent are similar, and all the homologous determinants may also be considered homodynamous. When therefore an ontogeny is directed by the ids of *one* of the parent species *only*, the type of this species is produced. The type of a human being, on the other hand, is constituted by ids of very many different kinds, no two of which are exactly alike, for each of them contains determinants of a somewhat different kind from the rest; and, speaking generally, the type is in all cases only the resultant of all these different components. The same type, or combination of characters, can only appear for a second time if the same components are again brought together. Except in the doubling of the fertilised egg in the case of 'identical' twins, this can, however, never occur, for a new combination of ids and idants which never existed before arises every time fertilisation takes place. Hence cases of complete reversion as regards individual characters to a previously existing form, can never occur; and only those instances are possible in which the reversion concerns more or less extensive groups of characters which already existed in an individual of a former generation.

We have now discussed the question of transmission from the parents, grandparents, and great-grandparents, to the child; and it will be as well, before going further, to summarise the various possible cases, and to see whether observation and theory contradict one another in any respect.

The most frequent case seems to be that in which the child is a mixture of the types of the two parents, the characters of both of which are either completely united, or else they alternate with one another in the different cells, parts, and organs, and even in the systems of organs themselves. In all such instances, the group of idants which predominated in the ontogeny of the parent, or rather of both parents, must have been contained in the germ-cell from which the child arose.

The question might here be asked as to how it can so frequently happen that only the dominant group of idants is present in the germ-cell destined to undergo amphimixis. We must, however, remember that a perfectly equal blending of the types of the two parents in the child occurs much more rarely than is generally supposed; and that it is difficult, and frequently in fact impossible, to say whether the maternal portion of a character is really derived from the type of the mother, and not from that of her father or mother. General characters only, of the mother's family are most usually combined with others derived from the family of the father. But those characters which have been peculiar to many members of a family for several generations are the most likely to occur in numerous ids and idants, and they therefore reach the germ-cells in larger numbers in many of the modes of 'reducing division.' Theoretically the dominant group of idants of the parent would not be necessary for such a transmission of a general family resemblance, but only a majority of the idants of this group.

The case, however, is different when the resemblance refers mainly to the type of *one* parent; and we must then assume that the dominant group of idants of this parent is present, and is opposed to another weaker group derived from the other parent, and provided with a smaller number of homodynamous determinants.

A third case may also occur in which *the type of an uncle or aunt* is more or less accurately reproduced in the child, or in which this type is blended with the characters of the parent on the other side. I know a man who closely resembles a maternal aunt, but nevertheless possesses many general characters of his father's family. This fact may in all probability be explained in terms of the idioplasm as follows: — the egg-cell from which this person was developed contained the group of idants which was dominant in the ontogeny of the mother's sister, and not that which predominated in the case of the mother. Theoretically this might very well be the case. Let us suppose that the primary germ-cells of the maternal grandfather (m p) contained the idants a, b, c, d, e, f, g, h, and those of the grandmother the idants i, k, l, m, n, o, p, q; and also that the fertilised egg-cell from which the mother was developed contained the idants a, b, c, d × i, k, l, m, and that from which the aunt was developed the idants *a*, *b*, *c*, f × *l*, n, o, p. We will further assume that the group of idants which was dominant in the ontogeny of the aunt were those indicated by the letters *a*, *b*, *c*, and *l*, printed in italics. It will then be seen that the same combination *a*, *b*, *c*, *l* can also be formed from the germ-plasm of the mother by means of the reducing division, for all four of them are present in this germ-plasm (*a*, *b*, *c*, d × i, k, *l*, m). It is doubtful whether such a case ever occurs so accurately, and I know of no instance which renders this hypothesis necessary: the resemblance is always an imperfect one.

We must also consider a fourth case, in which the child neither specially resembles its father or mother, presents a recognisable combination of the characters of both, nor bears a striking resemblance to one of the four grandparents, but displays an entirely new combination of characters. Such a child would probably always bear some resemblance to both, or at any rate to one, of the parental families, but it would not exhibit certain marked characters of the respective fore-fathers.

These cases do not contradict our theory, for by means of the reducing division it may possibly happen that none, or only certain of those very idants which were dominant in both parents are present in the germ-cells of the parents which undergo amphimixis.

If, for instance, the germ-plasm in the ontogeny of the father had the composition *a*, *b*, *c*, *d*, *e*, *f*, g, h, of which the idants a, b, d, f, were dominant, and the germ-plasm in the ontogeny of the mother consisted of the idants *i*, k, *l*, m, *n*, *o*. p, q, of which i, l. n, o essentially determined her type. If the reducing division then resulted in an egg-cell containing the idants k, m, p, q, and a spermatozoon was formed which contained the idants c, e, g, h, and these came together in the process of amphimixis, the resulting germ-plasm would certainly contain characters of both families, but it could not possibly produce the type either of the father or mother, or a mixture of both.

I have already discussed the question of reversion of the child to one of its four grandparents in detail, and need not return again to this point. I should, however, like to call attention to the fact that, theoretically, *a child can never exhibit a combination of the types of two of its grandparents*, simply for the reason that at least half of the idants of a germ-plasm determine the type of the child, the germ-plasm of which can only contain one quarter of the idants of *two* grandparents. An exception to this rule could only occur in cases of close interbreeding, in which both grandparents in part contained the same idants.

No case has, however, ever come before my notice in which a child bore no resemblance to its parents and yet exhibited a combination of the types of that of two of the grandparents.

3. *Reversion to the characters of ancestors far removed in animals and plants.*

We will now consider the question of reversion *to remote ancestors.* Cases of such reversion are so well known through Darwin's researches, that a more detailed description of the mere facts might almost be considered superfluous. But such a description is nevertheless necessary, and I must even express the opinion that the facts have not been so accurately investigated from all points of view as is desirable in order that a theoretical explanation of them may be arrived at. Darwin was the first to point out that in cross-breeding, either of species or of mere varieties, characters not infrequently appear in descendants which were not present in the parents, in some of which cases, indeed, it can be proved, and in others shown to be very probable, that they have been derived from remote ancestors. Mules, for instance, sometimes exhibit distinct stripes, like those of the zebra, on the fore-legs and shoulders, while such stripes are very seldom present in the horse and ass, and are even then only very faint; but we must nevertheless suppose them to be derived from the ancestral form of the two species. And, again, when certain races of pigeons are crossed. offspring are produced in which the plumage has the slaty-blue colour of the wild rock pigeon, although the races used for crossing were of quite another colour; in this case, moreover, the descent from the wild rock pigeon is certain. Similar instances also occur in plants. The hybrids of *Datura ferox* and *D. lævis*, in both of which the flowers are white, regularly bear blue (purple?) flowers, and Darwin * has shown that this is to be looked upon as a reversion to ancestors which possessed blue flowers, as, in fact, is the case at the present day in an entire group or species of *Datura*.

I will now attempt to explain these three instances in accordance with my theory. In them, as is universally the case, reversion must be attributed to the presence of old unmodified determinants in the germ-plasm, which take the place of the younger homologous determinants as regards obtaining the control of the cell or region of cells in question. Similar assumptions must be made in every theory of heredity. In his theory of pangenesis, Darwin makes use of old gemmules for this purpose, while de Vries assumes that reversion is due to the presence of old pangenes. Some unmodified portion or other of the hereditary substance must always form the starting-point in attempting to explain the problem; and the only question is, whether we are to remain satisfied with such a statement, and leave everything else in obscurity, or whether it is possible to obtain a certain insight, in principle at any rate, into the question as to why these minute parts can remain unmodified, and why and under what circumstances they can suddenly obtain control precisely in the region of transformed parts which are homologous to them.

A solution of the first of these problems has already been given above. It has been shown that, according to the principle of selection which controls the whole, a transformation such as is required by the vital conditions of a species never necessitates the transformation of *all* the determinants relating to the parts to be transformed, but that this process is only necessary in the

* Darwin, 'Animals and Plants under Domestication,' Vol. II. p 254.

case of a majority of them, which is sufficient to ensure the occurrence of the modification in question in every ontogeny, — *i.e.*, in every individual of the species. This is all that is required, and consequently *the processes of selection cannot accomplish more.* After every transformation of the body in the process of development of the species, the germ-plasm will thus contain some unmodified determinants in addition to those which have been transformed in any part, and these will only disappear very gradually in the course of the further history of the species.

The existence of the *material* by means of which reversions to all specific characters operate, can therefore be proved on the basis of the theory of selection: *every germ-plasm must contain a larger or smaller number of old determinants corresponding to the characters of the ancestral species.*

The solution of the second problem, as to why these 'ancestral determinants' always exert their influence at the right spot in the body, naturally follows from our theory, in which the mechanism of ontogeny is referred to the gradual disintegration of the germ-plasm.

The third problem then only remains: *how can the ancestral determinants, which are present in a minority, gain control over the majority of younger ones?*

We have seen that reversions to the ancestral form occur in the offspring of hybrid plants, — even if they are fertilised by their own pollen, — when those cells to which the *entire* group of idants of the ancestral species of the *same* parent has been distributed at the reducing division happen to come together in the process of fertilisation. Whenever the germ-plasm of the fertilised egg-cell contains idants of the species A *alone*, an organism of the species A can alone arise. But this occurrence is out of the question in cases of reversion to the characters of remote ancestors. The germ-plasm can then never consist entirely of idants of the ancestral species; and, in fact, we may doubt whether entire ancestral idants exist at all in the germ-plasm of any individual of a long-established species. For the number of idants (nuclear rods) is not extremely large in any species; and if, as in the instance given above, we assume this number to average sixteen, even the proportion of *one* unmodified to fifteen modified idants would be rather large, and would render an occasional reversion to the ancestral form possible in the ordinary reproduction of the species. Fertilisation need only occur eight times successively, two of the unmodified ancestral idants being added each time to those already present, in order to produce a germ-plasm consisting of pure ancestral idants, and the organism developed from it would then agree with the ancestor in question in all its characters. Under particularly favourable circumstances the process might even be completed in four generations, as is shown in the following table:—

Generation I.	a′, b, c, d	e, f, g, h = ♂	germ-cell.
	a′, i, k, l	m, n, o, p = ♀	"
Generation II.	a′, b, a′, d	c, i, k, l = ♂	"
	a′, q, a′, s	t, u, v, w = ♀	"
Generation III.	a′, a′, a′, a′	b, d, q, s = ♂	"
	a′, a′, a′, a′	α, β, δ, ε = ♀	"
Generation IV.	a′, a′, a′, a′, a′, a′, a′, a′ = ♂		"
	a′, a′, a′, a′. a′, a′, a′, a′ = ♀		"

The letters in the above table indicate the idants, which—with the exception of a′, which represents the unmodified ancestral idant, and only occurs *once* in the first generation—are only individually different from one another. The vertical line represents the reducing division of the mother-germ-cell, which in the most favourable case we are now considering, always causes this ancestral idant a′ to pass into the germ-cell about to undergo amphimixis. For the sake of simplicity, the conjugating germ-cells are assumed to be similar in respect to their contained ancestral idants. It would thus follow that germ-plasm consisting of ancestral idants only, would be produced in the fourth generation.

Although we must certainly admit that so favourable a combination of events can hardly ever occur, it cannot be doubted that an accumulation of ancestral idants may take place in *one* germ-cell in the course of a large number of generations, and that a *majority* of these ids may consequently come together in fertilisation. In this case a more or less complete reversion to the ancestral form must take place. As such reversions to the complete type of the ancestral species cannot be observed to occur in the normal reproduction of pure and long-established species, we must conclude that *all the idants have become modified in such old and pure species*, each of which produces the type of the species, and not that of an ancestor, in case it alone has to control the ontogeny.

The foregoing considerations seem to me to afford a good explanation of the *frequent occurrence of reversion in young species which are not yet definitely established.* Cultivated varieties of flowers, such as the Heartsease (*Viola tricolor*), always produce, among a large number of seedlings, certain plants which bear flowers more or less resembling those of the wild species. Evidently only a portion of the idants have become transformed in this modern form, another and smaller portion having remained unmodified. Since the reducing division separates the idants into two groups of all possible combinations, germ-cells, both male and female, containing a preponderance of unmodified ids, must also occur; and if two such germ-cells come together in fertilisation, reversion must result.

Let us suppose that, of the sixteen idants in the germ-plasm, ten were modified and six unmodified. Under favourable circumstances a case might then occur in which the majority of idants would remain unmodified, without further accumulation taking place in the course of generations — *i.e.*, at every fertilisation. This may be illustrated by the following table, in which the letters with a dash represent the unmodified, and those without one the modified idants, the line dividing the letters indicating the reducing division:

Maternal germ-plasm	a′, b′, c, d′, e, f′, g′, h	i, k, l′, m, n, o, p, q.
Paternal "	f, r′, s, t′, u′, v′, w′, x	y, z, a, b′, c, d, e, g.

Germ-plasm of offspring I.: — a′, b′, c, d′, e, f′, g′, h × f, r′, s, t′, u′, v′, w′, x.

Owing to the manner in which the reducing division takes place, the maternal and paternal germ-cells each contain five unmodified idants, so that in the germ-plasm of the offspring ten unmodified and six modified idants are present. If, on the other hand, we suppose that the other two cells produced by the reducing division from the same mother-germ-cell undergo fertilisation, the germ-plasm of the offspring II. will have the following composition: —

i, k, l′, m, n, o, p, q, y, z, a, b′, c, d, e, g.

It therefore only contains two unmodified idants, and must consequently give rise to the modified form of the plant.

A large number of cases of heredity which have not hitherto been understood can thus be explained in a very simple way. Take, for example, the *varying degrees of certainty with which varieties of cultivated plants transmit their peculiarities*. The extent to which such peculiarities are transmitted must depend on the number of idants in the germ-plasm which have remained unmodified, and the greater this number, the more easily will reversion occur.

These theoretical considerations will probably account for the first of the three cases which were mentioned above as types of reversion to more remote ancestors, viz., that of *Datura ferox* × *lævis*. The two species of *Datura* have white flowers, but produce hybrids in which the flowers are blue; and this is not only occasionally the case, but occurred in every one of the 205 plants which Naudin raised from this cross,* and also in all those cases which had been observed previously by Kölreutter and Gärtner.† If we assume that, in addition to their own specific idants, the two species of *Datura* contain a certain number of ancestral idants, the latter might be relatively increased in individual germ-cells in consequence of the reducing division; and when these met with germ-cells of the other species, which also contain numerous ancestral idants, a germ-plasm possessing a larger number, and perhaps even a majority, of these idants would be constituted. Individual cases of reversion to the common ancestral form might thus occur. But this assumption is evidently insufficient to account for the facts, for the blue colour of the flower appears in *all* the hybrids. The reversion in these cases must therefore be independent of a greater accumulation of ancestral idants which may possibly occur in individual germ-cells in consequence of the reducing division. These idants must, on the contrary, *come together in each fertilised egg-cell in a sufficient number to preponderate over the modified idants*, and to control the ontogeny. They cannot possibly, however, be stronger *numerically* in every case, and another factor must therefore take part in the process, which causes the ancestral primary constituents to preponderate in every case; and this is in all probability *the specific diversity of the modified idants*. We have assumed from the first that

---

* Darwin, 'Animals and Plants,' Vol. II., p. 254.

† *Cf.* Focke, 'Pflanzenmischlinge,' p. 269.

homodynamous determinants have a cumulative effect, while heterodynamous determinants tend to counteract one another. The same must be true of the *groups* of determinants, — the ids and idants; similar idants must increase the effect produced, while dissimilar ones will interfere with one another to a greater extent the more diversified their composition out of ids and determinants is. Let us confine our attention for the present to the groups of determinants for the flowers. The two species of *Datura* certainly contain determinants which tend to form *white* flowers, and we might therefore be disposed to infer that these are homodynamous, and that their effect must be cumulative. But this conclusion would be erroneous, for it is quite possible that these determinants only correspond as regards the production of a 'white' *colour*, and differ widely in respect of many other characteristics of the cells, such as those relating to their size and minute structure. On the other hand, the 'blue' determinants are actually homodynamous, and correspond not only with regard to the colour to which they give rise, but also in respect of all the other characters of the cells of the flowers, for they are derived from the common ancestral species. When the hybrid begins to develop flowers, the structure of the cells in the latter will depend on the determinants of the two white species and of the blue ancestral form. Although the 'blue' determinants are in the minority in the idioplasms of either parent, they may nevertheless, if they all combine, possess a greater power of transmission than the 'white' ones, if the latter are not homodynamous, *i.e.*, do not possess an exactly similar force of heredity, and consequently cannot produce an intermediate effect. They interfere with one another in their action, as they act in different directions to a greater or less extent. Many cases of reversion can be understood — even though only in principle — by means of this law. They can, moreover, easily be rendered comprehensible in individual cases if we have recourse to figures.

Let us suppose that the 'blue' ancestral determinants are not only contained in individual ids, but are present in all the ids of entire idants: a minority of old ancestral idants would then be opposed to a majority of modern ones, half of which, however, would correspond to the type of *D. lævis* and half to that of *D. ferox*. We assume that there are in all sixteen idants, six of which are 'blue' and ancestral, and ten 'white.' Since, therefore, five of the latter belong to *D. lævis* and five to *D. ferox*, and these are consequently dissimilar, the six similar ancestral idants, which have a cumulative effect, will preponderate, because the 2×5 different idants of *D. lævis* and *D. ferox* do not produce the cumulative effect of 10.

Hitherto I have assumed that we are in this case dealing with a *complete* reversion to the ancestral form, and not merely with reversion to *individual ancestral characters*. I cannot, however, judge with certainty in this respect from the facts known to me concerning these hybrids; and as this instance was not chosen for its own sake, but merely as an example in which complete reversion to more remote ancestors might be accounted for, it must remain undecided whether it really belongs to the above category, and whether such instances of complete reversion actually occur. The blue colour of the flowers is at any rate not the only apparently new character in these hybrids; the stem, for instance, is brown, while it is green in the two pure species. In only one of these species (*D. ferox*) is the stem brown at first, and this colour is only retained as a brown ring around the base of the stem. I must leave botanists to decide whether the shape of the leaves, and the structure of the stem or fruit, afford any reason for considering these parts as intermediate between those of the two parental species, or whether they are to be regarded as deviations from both, and presumably, therefore, as reversions to the ancestral species.

It may in all probability be assumed that the process is not a complete one in the above-mentioned cases in which reversion to the wild ancestral form in different races often occurs when the latter are crossed. Darwin certainly gives one instance of a pigeon which 'was hardly distinguishable from the wild Shetland species,' but which was, nevertheless, descended from four grandparents which differed very considerably from the wild species (*Columba livia*).* This bird, which was blue, and possessed the typical black bars on the wings and tail, was descended from a red spot, a white fantail, and two black barbs These breeds, as is well known, differ from the wild pigeon in colour as well as in many other details, such as the length of the beak and number of tail feathers; and it would therefore be interesting to ascertain whether these racial characters had all disap-

* 'Animals and Plants under Domestication,' p. 14.

peared in the grandchild, and had become retransformed into the corresponding characters of the wild species. Were this so, the reversion might be considered complete, and the same theoretical explanation could be given for it as in the case of the *Datura* hybrid. Unfortunately Darwin leaves this point untouched, as he devoted his attention chiefly to the coloration so characteristic of the species. It seems to me to be very probable, however, from several of his statements, that this was also essentially a mere case of reversion as regards the coloration of the plumage. I conclude this principally from the fact that the blue or original coloration of *Columba livia* occurs in all the principal breeds of pigeons, although these blue subvarieties are rare in some of them. The other racial characteristics do not at any rate exclude the possibility of a blue coloration; and thus, on the other hand, reversion to the blue colour is not necessarily accompanied by a reversion to all the other characters of the ancestral form.

It is perfectly certain that in most cases the reversion produced by cross-breeding is not complete, even as regards the coloration, but gradually becomes more marked, so that at first very faint and hardly perceptible indications of the wing- and tail-bars are seen, and these become more pronounced by degrees, so that a partial blue coloration with perfect bars, and finally the perfect slate-colour and complete bars of the ancestral form are produced. The greater number of reversions in pigeons must consequently be *incomplete*, — *i.e.*, they must refer to individual *characters or groups of characters* only, and we are here concerned with the theoretical explanation of such cases as these.

I take for granted the facts that all valuable races of pigeons breed true, that all the main breeds are characterised by differences in *form*, and that sub-breeds differ merely as regards *colour*. In my opinion this implies, in the first place, that the germ-plasm of the main breeds has become essentially modified from that of the rock-pigeon, and that only lesser portions of it correspond to that of the ancestral form; and also, that *all the determinants have not become modified to an equal extent*, — those for the coloration having undergone the least, and those for the whole body the most, alteration. I therefore suppose that the germ-plasm of one of the main breeds consists of a number of modified idants, none of which any longer correspond exactly to those of the ancestral form, so that no one of them, did it control the ontogeny, could result in the development of a rock-pigeon. This conclusion depends entirely upon the race breeding true, for a germ-plasm which still contained individual unmodified idants of the ancestral form would also necessarily, or at any rate probably, once in a way, contain a majority of ancestral idants at the same time, in consequence of the incessant recombination of the idants in the reducing division and in fertilisation; and this must result in reversion to the ancestral form. Such a reversion, however, never occurs in purely bred races, but only when crossing takes place.

The germ-plasm of a race of pigeons therefore consists, according to my view, of a certain number of idants, each of which represents the type of the race. The majority of the ids of which each idant is composed must consequently virtually contain this type; or, expressed more accurately, *the whole of the racial determinants — as compared with those unmodified determinants which may still be present — are in the majority in all the ids of every idant.*

The fact that races breed true can in this way be thoroughly explained.

As regards reversion to the coloration and markings of the wild pigeon, we must suppose that, in the process of artificial selection to which the different races owe their origin, only just as many ids have become completely transformed into racial ids as were required to ensure the desired object, viz., the preservation of the racial characters. A larger or smaller number of determinants in many or perhaps all the ids must have remained unmodified in all, or at any rate in many, of the idants. The determinants on which the coloration depends, must have remained unmodified in larger numbers than did those relating to any other characters, for the coloration is the most liable to revert.

Reversion in the coloration will therefore occur when the ancestral determinants for any particular region of the bird's plumage gain a predominance over the racial determinants in the course of development; and in cross-breeds it will take place *when the racial determinants are so different that their forces counteract one another instead of being cumulative.* Although the ancestral determinants are in the minority in the germ-cells of the two different breeds which unite at fertilisation, their controlling forces will nevertheless be cumulative, and if they are sufficiently numerous, they will determine the coloration in question, and thus produce reversion.

In this way we may, I think, account not only for the phenomenon of reversion in general, but also for many special details, and more especially for *the different degrees of reversion in different races of pigeons.* A sufficient number of experiments has certainly not been made with regard to this point, but it is nevertheless recognised that reversion occurs more easily and more markedly in some races than in others. Darwin, for instance, by crossing two black barbs with two red spots obtained dark hybrids, of which 'no less than six presented double-wing bars.'* On the other hand, the mongrels derived from two black barbs and two snow-white fantails showed no trace of reversion. This must have been due to the retention of a different number of unmodified specific determinants in the germ-plasm of the two races, as well as to the difference of the modified racial determinants; for the more marked the difference between the two crossed races, the more easily will the ancestral determinants gain the predominance over them.

The last-named experiment by Darwin was continued by pairing two of these hybrids, one of which was brown and the other black. The first brood (it is not stated how many there were) displayed wing-bars 'of a darker brown than the rest of the body.' This must have been due to the accumulation of a larger number of unmodified determinants in individual germ-cells in consequence of the reducing division, and to the subsequent union of two of these cells in the process of fertilisation. We should therefore expect that reversion would not occur in all the offspring of this pair, for the reducing division must also result in certain germ-cells containing a majority of the modified determinants. In the second brood of the same parents, in fact, a brown bird was produced which possessed no trace of wing-bars.

It is easy on the basis of our theory to account for the fact that a simple crossing of two species did not in many instances produce any traces of reversion, although reversion resulted from *the subsequent double crossing.* The most complete case of reversion obtained by Darwin was produced as follows. A

* Darwin, *l.c.*, Vol. I., p. 208.

mongrel female barb-fantail and a mongrel male barb-spot were paired. Neither of these birds 'had the least blue about them, nevertheless the offspring' (the number is not stated) 'from the above two mongrels was of exactly the same blue tint as that of the wild rock-pigeon from the Shetland Islands over the whole back and wings; the double black wing-bars were equally conspicuous.'* In this case the ancestral determinants evidently prevailed all the more strongly because they were opposed by racial determinants of three or four different kinds, the controlling forces of which could not simply be cumulative, like those of the ancestral determinants, but could only partially weaken and neutralise one another.

The fact which may be deduced from Darwin's observations, viz. that the offspring of simple crosses display practically a very similar tendency to reversion, can also easily be explained theoretically. For the germ-plasm of a well-established race will contain a certain percentage of ancestral determinants, and the germ-cells of an individual will be liable to but few fluctuations in this respect. In such simple cases of crossing, an almost similar number of ancestral determinants, as well as of racial determinants, must therefore come together at each fertilisation; and the struggle between these different kinds of determinants must always produce approximately the same result. A perfect uniformity in the offspring of the same cross cannot be expected; but if reversion does occur in some cases, it will be absent in others, and if it occur partially in some, certain parts in others will revert. An instance of the former kind is seen in the above-mentioned cross between a black barb and a white fantail, and of the latter in the case described by Darwin on p. 207 of his 'Animals and Plants under Domestication,' Vol. I. He crossed a white nun with a red tumbler, and reared five young, all of which presented traces of reversion. One possessed a blue tail; the second and third 'presented a trace of the bar at the end' of the blue tail; the fourth 'was brownish, and the wings showed a trace of the double bar'; and the fifth 'was pale blue over the whole breast, back, croup, and tail, but the neck and primary wing-feathers were reddish, and the wing-bars presented two distinct bars of a red colour.' Thus all the five young had reverted — some more and some less markedly —

* *Loc. cit.*, Vol. I., p. 209.

in one part or another. The differences between the young support the theoretical assumption that the number of ancestral determinants, which determined the type of the five young ones, was by no means the same in the different germ-cells of the two parents. Their dissimilarity, moreover, admits of a further conclusion, which could likewise be foreseen theoretically, viz., *that the determinants for the coloration of the different regions of the body are represented in the different germ-cells in fluctuating numbers*. Not only should we naturally suppose this to be the case from the structure of the germ-plasm as here assumed, but we can hardly even imagine it to be otherwise. If the various idants in the germ-plasm of a hybrid together contain twenty ids with ancestral determinants, it does not follow that the latter *only* are present in each id. There is no reason why in any one id the determinants for the colours of the wings, for instance, should not be ancestral, while those for the colours of the tail are racial, or *vice versa*. The bird would display reversion in the tail or wings, according to whether the former or the latter kind of determinants is in the majority in the germ-cell: for our theory shows that the struggle of the determinants takes place independently in every cell, and consequently also in every part of the organism; and the result will therefore be independent in each part, and will depend solely on the combination of homologous determinants which contend with each other in the cell or group of cells under consideration. It can thus easily be explained how it comes about that though all the offspring of a certain cross show reversions, the latter are nevertheless combined in very different ways with the racial characters of the parents.

We must now consider the third type of reversion mentioned at the beginning of this section, viz., that which concerns *individual characters of very remote ancestors, and is seen, for instance, in the stripes on the legs of horses and asses, and more especially of mules*.

It will not be necessary to repeat the arguments which Darwin used to show that the horse was primitively dun-coloured and striped. Indications of such a coloration are seen at the present day in horses of different colours in all parts of the world, in the form of a dark stripe down the back, and of transverse markings on the legs and shoulders; such markings, however, are not common, and most frequently occur in dun-coloured horses.

The ass has been proved to be descended from a species possessing stripes on the legs, and corresponding to the existing wild species (*Asinus tæniopus*) of Abyssinia The domesticated ass has, as a rule, only retained the crossed stripe on the back and shoulders, but occasionally transverse markings occur on the legs, and this is most often the case in animals of a pure grey ground colour. The stripes on the legs are very rare both in horses and asses, as I know from personal observation, and even when present they are very faint.

In mules these stripes occur much oftener and much more distinctly — particularly in light grey animals, and are sometimes seen on the hind-legs as well as on the fore-legs and shoulders. Thus reversion occurs to the ancestral form of both sides.

The theoretical explanation of this fact must be based on the assumption that a certain number of unmodified ancestral determinants for the hairy coat have been retained in the germ-cells of the two species; and that these, when they are brought together in the germ-cells of the two parents, may predominate over the modified determinants of the parents. The fact that such reversion does not by any means always occur in mules, points to the conclusion that the number of ancestral determinants varies very considerably in the germ-cells of individuals, and that these may even be entirely absent, or only sparsely represented, or, on the other hand, may be present in large numbers. In the latter case, when two of these germ-cells come together in fertilisation, partial reversion will occur, and will be more marked and extensive the greater the number both of the ancestral determinants which meet together and the parts of the body to which these belong. The determinants for the stripes on the hind-legs are evidently much less numerous in the combined germ-plasm of the two existing species than are those for the marks on the fore-legs as is shown by the manner in which these stripes have been relatively retained by several existing wild species of *Equus*. All these determinants must, however, be present in very varying numbers in the germ-plasms of different individual horses and asses, for, as already stated, reversion does not by any means always occur in mules; in Italy, where these animals are used in large numbers, it is not common, and, as far as my experience extends, is perhaps only exhibited in one or two animals out of a hundred Gosse, however, states that in the United States of America nine mules out of every ten are striped.

Before concluding this section, I will analyse a case of reversion to remote ancestors in plants in greater detail. My reason for so doing is not because different principles are necessary for its explanation, but because we possess the results of experiments which render a closer examination of the theory possible.

I will select as an example the case of the reversion of *irregular or unsymmetrical flowers to a regular or peloric form.* Many instances of reversion of this kind have been described, but they are very exceptional: their origin cannot be connected with any external causes, and must evidently be due to purely internal ones, viz., to the composition of the germ-plasm.

If my opinion concerning the transformation of the germ-plasm in the course of phylogeny is correct, it follows, as has already been shown, that individual unmodified determinants of an old character must always appear here and there in the germ-plasms of the modern species, even after an enormous number of generations. Such ancestral determinants — those of the original regular flowers, for instance — need not by any means be contained in the germ-plasm of every individual plant; and the older the modern species is, the fewer will be the number of these determinants, which we will call 'peloric.' They must gradually be displaced by the 'asymmetrical' determinants; for the latter, being better adapted to the existing conditions, have a better chance in the struggle for existence. A large number of plants, such as *Corydalis tuberosa*, for instance, will thus no longer contain any 'peloric' determinants, and this makes it apparent why reversion so seldom occurs in these cases. The fact that it can occur at all, is to be accounted for by the processes of reducing division and amphimixis in two germ-cells, the latter of which always follows on after the former. For if a small number of 'peloric' determinants still remained in various idants of individual plants of the species in question, they might occasionally come together in one germ-cell owing to the reducing division; and if two such germ-cells meet one another in the process of fertilisation, this group of determinants may predominate, and reversion will then occur — provided that the combined power of these 'peloric' determinants is sufficient to predominate over the 'asymmetrical' determinants.

The experiments which have been made with respect to the transmission of characters in such abnormally peloric flowers, prove that this explanation must be correct in principle. Darwin crossed the peloric snapdragon (*Antirrhinum majus*) with its own pollen, and from the seeds thus obtained raised sixteen plants, which 'were all as perfectly peloric as the parent plant.'* We need not be surprised that the pelorism was inherited, for the 'peloric' determinants were in the majority in the parental germ-plasm on either side; it may, however, have been due to chance that *all* the sixteen plants which were reared proved to be peloric. If a larger number of offspring had been raised, some of them would certainly have produced asymmetrical flowers, for the reducing division would in most cases divide the 'peloric' determinants unequally amongst the two resulting cells, and consequently two germ-cells containing no, or only a minority of, 'peloric' determinants might meet together in fertilisation. Reversion to the ordinary form of the flower must then occur.

The result of Darwin's counter-experiment is particularly interesting. The peloric snapdragon was crossed with the common form, and 'two great beds of seedlings' † raised, '*not* one of which was peloric.' And in ninety plants which were carefully examined there was not a trace of pelorism, 'except that in a few instances the minute rudiment of the fifth stamen, which is always present, was more fully or even completely developed.' Darwin attempted to explain this fact by the assumption that in this case the common form of the flower possessed a 'prepotent force of transmission;' but apart from the fact that this statement is merely another formulation of an observed fact, and can hardly be looked upon as a real explanation, it does not hold in the case of Darwin's subsequent experiments. For the plants obtained by crossing the common snapdragon with the peloric form, 'which perfectly resembled the common snapdragon, were allowed to sow themselves; and out of a hundred and twenty-seven seedlings, eighty-eight proved to be common snapdragons, two were in an intermediate

---

* 'Animals and Plants,' &c., Vol. II., 1888, p. 46.

† *Loc. cit.*, Vol. II., 1888, p. 46.

condition between the peloric and normal state, and thirty-seven were perfectly peloric.'* If now, the character of asymmetry possessed a 'prepotent force of transmission,' we should expect it to become more apparent when the flowers of both parents were asymmetrical, than when this was the case as regards one parent only. In the chapter of the same book which treats of pangenesis, Darwin has attempted to give a special explanation of the fact 'that a character gains strength by the intermission of a generation' in which this character is not present.

My own explanation of the above-mentioned facts follows almost directly from what has already been said. The production of so many individuals of the ordinary kind by the crossing of the common snapdragon with the peloric form, and the fact that a large number of 'perfectly peloric' offspring, as well as of the common form, were produced when those obtained by the first cross were fertilised with their own pollen, is due simply to the reducing division of the germ-mother-cells occurring in different ways, so that sometimes only 'ordinary' and sometimes only 'peloric' determinants reach the germ-cell, and sometimes, again, a combination of both, in which either the 'peloric' or the 'ordinary' determinants preponderate. The daughter-plant will produce flowers which are either quite of the common form or more or less peloric, according to whether the determinants which are brought together by the two germ-cells in the process of fertilisation are chiefly 'peloric' or chiefly of the 'ordinary' kind. It is evident, however, that most of the parents of this generation, which possessed flowers of the ordinary form, must have contained 'peloric' as well as 'ordinary' determinants in their germ-plasm before the reducing division, for they were all derived from a peloric father or mother. The predominance of the ordinary form in the next generation may be explained as being due to the peloric grandparents possessing only a slight majority of 'peloric' determinants in their germ-plasm, in addition to a considerable number of 'ordinary' determinants. The whole of the germ-plasm of the germ-cells from which the parent generation arose must have contained very many more 'ordinary' than 'peloric' determinants.

---

* *Loc. cit.*, Vol. II., p. 46.

4. *Reversion to Rudimentary Characters.*

It is well known that organs which have lost their value for the preservation of the species become rudimentary in the course of generations: they diminish in size, become stunted, and ultimately disappear altogether.

This may be expressed in terms of the idioplasm as follows. The group of determinants in the germ-plasm for the organ in question becomes reduced in one id after the other, first in one determinant and then in another, until finally it disappears completely; and the process is repeated in an increasing number of ids, until eventually this group of determinants is no longer contained in any of them. It cannot be stated how long a time and how many generations are required for this process, but it may at any rate be asserted, and even proved, that individual ids still contain determinants for the organ in question long after its disappearance from the mature individual. The fact that the organ occasionally reappears, and that consequently *reversion may take place*, proves that this is the case.

The supernumerary nipples which occasionally occur in human beings are an interesting example of this kind. The two normal nipples generally occur in a rudimentary form in men, but, in addition to these, very diminutive ones are occasionally met with in parts in which they are normally present only in the lower orders of mammals, such as carnivores, rodents, and lemurs:— a pair may be present above the normal teats near the axillary region, and two or three others lower down on the abdomen. All of them certainly never occur in the same individual, but usually only a single one or a pair are present: these, however, are found both in men and women. They are undoubtedly to be looked upon as reversions to extremely remote characters possessed by our lower mammalian forefathers. We owe an accurate account of their occurrence in the male sex to the numerous and detailed investigations of Otto Ammon,* who found them in three per cent. of our recruits.

As Ammon's researches only extend over two, or at most three generations, and only refer to individual cases, we cannot form a precise estimate as to the degree and extent to which these structures are transmitted. We may, however, attribute the

* I am indebted to Mr. Ammon for these details, which are not yet published.

stimulus to reversion in these instances to interbreeding; that is to say, to the amphimixis of such germ-cells as contain a certain number of ancestral determinants in their germ-plasm on either side for those regions of the skin in which the nipples were situated in remote ancestors. Merely in consequence of the reducing division, these determinants may possibly accumulate in one germ-cell in sufficient numbers to produce the character in question. The same may be said with regard to many small individual marks, and such very ancient hereditary parts as supernumerary nipples usually follow the same rule as do individual characters; they have to a certain extent degenerated so as to come under the same category, and for a long time past have not been contained in the germ-plasm of every individual; their determinants, on the contrary, are entirely absent in most cases, and are only found in a certain number of ids in certain individuals. These determinants, like those of individual characters, may be transmitted for several generations without attaining development, and may then suddenly become manifest in consequence of a favorable combination of two conjugating germ-cells. The only difference between these ancestral characters and ordinary individual peculiarities is, that the determinants of the latter are contained in a larger number of ids, and we must therefore conclude that they become developed far more frequently and regularly.

Although in the case of the supernumerary nipples in the human race we cannot definitely indicate the ancestor from which they were derived, or the length of time during which they have been transmitted, it is at any rate possible to do so approximately in the case of *the supernumerary toes of horses*. For, thanks to the excellent researches of Kowalewsky, followed by those of Marsh, we are well acquainted with the phyletic development of the horse: we now know that horses belonging to the genera *Mesohippus*, *Miohippus*, and *Protohippus* or *Hipparion*, which possessed two smaller lateral toes in addition to the large median ones, existed in the middle Tertiary period. When horses are occasionally born at the present day in which one or two such accessory toes are present on two or even on all four feet, we are perfectly right in considering the development of these toes to be due to reversion to an ancestor of the Miocene period. We must therefore assume that in certain series of generations of the existing horse, some idants have been retained in which ancestral determinants of the fore and hind-feet are present, but that these are in the minority, and a large number of them can only accumulate in one germ-cell when a particularly favourable reducing division occurs. This, however, would not in itself be sufficient to produce the character in question: chance must also play a part, in order that such an egg-cell, containing an abundance of ancestral determinants, may be fertilised by a spermatozoon in which a certain number of these determinants are also contained. Then, and then only, will there be a likelihood that the entire number of the latter will be sufficiently large to preponderate over the modern determinants of the foot in the process of ontogeny.

Cases of reversion of this kind occur very rarely. Marsh has, however, brought forward a small series of such instances, the oldest of which relates to a horse belonging to Julius Cæsar, and the most recent was observed by him in a living animal.*

5. *Preliminary Summary of Sections 1–5.*

All the phenomena of reversion which have so far been considered can be explained on the supposition that every germ-plasm is composed of a large number of equivalent units or ids, each of which possesses all the determinants required for the development of an organism; every character is therefore produced by the co-operation of many determinants of the same region (homologous determinants). The transformation of a species or race into a new one, moreover, never depends primarily on a simultaneous modification of all the ids and determinants: when the modification begins to take place, even entire groups of ids (idants) may remain unmodified, and this will subsequently be the case as regards a minority of ids, and still later at least with regard to certain of the determinants in individual ids. The characteristic form of every individual cell which takes part in the process of ontogeny is the result of the struggle of the ids which occurs in this cell; and as some amongst the mass of determinants which constitute every id in the germ-plasm may become modified and others not, and the proportion of modified to unmodified determinants may vary from id to id, we can understand that in crosses between species

---

* O. C. Marsh, 'Recent Polydactyle Horses,' *American Journal of Science*, Vol. xliii., April 1892.

or races, reversion is sometimes more and sometimes less marked — in some cases restricted and in others very extensive. In those cases, therefore, *in which rudimentary and insignificant characters only are concerned, such as stripes in mules, pelorism in flowers, and rudimentary nipples in the human race, individual variations must frequently occur in all degrees, from the complete absence of reversion to the most marked form of it.*

Conversely, it is just as easy to explain why reversion occurs *without exception* in other cases, such as in that of *Datura*-hybrids, if we assume that entire groups of ids or idants, as well as certain determinants in individual ids, have remained unmodified. For if the former are present in such numbers as to preponderate in a hybrid over the two kinds of modified ids, which counteract one another, they must be contained in nearly equal numbers in the germ-plasm of every individual.

We can in principle understand the whole series of the phenomena of reversion, if we represent the transformation of the germ-plasm as being due to the majority of the ids in most of the idants becoming modified first, while a minority of them remain unmodified: in the course of generations the number of these unmodified ids then gradually becomes smaller, owing to the action of natural selection, until the ancestral species is only represented by a number of scattered ids; and finally, by continued selection, these ids also become modified in the same direction, to such an extent that eventually only the determinants of those individual characters escape modification which are less important or quite meaningless for the life of the organism. We know that complete reversion to the ancestral form may occur in young species in the case of favourable crosses between allied species (*Datura*); and that in the further course of phylogeny, that is, in older species, total reversions can no longer occur, although reversion to single characters, or even to entire groups of characters, may still take place, and does so with certainty under certain conditions — as is exemplified by the reversion of *certain* hybrids between different races of pigeons. In its last stage, then, reversion occurs as an entirely uncertain and apparently capricious re-appearance of an individual ancestral character, such as, for instance, that of the occasional striping in mules.

By means of our theory much may also be deduced and rendered comprehensible in principle concerning the *external causes* of reversion. It can easily be understood that the crossing of different species and races is liable to lead to a preponderance of those ancestral idants, ids, or determinants which are common to the two parents. We might, however, also infer from the theory that certain of the offspring of hybrids are very liable to revert in the next generation to one of the ancestral species; for the germ-cells of the hybrids are very dissimilar as regards the germ-plasm they contain, in consequence of the reducing division which every germ-cell undergoes at its formation: most of them will contain idants of both parents in every conceivable proportion, while in others, idants of only one of the parents will be present. *The 'reducing division' is therefore one of the most efficacious of the primary causes of reversion.* for it renders possible the uneven distribution of the different qualities of the idants which were contained in the germ-plasm of the germ-cells of this parent. This principle applies equally to the cases in which not a single entire idant remains untransformed in the germ-plasm, which then only contains a minority of unmodified ids or even merely of determinants, scattered amongst several idants.

Reversion in any degree therefore depends, in the first instance, on the process of amphimixis, for without sexual reproduction the reducing division would not have been introduced into the organic world, and the second extremely important factor in reversion — viz., the crossing of different germ-cells — would also not exist. Reversion is, however, not of necessity connected with *actual* amphimixis, but may, as I shall show later on, occur in connection with parthenogenesis and gemmation; although this is only true in the case of those organisms in which amphimixis formerly occurred, and the germ-plasm of which may therefore contain ancestral ids or ancestral determinants. It is evident, however, that the chance of reversion occurring must be much greater when amphimixis takes part *directly* in the process of reproduction; for the relative proportion of modified and unmodified units of the idioplasm of this order may be quickly or suddenly altered in favour of the unmodified units, and thus the ancestral units which were originally present may undergo accumulation, and predominate over the somewhat dissimilar homologous units of the modified kind.

Our comprehension of the problem just discussed is certainly increased by the above considerations, and the following passage from Darwin's 'Animals and Plants under Domestication'* will best show how much our insight has since then advanced:—'In purely-bred races, occasional reversion to long lost characters of the ancestors often occurs without our being able to assign any proximate cause.'

6. Reversion in Asexual Reproduction

(a) *Reversion in the Process of Gemmation*

As already stated, *reversion does not entirely depend on amphimixis*, and may in fact, also occur apart from the crossing of two individuals. The *bud-variations* of plants form a well known instance of this kind.

For many years I possessed in my garden a maple (*Acer negundo*) with variegated leaves which were almost entirely white, and one branch of this tree bore ordinary green leaves, flowers, and seeds. Owing to the greater amount of chlorophyll, this branch grew and bore flowers and seeds far more luxuriantly than the main branch from which it arose. If we look upon the offshoots of a tree as persons, this would be an instance of a person of the plant-stock produced asexually, which reverted to the ancestral form.

We must go back to the origin of variegated species in order to find a theoretical explanation of this phenomenon. Like most similar varieties, indeed, of our trees and shrubs, this form must have arisen by *bud-variation*; in other words, a normal maple, from certain unknown causes, gave rise to a branch bearing variegated leaves. The cause of this modification, traced to the idioplasm, must have been due to the determinants of the leaves and other green parts of the shoot becoming modified in such a way as to result in the production of organs deficient in chlorophyll. If, however, only the majority, and not all the ids in the apical cells of the first variegated shoot became modified in this manner, a reversion of the variegated variety to the green ancestral species would become possible.

Another assumption, however, which cannot yet be proved to be true, is required in order to account for the appearance of a green branch upon a variegated tree: we must suppose that even in ordinary cell and nuclear divisions, the division of the idioplasm may take place in an irregular manner, so that all the

* Vol. II., p. 25.

ids do not reach both the daughter-nuclei; but that both the halves of some of them which result from the division pass into the *same* daughter-nucleus. On the other hand, it might even perhaps be possible for an entire idant to be transmitted to only *one* of the daughter-nuclei. It must be possible for some such irregularity to occur in the distribution of ids during nuclear division, for otherwise the occurrence of a different combination of the primary constituents in the course of growth — such as, indeed, actually occurs in cases of bud-reversion — would be inexplicable.

Let us distinguish between those ids which correspond to the original form, and those which have become modified, by describing them respectively as 'green' and 'variegated' ids. The reversion of a shoot must then be due to the unequal distribution of the ids amongst the daughter-nuclei during the divisions of the apical cells of the shoot, so that a majority of 'green' ids were distributed to a cambium cell containing 'blastogenic' germ-plasm, or even also to an apical cell of a young lateral bud. Conversely, a variegated shoot might subsequently originate from a green one, as in fact actually occurs.

I should be inclined to offer a similar explanation of the reversions to the ancestral form which so frequently occur in all the numerous varieties of our trees and shrubs, such as, for example, the oak-leaved heterophyllous hornbeam, the fern-leaved oak, the varieties of the maple and birch which possess greatly subdivided leaves, and the copper-beech and copper-hazel. The tendency to revert is very varied even in the different varieties of the same species. The golden-striped variety of *Euonymus japonica*, for instance, is very liable to revert, while in the silver-striped variety of this plant reversion rarely occurs.

This difference would simply depend on the relative minority of ancestral ids in the variety in question. Reversion will never, or only rarely, occur when the idioplasm only contains a few ancestral ids; if, however, there are so many of these that the unmodified ids only form a small majority, reversion can easily take place.

In this connection I must mention another instance, which has very frequently been discussed since Darwin's time, viz., that of the peculiar 'graft-hybrid' *Cytisus adami*. The combination of the characters of the two ancestral species continually varies in this plant, which sometimes bears the yellow clusters of flowers characteristic of the common laburnum (*Cytisus laburnum*), and sometimes purple flowers like those of *Cytisus purpurea*, or, again, both these colours may be present in the same flower. I will here quote Darwin's description of the plant*:—'To behold mingled on the same tree tufts of dingy-red, bright yellow, and purple flowers, borne on branches having widely different leaves and manner of growth, is a surprising sight. The same raceme sometimes bears two kinds of flowers, and I have seen a single flower exactly divided into halves, one side being bright yellow and the other purple, so that one half of the standard-petal was yellow and of larger size, and the other half purple and smaller. In another flower the whole corolla was bright yellow, but exactly half the calyx was purple. In another, one of the dingy-red wing-petals had a narrow bright yellow stripe on it; and lastly, in another flower, one of the stamens, which had become slightly foliaceous, was half yellow and half purple.'

The result of the struggle of the parental idants evidently cannot depend in the case of *Cytisus adami*, as it does in that of the individual characters of the human race, upon the fact that the number of homodynamous determinants varies in the parental idioplasm according to the part concerned; for were this the case the same parts of the flower could not sometimes be yellow in some instances and red in others,—all the flowers, on the contrary, would display the same composition as regards the parental hereditary parts, even though slight variations might occur, such as would be produced by a dissimilarity in the conditions of nutrition. As in the case of the hybrids of *Oxalis* already mentioned, the flowers would at least display a certain combination of parental characteristics which would be uniform in one and the same plant. The fact that this is not the case, seems to me to afford a decisive proof that *Cytisus adami is a real graft-hybrid, and not an ordinary seminal hybrid*, as in fact was stated to be the case by the nurseryman Adams, who first produced it. I therefore consider the controversy ended as to whether graft-hybrids exist at all, and offer the following explanation of the hereditary phenomena concerned.

---

* *Loc. cit.*, Vol. I., p. 414.

Cytisus adami was obtained from a young bud on a portion of the bark of *Cytisus purpurea*, which was grafted into the stem of *C. laburnum*. This bud developed into a shoot which exhibited an intimate combination of the parental characters. The shoot was afterwards propagated, and the plants raised from it exhibited 'reversions to both of the parental forms,' as well as dingy-red, *i.e.*, mixed blossoms, so that the pure characters of the parents were displayed in more or less extensive regions of the hybrid.

From a theoretical point of view, it can obviously be granted that a mongrel-plant may originate by contact of the living tissues of the parents, only if the transformation of the rudiment of an existing shoot is out of the question, and if the rudiment then appears for the first time. An existing dormant bud, which contains all parts of the shoot, cannot become modified as regards the idioplasm which it contains by the stock of another species which nourishes it: its apical cells, from which further growth proceeds, cannot receive a supply of extraneous idioplasm from the supporting stock; for the nuclear rods alone contain idioplasm, and this is a solid substance, and can only undergo combination by the fusion of two cells and their nuclei. It is therefore also worthy of notice that Adams did not observe that a hybrid was developed from the *single* dormant bud which was from the first present on the graft, but that the hybrid arose from *one of the later buds which were formed in the second year*; moreover, only *one*, and not *all*, of these buds produced bud-hybrids. The formation of this single hybrid-bud must have been regulated by an unusual and accidental occurrence, for all efforts to produce the hybrid a second time have so far been in vain. This accidental occurrence must have been that the cambium-cells of the two species came to lie close together, so that they could both enter the same bud arising from the cambium. Botanists must decide whether two cambium-cells, belonging to different species, can conceivably become united into one by a process of conjugation similar to that which occurs in the union of the male and female cells in fertilisation, and whether the foundation may in this way be laid for a new growing point.* If such a process

* A *fusion of nuclei*, apart from that seen in the process of fertilisation, actually occurs in plants in the case of the embryo-sac. Guignard (*loc. cit.*) describes this process in detail as follows:—The 'upper and lower pole-nuclei' of the embryo-sac approach one another, each accompanied by

occurred in this case, *the number of idants of Cytisus adami must be as large as that of the two ancestral species taken together*; for, as far as we know, a process of reducing division only occurs in the formation of sexual cells. The correctness of my assumption can therefore be controlled by observation.

It is hardly conceivable that two young plant-cells could, without fusing, have formed the growing point of the hybrid-shoot; for probably only one of these cells could have performed the function of an apical cell, and consequently the hereditary influence of the other could not have extended to countless daughter-shoots, as was actually the case. In the course of growth, every cell below the apical cell must necessarily have gradually come to be situated further away from the growing-point. Such an intimate combination of characters as actually occurred could not have been effected in this way.

I am therefore inclined to suppose that the unusual phenomena exhibited by *Cytisus adami* were due to an abnormal kind of amphimixis, so that the idants of both species were combined in the apical cell of the first shoot; but that in the subsequent cell-divisions an unequal distribution of the two kinds of parental idants amongst the daughter-nuclei took place, thus producing the variations in the combination of the characters.

Such an unequal distribution of the superabundant idants might also occasionally occur in an apical cell itself. This process may, moreover, be connected with the frequent complete reversion of an entire branch to *one* of the ancestral species, as well as with the fact that a modification tending to make the parental characters more and more distinct has occurred in the course of time in many examples of hybrids. Shortly after the first appearance of the *Cytisus*-hybrid, the colour of all the flowers was a dingy-red, — that is, an intimate mixture of the two ancestral colours, yellow and purple; but by degrees this mixture became less perfect, until eventually pure yellow and pure purple flowers, and even entire clusters of flowers and entire branches possessing almost the pure character of one or other of the parent-species,

---

its centrosome, which has already become doubled by division. The nuclei then come to be situated close together, the centrosomes uniting in pairs, just as in the ordinary process of amphimixis; and finally the two nuclei fuse together comp ete y. A spindle of division is then formed, and several divisions follow c ose aft r one another.

were formed. It seems only possible to explain this circumstance by supposing that the connection of the two idioplasms was easily severed, and that differential nuclear divisions occurred in such a manner as to cause a larger number of idants of *C. purpureus* to pass into one daughter-nucleus, and a larger number of those of *C. laburnum* into the other; or, at any rate, to cause the one or the other idant to pass completely into one of the daughter-nuclei, instead of dividing longitudinally and one of the resultant halves entering each daughter-nucleus. Though this is certainly only a conjecture, it is, however, not altogether an unjustifiable one, for the apparatus for division in each of the two species is certainly concerned with a smaller number of idants than that which must be present after the fusion of two nuclei; irregularity might therefore easily occur at the division. It is possible that unknown forces of attraction may also play a part in the process: the idioplasms of the two species do not at any rate exhibit a marked attraction to one another, as might perhaps be conjectured from the negative results obtained by Darwin, Reisseck, and Caspary in ordinary experiments on hybridising. By fertilising *C. laburnum* with the pollen of *C. purpureus*, Darwin obtained pods which dropped off 'in sixteen days after the withering of the flowers,' and the reverse cross resulted even less successfully.

However this may be, — and the point can be settled by determining the number of idants, — the phenomena of heredity at any rate indicate that the idioplasms of the two parents can easily again become separated in the course of cell-divisions. This separation might perhaps begin with the passing over of *one* idant only from one side, which would result in the preponderance of one parent in many of the blossoms, &c. This preponderance would have increased considerably in the course of growth, so that far larger groups of cells now frequently contain pure idioplasm of *C. laburnum* or *C. purpureus*, and new shoots are formed which apparently contain idioplasm of one or the other species only. The fact that the plants raised by Herbert from seeds of pure yellow flowers of *Cytisus adami*, which, although they bore yellow flowers, showed a purple tinge on the flower stalks, proves that even these shoots may still contain some individual idants of the other ancestral species. The idants of *C. purpureus* seem, however, to have disappeared entirely from certain shoots; for Darwin raised plants from the seeds of yellow flowers, which resembled *C. laburnum* 'in every character, with the exception that some of them had remarkably long racemes.'

My explanation of the fluctuation in the combination of parental characters in *Cytisus adami* rests on the fact that we are not here concerned with the subtle differences in the forces which, in the reproduction of the human race, cause either the maternal or the paternal idants to predominate, and which we have attributed to the number of homodynamous determinants, which differs according to the characters concerned: this fluctuation depends on the grosser differences, and on the number of idants on the two sides. Sometimes the idants of *C. purpureus*, and at others those of *C. laburnum*, predominate in the idioplasm of a cell; and, under certain circumstances, even only the *one* kind of idant may be represented in a cell, in which case it alone will also be present in all the offspring of this cell.

My theory is not therefore rendered less likely by the case of *Cytisus adami*; we can, on the contrary, even explain the most minute details of such phenomena by its means, although it was not formulated for this purpose.

6. *Reversion in Parthenogenesis.*

Reproduction by means of unfertilised ova occurs regularly in some plants and in many animals — more especially in crustaceans and insects. We might be inclined to expect, *a priori*, that no great degree of deviation between the mother and daughter could occur at all in this monogonic form of reproduction, and that at most a reversion to more remote ancestors could take place.

Such a supposition is not, however, borne out by experience. A series of experiments with parthenogenetic species, which I have made during the course of the last eight years. have shown that although the expected uniformity between parent and offspring generally results in a very high degree, exceptions occur occasionally, and that these must be regarded as exhibiting reversions to an ancestral form many generations removed.

The facts are briefly as follows. Two varieties of a small Ostracod (*Cypris reptans*), possessing a very marked coloration, occur in certain ponds in the neighbourhood of Freiburg. The colour of one variety A, is light yellow ochre, five green spots being present on either side of the shell; the other variety B, appears dark green, owing to the yellow-ochre ground colour being reduced in extent by the presence of six large green patches. These patches exactly correspond as regards position in the two varieties, and are merely much smaller in A than in B, the sixth spot being wanting in A. Both varieties are reproduced parthenogenetically in the neighbourhood of Freiburg, and males never appear.

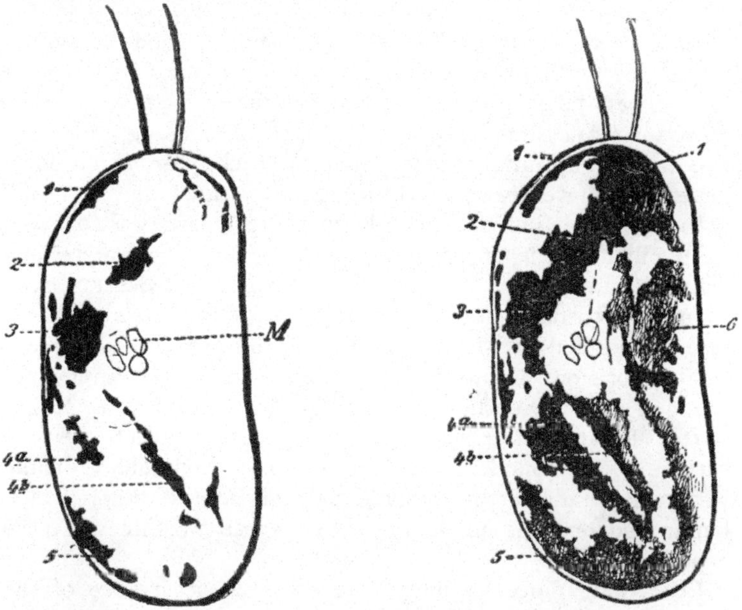

FIG. 23.—The two varieties of *Cypris reptans*. 1–6, indicate the six main patches of colour on the shell.

My experiments were made by isolating one female of each variety in a small aquarium, feeding them well, and allowing them to multiply until each vessel was filled with their mature descendants, which in their turn produced eggs. The individuals of the colony were then examined, and the greater number killed and preserved, one or more of them being kept alive and placed separately in fresh aquaria for breeding purposes.

These animals breed very rapidly throughout the year, and thus in the course of the last eight years many thousands of individuals have passed through my hands.

The first result obtained by these experiments confirmed

my expectations: *the descendants of the same mother resembled one another as well as the parent with which the experiments began, even as regards minute details of the markings.* The differences were mostly as small as those which may be observed in identical human twins: it cannot be stated whether these were due to a want of similarity in the germ-plasm, or whether they were to be accounted for by accidental differences in nutrition.

Apart from the exceptions about to be mentioned, no modifications occurred even in the course of many generations. I now possess colonies of A and B which cannot be distinguished from their ancestors in the year 1884. Reckoning that from five to six generations were produced each year, about forty generations have been passed through since then.

In 1887, some individuals of the dark-green variety B appeared for the first time in an aquarium containing the typical yellow-ochre-coloured variety A; and since then I have twice observed a similar occurrence in other broods of A. In the last of these cases, examined in May 1891, it could be proved that only a single Cypris out of 540 adults contained in one aquarium had changed into the dark variety *suddenly, and without apparent cause.* In another case, intermediate forms between the two varieties were found, as well as ordinary individuals of the variety B; and this fact is not only of interest theoretically, but it also removes all doubt as to the trustworthiness of the experiments.

For a long time I waited in vain for the occurrence of the reverse transformation from the dark variety to the light-coloured one, and was inclined to consider that the former was the ancestral form. But in the winter of 1890–91, a colony of B appeared in which a few typical individuals of the variety A were found, together with typical specimens of B which had been bred in this aquarium for many years.

It is out of the question that these sudden transformations were due to external influences, for *both* forms made their appearance together in the same small aquarium, under precisely the same conditions. We can only suppose that *modifications in the composition of the germ-plasm* must have taken place, and I think it is possible to prove that this was the case.

Parthenogenesis, using this term in the strict sense, has in all cases been derived from sexual reproduction, as is proved in this case by the fact that the unmated females retain the receptaculum seminis, which is unused and always empty.

The two varieties must have originated at a time when sexual reproduction occurred — at any rate periodically: were this not so, primary constituents of A could not be present in the germ-plasm of B, and *vice versa*. The co-existence of both kinds of primary constituents in the same animal, can only be understood if we suppose that sexual reproduction had occurred at no very distant period.

The explanation of the process of reversion naturally follows from the fact that *in species in which parthenogenesis regularly occurs, a reducing division nevertheless takes place*, but only a *single* one: *one* polar body is separated off from the egg, and not two, as in sexual reproduction. This single halving of the idants in the ovum must undoubtedly be preceded by a doubling, just as occurs in sexual reproduction; for a reduction of the number of idants to one half would otherwise take place from generation to generation, so that ultimately only a single one would be left. If, however, a reducing division preceded by a doubling of the number of idants takes place, reversion becomes possible.

Let us take a case of the simplest possible kind, and suppose that there are four idants in the germ-plasm, three of which are entirely composed of ids of the type A, and the fourth of ids of the type B. The four idants a, a, a, b, first become doubled, so that eight idants, a, a, a, a, a, a, b, b, result. Let us assume the most favourable case for reversion towards the variety B to occur, the reducing division resulting in the separation of these idants into the groups a, a, a, a, and a, a, b, b, the latter of which forms the nucleus of the egg-cell: the daughter individual arising from this egg would then produce primary ova containing the group of idants a, a, b, b. After the process of doubling in the ripe ovum, this group would then have the composition a, a, a, a, b, b, b, b; and if the reducing division occurred in such a manner as to result in the combination a, a, a, a, passing into the polar body, and the combination b, b, b, b, remaining in the nucleus of the ovum, an individual of the variety B would undoubtedly arise from the egg, and reversion would ensue.

These processes would in reality be much less simple, and would take place much more slowly. For the number of idants is doubtless in most cases much larger, and the addition of a single idant of the other form would result in a much smaller percentage. I have been able to prove the existence of 24–26 idants in the case of *Artemia salina*, a crustacean which multiplies parthenogenetically. If even several of these belonging to one of the two ancestral varieties were present in a minority, it is questionable whether they would ever become accumulated in one and the same germ-plasm in the course of generations and the corresponding reducing divisions, so as to form a preponderating majority: such an accumulation is, however, conceivable. This would partly depend on chance, for the majority of the ova produced by an individual always perish, and rare combinations which would produce reversions are therefore also usually lost.

We may thus account for the fact that reversion only occurred rarely in my experiments and only in certain individuals of a colony at the same time, as well as for its appearing either suddenly, or else after the production of intermediate forms. The latter case is to be explained theoretically by supposing that a balance of the two kinds of idants was first produced, and that this then in the offspring partially led to the preponderance of the idants on which reversion depended.

The possibility of reversion in parthenogenesis therefore depends upon two factors: firstly, upon the composition of the germ-plasm out of different kinds of ids and idants, — *i.e.*, upon the occurrence of sexual reproduction in a previous generation; and, secondly, upon the 'reducing division' which always takes place at the formation of germ-cells.

7. *Proof that the Determinants become Disintegrated into Biophors*

I will conclude this chapter with some remarks which might have been more suitably inserted in the chapter on *the control of the cell by the idioplasm*, but in that place they would have been unintelligible, as they depend upon a knowledge of the phenomena of maturation of the ova and spermatozoa.

The fact that actual germ-plasm is removed from the animal-ovum by means of the cell-divisions occurring before maturation, — that is, in the process of *the separation of the two polar bodies*, — was not by any means easily determined, and was only proved after at least ten years of prolonged and difficult investigation and reasoning. I have not thought it advisable to give an account of the whole history of the development of our knowledge of this subject,* as this is unnecessary for understanding the phenomena of heredity. I must, however, refer to a period of this history, which is instructive as regards the method by which the cell is controlled by the emigration of nuclear matter into the cell-body as enunciated by de Vries and accepted by myself.

After the so-called 'extrusion' of the polar bodies of the egg had been shown to correspond merely to a very unequal cell-division, and convincing proof had been given that the controlling substance must be situated in the chromatin of the nucleus, it followed as a logical postulate that the ovum, like every other cell, must be supposed to contain a special controlling substance, or *specific idioplasm*, the function of which is to produce the special histological nature of the cell in question. This conclusion I arrived at, and assumed that the ovum, from its earliest stage to the attainment of its full size and specific nature, is controlled by a special idioplasm, which differs entirely from the idioplasm which becomes active after the completion of maturation. If the nature of the cell is determined at all by its idioplasm, the ovum, while still growing and undergoing histological development, cannot possibly be controlled by the same idioplasm as that which serves for embryonic development. I consequently assumed the existence of an '*oogenetic*' idioplasm in the egg during the period of its histological differentiation, and also that after maturation, this substance gives up the control of the cell to the germ-plasm.

The question then arises as to what becomes of the oogenetic idioplasm when this change in the control takes place.

My answer to this question was, that the oogenetic idioplasm is removed from the ovum by means of the polar divisions, and that it was thus rendered possible for the germ-plasm — which was already present in the nucleus of the ovum, and had in the meantime increased considerably in bulk — to obtain control of the cell.

This conjecture has since turned out to be erroneous. Investigations which I subsequently made soon showed that at least *one* of the two polar divisions has a totally different significance,

* Such an account is given in my essay on 'Amphimixis,' and in still greater detail in the previous Essays upon Heredity.'

inasmuch as it results in the halving of the germ-plasm itself. It became apparent subsequently that *both* divisions serve this purpose, and each of them causes the removal of germ-plasm, and not of oogenetic idioplasm from the ovum.

My hypothesis must therefore be given up, but I nevertheless believe that the conclusion on which it was based was a correct one, and that it may be further utilised in the light of the theory of heredity here developed. The oogenetic idioplasm must exist, and, using the terminology I have now adopted, it may be spoken of as the oogenetic 'determinant.' This determinant will consequently be the first to become separated from the mass of germ-plasm of the young egg-cell, to disintegrate into its constituent biophors, and to migrate through the nuclear membrane into the cell-body. In this way alone can we account for no trace of it remaining in the nucleus, and for embryonic development not being subsequently impeded by its presence. *This determinant is used up, and disappears as such;* and the fact that it is not expelled from the egg strongly indicates, if it does not prove, that the control of a cell by a determinant is accompanied by the absorption of the latter, and a further support may thereby be obtained for the hypothesis of emigration.

A precisely corresponding process must be assumed to occur in the formation of the sperm-cells, in which also the function of the idioplasm during the histological differentiation of the cell differs widely from that of the germ-plasm of the mature spermatozoon. The necessity for assuming the existence of 'histogenetic' determinants is perhaps rendered still more evident in the case of the egg-cell, as in some animals two kinds of eggs are produced which are very different as regards size, the relative quantity of food-yolk, colour, and nature of the shell. The assumption of *two kinds* of oogenetic determinants cannot be avoided in this case, for we cannot suppose that the same germ-plasm can have such different effects on the cell. In the section on alternation of generations, it was shown to be necessary to assume that such species contain two kinds of germ-plasm, containing determinants which are in part similar and in part dissimilar. Thus the germ-plasm from which the winter-eggs of the *Daphnidæ* are developed must contain an oogenetic determinant which is quite different from that of the germ-plasm in the summer-eggs, for these two kinds of ova are entirely dissimilar.

I know of no instance in which there is such a wide difference as regards the activity of the idioplasm in successive cell-generations as is the case in the germ-mother-cells and the mature germ-cells arising from them. If, however, even in this very striking instance of a sudden change of function of the idioplasm, the idioplasm which was active at first is not removed from the cell, such a process cannot occur in any other case; and we are consequently justified in applying to all other cells the conclusion derived from the behaviour of the germ-cells, and in considering it as proved that the *active idioplasm of a cell becomes used up in consequence of its activity*.

CHAPTER XI

DIMORPHISM AND POLYMORPHISM

1. Normal Dimorphism

The phenomena of reversion discussed in the last chapter depend on the capacity possessed by organisms of conveying in their idioplasm characters which they do not themselves possess in the form of 'latent' primary constituents, and of transmitting these to their descendants, in which they may, under favourable circumstances, undergo development.

It has hitherto been supposed that all the individuals of a species possess these latent primary constituents in the same degree; and that, consequently, characters which are capable of becoming occasionally manifest at all in any organic form exist in a latent condition in all individuals of the species, their subsequent appearance or non-appearance depending solely on certain developmental conditions. Even Darwin was of this opinion, as is shown by many passages in his works. He imagined that latent 'gemmules' existed, for instance, of stripes like those of the zebra in every horse, of the slate-blue coloration of the rock-pigeon in every domesticated pigeon, and of the two parental species in every hybrid. In the last chapter I attempted to show that this may be true in many cases,— such as, for instance, in the races of pigeons, — but that it is by no means necessarily so always, and that in numerous instances certain latent ancestral characters are not present in all, but only in a larger or smaller number of individuals of the form in question. We have seen that the reducing division may, indeed, from one generation to another, divide the germ-plasm of the parent in such a manner that the germ-plasm of some of the offspring receives no portions of the idioplasm of one of the grandparents at all. The most striking example is seen in plant-hybrids, in which reversion to one of the parental forms may even occur amongst the offspring of the hybrid. In spite of the nearness of the generations between such descendants and the parental forms, none of the characters of one of the parents are present in a latent condition.

Such a rapid removal of characters from the germ of individual descendants is, however, only possible if — as in the case of these hybrids — they belong to one parent only. Thus in the human race, any individual traits of the mother or father may not only be absent in certain of the children, but the corresponding determinants may even be wanting in their germ-plasm, so that these traits cannot reappear in the grandchildren or great-grandchildren. The case, however, is very different with regard to those characters which are possessed by both parents. These cannot disappear from the germ-plasm of individual descendants from one generation to the next, for their determinants constitute a majority in the maternal as well as in the paternal half of the germ-plasm. As long as these characters are present in *all* the individuals of the group of forms in question, the determinants for these characters will predominate in most of the idants, and it will then hardly be possible for them to dwindle into a minority in consequence of successive reducing divisions. But let us take the case of a gradually disappearing character, and select as an example the wing-bars characteristic of the rock-pigeon which are present in the domesticated breeds. The absolute number of the 'wing-bar' determinants in the germ-plasm must have been diminished gradually in the course of the processes of selection which led to the formation of the various races, owing to many of them becoming transformed into differently constituted or 'modern' determinants. The smaller the number of 'ancestral' determinants, the more liable are they to be totally eliminated from *one* half of the germ-plasm in the reducing division. When, however, they are entirely wanting in individual germ-cells, two of these cells may possibly come together in conjugation; and an animal would then be produced which possessed no 'wing-bar' determinants, or, in other words, *which no longer contained this character in a latent condition*. In proportion as such individuals became more frequent, the average number of 'bar' determinants in the germ-plasm of the race must still further decrease, owing to the constant interbreeding of these individuals with others, until finally only a small percentage of individuals would contain such determinants at all.

I do not think that the above argument can be strictly applied in the case of this particular character of the pigeon. The frequent appearance of bars in crosses of different races of pigeons indicates, on the contrary, that most individuals still contain a number of these 'bar'-determinants. We may, however, conclude with regard to the occurrence of stripes like those of the zebra on horses and asses, that the infrequency with which reversion occurs, even in the case of crosses, is probably due to the fact that the character in question has long ceased to be present in a 'latent' condition in all individuals of both species, and that the germ-plasm usually no longer contains 'zebra'-determinants.

We certainly cannot ascertain whether some few ancestral determinants may not nevertheless still be present occasionally, for even in crosses their cumulative effect might no longer be apparent; but general considerations compel us to assume that even these sporadic old determinants must ultimately disappear. They will be contained in the germ-plasm of a gradually decreasing number of individuals, and as these derive no benefit from their presence, they will presumably become less and less frequent.

We must therefore take into consideration, on the one hand, the dwindling specific or racial characters, which may be present in a latent condition, though all of them need not by any means necessarily occur in all individuals; and on the other, the individual characters, which may exist in a latent condition in a varying number of individual descendants.

The actual specific characters are, however, transmitted to every individual, though they also do not always become manifest, for some regularly remain latent when another opposing group of characters becomes apparent. I am here alluding to *primary and secondary sexual characters:* but all characters — even the non-sexual ones due to *dimorphism* and *polymorphism* — come under this category. I must now attempt to explain these phenomena in accordance with my theory.

In considering this problem, we must naturally begin with the simplest case of sexual differentiation, in which the originally monomorphic germ-cells become differentiated into male and female. The question then arises as to *the origin of this differentiation in the idioplasm, and how it has arisen phyletically*.

Let us take a definite example. The *Volvocineæ* are lowly

organised multicellular algæ of a globular form, which rotate and are propelled through the water by the movement of their cilia. In addition to multiplication by means of asexual germ-cells, several genera of this order, such as *Eudorina* and *Pandorina*, exhibit a sexual mode of reproduction, consisting in the conjugation — *i.e.* complete fusion — of two germ-cells which are apparently quite similar to one another.

As long as this is the case in all individuals, we might suppose that the reproductive-cells are controlled by the idioplasm which directs the development of the species in general, — that is, by the germ-plasm, which is composed of a varied number of similar determinants. This will, however, no longer be the case as soon as the conjugating germ-cells become differentiated into male and female, as has occurred in the allied genus *Volvox*. The utilitarian motive for this differentiation is not far to seek, for it must be advantageous for the germ-cells to contain the greatest possible accumulation of nutritive material; and this could only occur to a very slight extent as long as the two germ-cells destined to undergo amphimixis retained a marked power of movement, as is the case in *Pandorina*. They consequently become differentiated into the stationary egg-cell, containing a large quantity of nutritive material, and the small motile sperm-cell, in which very little food-material is present. On what process in the idioplasm, then, does this differentiation depend?

The idioplasms of the egg- and sperm-cells evidently cannot be precisely similar. They cannot simply consist of germ-plasm; but the egg-cell must contain a determinant which gives it the histological character by which it is distinguished from the sperm-cell, which latter must also contain a determinant controlling its histological development. The germ-plasm of *Volvox* must therefore contain *spermatogenetic* and *oogenetic* determinants besides those for the ciliated somatic cells, only one or the other of which, however, becomes active, and impresses the male or female character on the germ-cell. I imagine that these sexual determinants are *double*, each of the two parts always occurring together and being closely united, but so regulated that only one of them becomes active at a time. We might represent this figuratively by supposing that each id of the germ-cells consists of a central globular mass of germ-plasm, surrounded by a layer composed of this sexual double-determinant, either the male or female part of which may be external, and consequently dominant.

This is only a metaphor, and is not intended to represent the actual occurrences. We are ignorant of the forces and substances which are here concerned, but we at any rate know that the idioplasm of the primary germ-cells in the higher animals is still capable, in by far the most cases, of giving rise to *either* kind of germ-cells, and that the decision as to whether the germ-cells will develop into ova or spermatozoa occurs at some early stage in embryogeny: in the eggs of the bee it takes place at the beginning of embryonic development, long before the first primary germ-cell is differentiated, while in other animals it perhaps occurs at a later stage. The well-known researches of Siebold and Leuckart, prove that at any rate in the case of bees, this decision undoubtedly rests upon the occurrence or omission of fertilisation, — it occurs, that is, at the time when the germ-plasm which controls the new organism is constituted; and this fact seems to me to be of great importance. If fertilisation takes place, the organism develops into a female, and if not, a male will result. This at least proves that the decision *may* take place at such an early stage, and I doubt whether it can in any case occur later: in some animals, at any rate, it occurs still earlier, during the period of maturation of the egg. The *Phylloxera* lays large eggs which produce females, and small ones from which males arise. Both become fertilised, so that in this case fertilisation takes no part in the determination of sex.

These questions cannot be discussed more fully here. We are only concerned in making it clear that sexual determinants in the sense indicated must be assumed to exist, and that both kinds are contained in the primary germ-cells. The following considerations will render it apparent why we have assumed these determinants to be double, — *i.e.*, to consist of two groups of biophors of a common origin lying close to one another. Even apart from such low organisms as *Volvox*, it is well known that the male and female individuals only differ from one another as regards the kind of sexual cells they produce in many of the lower Metazoa — viz., the sponges and Hydromedusæ. In these forms the sexual determinants alone are double. In most other animals, however, the difference of sex is not confined to the germ-cells, but affects the soma itself to a greater or less extent. Hence in all *sexually dimorphic* organisms the germ-plasm must contain a varied number of double primary constituents of somatic characters, — namely, those which represent characters which differ in male and female individuals. These in the first instance concern the organs in which the sexual cells are developed, nourished, stored, and removed, — that is to say, the so-called sexual glands and their ducts; then follow the active and passive copulatory organs, and the structures connected with oviposition; and, finally, special sexual characters arise with regard to the organs for supporting and nourishing the offspring — such as mammary glands, teats, and uterus, — or they may refer to the instinct of carrying the eggs in the mouth, as in the male of a tropical species of frog, or to that in the female butterfly, which deposits its eggs in a definite manner on a certain plant. In the last two instances, the structure of the body and the nerve-centres must also be different in the two sexes, and the male and female type of these parts must exist in a latent condition in every germ-plasm. Under this head, moreover, '*secondary*' *sexual characters* must be included, such as the various tracking and alluring organs of the males, the gorgeous colours of male birds and butterflies, the scent-producing organs of the latter, which exhale perfume, and the song of male birds and insects.

In the human race we know that all the secondary sexual characters are transmitted by individuals of the opposite, as well as of the corresponding sex. A fine soprano voice, for instance, may be transmitted from mother to granddaughter through a son, and a black beard from the father to the grandson through a daughter. And in other animals, the sexual characters of both sides must be present in every sexually differentiated organism, some of them becoming manifest and others remaining latent. This fact can only be proved in certain cases, for we seldom notice the individual differences of these characters with sufficient accuracy; it can, however, be shown to be true, even in tolerably simply organised species, and we must therefore suppose that *latent characters belonging to the other sex* are always present *in each sexually differentiated organism*. In bees, the males developed from unfertilised eggs possess the secondary sexual characters of the grandfather; and in the water-fleas, in which several generations of females arise from one another, the last of these generations produces males with the secondary sexual characters of the species which must consequently have been present in a latent condition in an entire series of female generations.

The germ-plasm of the fertilised egg-cell must therefore contain the primary constituents of all the secondary male and female sexual characters, as well as those of the male and female germ-cells. We might, then, suppose that this would be accounted for by the assumption that the determinants of both kinds of characters are contained in the germ-plasm, and that the decision as to sex is not only determined with regard to the sexual determinants, — so that the germ-cells take on a male or a female character, — but also as regards the somatic determinants, in order that the secondary sexual character of the male or of the female may take the lead in the building up of the body.

This assumption is certainly indispensable, and it suffices as far as the latent transference of secondary characters of both kinds from the germ-plasm of one generation to that of the next is concerned, but it still requires an essential addition. A number of facts indicate that the latent primary constituents of secondary characters of both kinds are not only present in the germ-plasm from which the organism arises, *but also in the fully developed body of the organism.* The fact that the characters of both sexes can be transmitted to descendants, proves that the germ-plasm must contain the corresponding primary constituents of both; and in the sixth chapter it has already been shown how this transmission may be accounted for by a continuity of the germ-plasm from one generation of germ-cells to the next. We are now concerned with the fact that mature individuals may also contain the primary constituents of secondary sexual characters of both kinds, and that those of the female may be present in the body as well as in the germ-cells of a male organism, and conversely those of the male may exist in the female. The well-known facts to which I refer have been carefully collected and fully discussed by Darwin, and are briefly as follows: — the secondary sexual characters of one sex may, under special circumstances, become developed subsequently in fully-developed individuals. This results in both sexes, especially in the case of castration. The removal of the sexual 'glands' from young mammals and birds prevents the development of secondary male sexual characters. Castrated cocks, for example, retain the appearance of hens, and do not develop the beautiful tail or the large comb and spurs of the male bird, nor do they crow; and conversely, when hens become sterile from age, or if their ovaries become degenerated, they take on the external sexual characters of cocks. I possess a duck which no longer lays eggs, and has assumed the coloration of the drake. Men who have been castrated in their youth retain a high voice like that of the other sex, and the beard does not become developed.

These facts obviously compel us to assume that the capacity for the development of secondary female characters exists in a latent condition in the body of the male, and *vice versâ*, and that these are ready to undergo development under certain conditions. Darwin also arrived at the same conclusion. The only argument which might be advanced against its correctness is that a change of secondary sexual characters in any particular individual has only been observed in rare instances, and in very few species of the higher animals, such as birds and mammals. It might, therefore, be doubted whether it is possible to draw a general conclusion from such isolated observations. These instances nevertheless remain to be accounted for, and we must attempt to explain them in accordance with our theory.

Let us take a very simple example. In many butterflies of the family *Lycænidæ*, the upper surface of the wings is brown in the females, and blue in the males; and it seems very probable that brown was the original colour, as species of *Lycæna* exist at the present day in which both sexes are of this colour. The process which took place in the idioplasm in order that the change of colour may have occurred, must have consisted in the primary determinants of those cells which decide the colour of the wings — which we will speak of as 'brown' determinants — becoming transformed into 'blue' determinants in the germ-plasm, this change only occurring after they had become doubled and in such a manner that only one of the twin-determinants in each case remained brown, an arrangement also taking place which only allowed each to become active alternately. We thus arrive at the assumption of *double determinants*, just as in the case of the determination of sex in the germ-cells. I at first believed that it was indispensable to assume the existence of determinants with different halves, merely because the presence of inactive determinants in the corresponding part of the body could not be otherwise explained. As the adoption of the characters of the opposite sex cannot possibly be of any use to the species, natural selection can have taken no part in the addition of male determinants to the somatic cells of the female body, or *vice versâ*. Such a transference must therefore depend on an unintentional secondary effect of existing arrangements and forces. I soon, however, recognised that such arrangements actually exist, and that the assumption of mechanically unseparable double-determinants is not necessary in order to explain the presence of the *two* sexual determinants in the region where one of them undergoes development. I therefore attach no value to the idea of the material connection of the two dimorphic halves of the primary determinants in question. In fact I shall have occasion to show presently that these halves *must*, in any case, sooner or later become separated as independent determinants in the course of phyletic development.

The reason that such double determinants must, however, always remain close together, even after their separation, results simply from the mechanism for the ontogeny of the idioplasm, which we suppose to consist in the gradual disintegration of the mass of determinants of the germ-plasm into smaller groups. They divide according to a definite law into smaller and smaller groups in the course of the embryonic cell-divisions. None of them remain unused or undergo destruction, but each passes through a definitely prescribed course, and the determinants for any particular part or region of the body *must* necessarily remain together, even when they are not inseparably connected mechanically. In a *physiological* sense, therefore, they are still *double* determinants,—*i.e.*, each half controls the same region,—and in this sense I shall now use the term.

A transfer of the secondary sexual characters, such as occurs in birds, cannot take place in the *Lycænidæ*, because the wing-scales are never formed more than *once* in the course of life, and consequently we have no means of proving the presence of double determinants in the cells of the wing. Other observations on insects indicate, however, that their idioplasm is nevertheless capable of producing such a sexual transference.

This is especially shown in the case of the occasional hermaphrodite forms of insects, the most instructive instance—which has been very accurately investigated by Leuckart* and von

* R. Leuckart, 'Sitzungsberichte der deutschen Naturforscherversammlung,' 1864.

Siebold,* and more recently by Kraepelin — being furnished by bees. The male and female secondary sexual characters are combined in the most wonderful manner in these hermaphrodite forms: in some bees, the *right side was female and the left male*; in others, the anterior half of the body was male and the posterior female; while in others, again, the entire trunk was male and one side of the head female. As Leuckart remarks, 'the male and female characters' in these hermaphrodite bees 'have been intermingled in the most varied and unsystematic manner, so that it is difficult to discover two individuals with perfectly similar characteristics.'

We are indebted to Kraepelin for an excellent account of the external sexual parts in these hermaphrodite bees,† including the copulatory apparatus. His description shows that the blending of the male and female characters even affects very small parts. It not only often happens, for instance, that half the entire stinging apparatus of the left side is female, while an intromittent organ is developed on the right, but certain chitinous plates on the ventral side of the last abdominal segment, which is almost male in character, also display a distinct tendency to take on the form of the corresponding plates of the female stinging apparatus; in other words, *these chitinous plates are intermediate in form between those of the male and female*. Their formation must therefore have been controlled by a combination of 'male' and 'female' determinants. It would be incredible that these harmonising determinants could have met together at the right point in the extremely complex combination of determinants in the germ-plasm if they had not been arranged together from the first, and if an arrangement of male and female double determinants had not previously existed in every such region of the germ-plasm, so that they reached the corresponding part of the body *together* in the course of ontogeny, either the male or the female half then becoming active.

In the determination of sex in the normal development of bees, all these somatic double determinants must be correspondingly determined. We do not know to what factor the prevention of the determination of sex in a similar manner in the formation

* C. Th. von Siebold, 'Zeitschrift fur wissenchaftliche Zoologie,' Bd. xiv., 1864, p. 73.

† Kraepelin, 'Zeitschr. f. wiss. Zoologie,' Bd. xxiii., 1873, p. 326.

of the hermaphrodites is due. Siebold attempted to explain this difficulty by supposing that the eggs which produced hermaphrodites were imperfectly fertilised. He found that in the queen, after producing hermaphrodites, the receptaclum was almost empty; and as drones arise from unfertilised, and females from fertilised eggs, the view that imperfect fertilisation must produce hermaphrodites appeared to be a plausible one. It was not known at that time that a single spermatozoon suffices for fertilisation, and at the present day we are no longer justified in using such an expression as 'imperfect fertilisation.' Whenever a living spermatozoon enters an egg, the latter becomes fertilised; and an imperfect fertilisation could only be supposed to occur if the spermatozoon is abnormal—if, for instance, it contains too few idants. But even if such a case occurred, it would be of very little value theoretically, and we could only state that the determination of sex did not take place in all the double determinants at once, but only in certain larger or smaller groups,—in the case of the germ-cells as well as of the dimorphic parts of the body. Besides the ordinary case, in which the sexual gland of an individual was developed on the female type on the right side and the male type on the left, other instances occurred in which female and male germ-follicles were formed on the same side, seminal tubules and ovarioles being present close together. With reference to this fact, von Siebold remarks that 'the hermaphroditism of the sexual apparatus hardly ever corresponded to that observable in the external form;' and this seems to me to be of special theoretical interest, as it allows us to conclude with certainty that *the harmony of the normal condition is due to a simultaneous decision respecting the double determinants of the germ-cells and those of the body*, and not to a primary determination of sex in the sexual glands, from which the somatic male or female sexual characters would only be determined secondarily. The existence of double determinants in the germ-plasm can be actually proved in the case of bees. For if any egg can develop into a male or female according to whether it is fertilised or not, it *must* contain both kinds of determinants.

Although this assumption is undoubtedly a correct one, it alone is insufficient, because the secondary differences of sex do not always only concern single cells or groups of cells which correspond exactly in the two sexes, such as, for example, the brown and blue scales of the *Lycænidæ; and becau e in many, and perhaps in most cases, the dimorphic parts correspond only partially or not at all.*

The degree of sexual differentiation is very different in the various groups of the animal kingdom. In the lower and higher Crustacea the males commonly possess more 'olfactory setæ' on their antennæ than do the females. In the large water-flea, *Leptodora hyalina*, for instance, the anterior antennæ are represented in the female by short stumps provided with five olfactory setæ, while the male has long rod-like feelers, which bear about eighty setæ. This difference evidently cannot be referred to one double determinant. Each olfactory seta must be represented by a special determinant in the germ-plasm; even if the first five corresponded, and the differences between them could be attributed to the double determinants, more than seventy determinants for these setæ — which occur in the male sex only — would still be left, quite apart from those for the feeler itself. These seventy determinants are not double, because the corresponding parts are absent in the female, and consequently two *groups* of determinants must exist side by side in the germ-plasm, — those, namely, for the antennæ in the male and in the female, — only *one* of which becomes active in each case. We might imagine that the two groups pass, in close proximity to one another, through the cell-series of embryogeny up to the formation of the rudiments of the antennæ, and then become separated, the inactive group remaining in an 'undifferentiated' cell at the base of the feeler, while the other causes the development of the antennæ of the sex in question by continued cell-division.

The same must be true in numerous other and much more complicated cases. If a tail feather in a male humming-bird, for instance, is six times as long as the corresponding feather in the female, the colour of the former being of a brilliant ultramarine, and that of the latter greyish, we must assume that two groups of determinants are present, which differ in number and in nature. The two groups are situated close together in the germ-plasm, pass through the same cell-series in ontogeny, and ultimately reach the same part of the skin covering the last caudal vertebra. Here, however, one of them remains inactive, the other alone causing further cell-divisions and the consequent development of a feather to take place.

The greater the extent to which sexual dimorphism occurs, and the larger the parts which it affects, the larger the two groups of determinants must be, and the earlier in ontogeny will one of them remain inactive in a cell and cease to undergo further division, while the other gives rise to further cell-divisions.

Sexual dimorphism largely consists in *the arrest of development in an organ in one sex.* In many female butterflies, for instance, wings are wanting. This must be due to the group of determinants for the wings, which existed in a double condition, — *i.e.*, were male and female in the earlier phyletic stages, — becoming arrested as regards their male portion. Such females commonly possess rudiments of the wings, and in this case the two groups of wing-determinants must pass through ontogeny together up to the stage in the caterpillar in which the formation of the imaginal disc of the wing arises from a cell of the hypodermis. If the animal is a female, the arrested group, and if a male, the perfect group of wing-determinants will then become active. It is, however, also conceivable that the development of the group of determinants in the female might continue to be arrested until they disappeared completely, as in the case of female *Psychidæ*, in which the wings are altogether wanting.

But the highest degree of sexual dimorphism is not reached even when certain parts disappear completely. In various groups of the animal kingdom *species exist in which the males differ from the females in nearly all their characters.* In many Rotifers the males are very much smaller than the females, and exhibit an entirely different structure; the alimentary canal, moreover, is entirely wanting. In *Bonellia viridis*, a marine gephyrean worm, the male differs so much from the female that one might be tempted to class it with an entirely different group — the Turbellaria. The difference in the sizes also of the two sexes is still more marked in this instance, the length of the male being 1–2 mm., and that of the female 150 mm.; the former, moreover, lives as a parasite within the latter. In this case the eggs which give rise to males cannot be distinguished from those which develop into females, even in size; and the relatively enormous bulk of the female simply results from subsequent growth. Even the young male and female larvæ cannot be distinguished from one another, and their development only begins to differ when metamorphosis of the larva into the sexually mature animal takes place. The larva is of an elongated oval shape, is furnished with two bands of cilia by means of which it swims about, and possesses an alimentary canal with a mouth and anus. The transformation into the definitive form of the species begins with the loss of the posterior band of cilia,

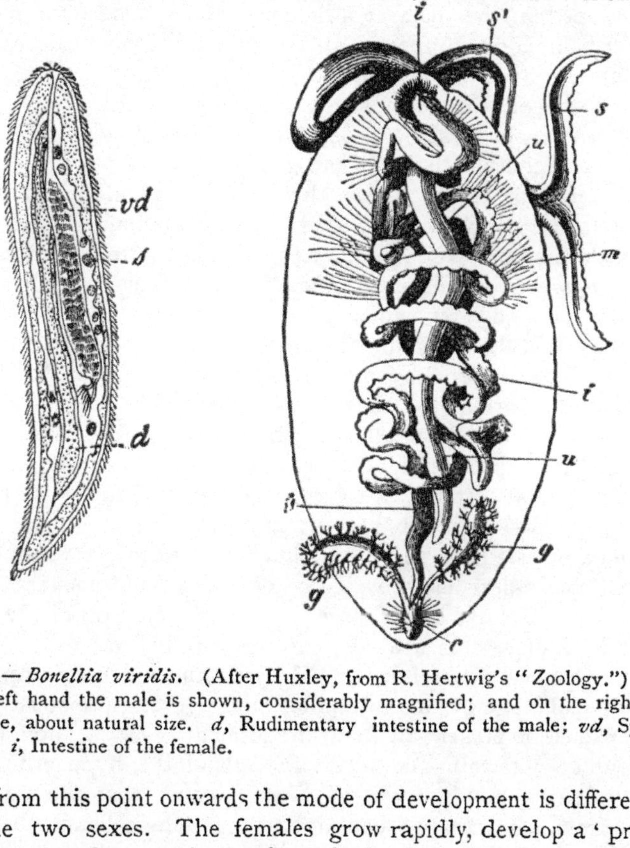

FIG. 24.—*Bonellia viridis.* (After Huxley, from R. Hertwig's "Zoology.") On the left hand the male is shown, considerably magnified; and on the right the female, about natural size. *d*, Rudimentary intestine of the male; *vd*, Sperm duct; *i*, Intestine of the female.

and from this point onwards the mode of development is different in the two sexes. The females grow rapidly, develop a 'proboscis' at the anterior end, and the intestine increases in length; the males, on the other hand, become entirely covered with cilia, and the mouth and anus, as well as the fore- and hind-gut disappear, the mid-gut, filled with yoke-granules, alone being retained. Notwithstanding their great diversity, the two sexes are formed on the same plan; the male, however, may be said in general to remain stationary at a certain stage of organisation, while the female continues to develop, and reaches a much higher stage of organisation, at any rate as regards many organs, such as the nervous and vascular systems, which are altogether wanting in the male. But neither is this the full extent of the difference between them, for the testes, as well as the skin and certain hooked organs of attachment are only developed in the male at a later stage in a peculiar manner. A certain correspondence still remains in the most essential points of structure of the body, in spite of the great difference between the adult sexual animals; so that, as Sprengel says, the male is also 'a Gephyrean possessing all the known structural characteristics of the group.'

In terms of the idioplasm, this course of development may be described somewhat as follows: the determinants which direct the development of the larvæ are single, and consequently *monomorphic* larvæ are produced. The idioplasm of all or most of the cells, however, which constitute the organs of these larvæ, contain double determinants or double groups of determinants, of which those for the female are, in most cases, far the larger: in fact, the group for the female will usually not be opposed by any for the male at all, — in the case, that is, of such organs as the long proboscis, of which there is no homologue in the male. It is certainly very remarkable that these groups of determinants, although present in the male and unopposed by others, do not become active; but, even although we do not in the least understand how it comes about that this inactivity is enforced, the case is not more surprising than that of the determination of sex as a whole, and the inactivity of existing individual primary constituents. Why do the wings first appear at the pupal stage of the caterpillar, and not long previously, since they must be present all along in rudiment — *i.e.*, in the form of a group of determinants in certain cells of the hypodermis? Or why does a boy not grow a beard, seeing that the necessary determinants must be contained in certain cells of the skin? We can no more account for all these cases than for the inactivity of sexually differentiated determinants. All that can be said is, that these determinants have the peculiarity of only becoming active at a certain ontogenetic stage; but this scarcely gives any further insight into the matter than does the statement that sexually differentiated determinants become active or remain inactive, according to the sex of the organism in question.

The ids of the germ-plasm must contain more determinants in sexually dimorphic than in monomorphic species, the number increasing in proportion to the increase of difference between the sexes. They must also increase in size; and the question therefore arises as to whether dimorphism may not perhaps increase to such an extent and finally involve all parts of the body, that *double ids* arise; that is to say, each id of the germ-plasm comes to consist of a male and a female half, in which all the determinants are different. This seems to be practically the case in some animals: in the *Rotifera*, for instance, the males commonly differ so much from the females that they, like the sexual form of *Phylloxera*, arise from special eggs, smaller than those which develop into females. But it would nevertheless be incorrect to suppose that each of these two kinds of eggs contained either male or female ids only. The number of common determinants present must certainly be small, but even here the germ-plasm of each egg must nevertheless contain all the male and female determinants. This is proved by the interpolation of generations consisting of females only in the cycle of generations which is passed through each year, the parthenogenetic females eventually producing males.

We must now return to the question of *sexual reversion*—if I may so call it—which has already been referred to; that is to say, the appearance of characters of the opposite sex after castration or degeneration of the sexual glands. Hitherto this has always been considered a universal phenomenon, but I do not think that such a conclusion is justified. As already mentioned, observations with regard to such cases of 'sexual reversion' have practically been confined to birds and mammals, and even in these do not always refer to all the parts which are sexually dimorphic in the species in question. Cases have certainly been observed in which, for instance, an old 'hen which had ceased laying, assumed the plumage, voice, spurs, and warlike disposition of the cock.'* This proves that in these birds all the secondary sexual characters of the male are present in a latent condition in the soma of the female We might, however, suppose that this form of reversion only takes place when the characters concerned are completely homologous in the two sexes; that is

* Darwin, 'Animals and Plants under Domestication,' Vol. II., p. 26.

to say, when they exactly correspond with regard to the time and place of their ontogenetic origin and the number of their determinants: but if this is not the case, such a 'reversion' to the characters of the other sex can hardly be possible, because the foundation is wanting from which the reverted organ could arise. Let us suppose that the same conditions as apply to fowls hold good in the case of butterflies:—that is to say, that the secondary sexual characters are present in a latent condition in the soma of the other sex, and undergo development on removal of the sexual glands. The male of *Lycæna alexis*, for example, which has blue wings, would then develop brown ones on being castrated. This would take place by the shedding of old scales and the growth of new ones; or if castration had been effected in the caterpillar stage, the scales in the growing wing would be brown from the first. The scales are homologous structures in the male and female, and each of them is controlled by a single determinant. If, therefore, a cell containing the determinant for a brown scale of the female is situated at the base of the already developed scale of the male, a reversion to the colour of the scales in the female might occur under such circumstances, which are of course purely imaginary.

The matter would, however, be entirely different if the female *Lycæna* had no wings at all, as is the case in females of some *Bombycidæ*. The character of the blue scales in the male would then have no homologue in the female, and an inactive cell, with the determinant of a brown scale, could not possibly be situated at the base of the blue scales in the male. This may be expressed in general terms as follows:—*double-determinants, possessing a definite male and female character, can only be present up to the phase and point in ontogeny in which the development of the two sexes is exactly homologous.* We can therefore only expect a reversion to the secondary characters of the other sex to occur when this point remains permanent. In *Lycæna* the divergence would occur at the formation of the wing-scales; in *Psyche* (the female of which is wingless), in a certain group of cells in the hypodermis of the thorax; in *Bonellia*, in all the cells of the larva; and in the Rotifera, in the egg itself. If our view is a correct one, a female *Bonellia* would consequently be incapable of developing male characters in consequence of castration, because it has long since passed that stage of ontogeny in which the divergence into a male or female occurs; and in the case of the Rotifera, it is not to be expected that any influence could cause a male to produce female characters.

In the part treating of sexual reproduction — including the section on the struggle of the paternal and maternal hereditary tendencies which takes place during the development of the offspring — no mention was made of a very general, and, in my opinion, erroneous conception of sex; and it will be as well to explain this omission before proceeding further.

The transference of sex has hitherto usually been looked upon as an act of transmission. This cannot be the case, inasmuch as every germ-plasm contains the primary constituents for both sexes, and the process of transmission itself has evidently nothing to do with the determination of sex. As already mentioned, it does not by any means follow that because a child is a female, its secondary or primary sexual characters will resemble those of its mother. This, indeed, has long been known, but has not led to a general recognition of the fact that sex is not transmitted at all; and that, on the contrary, the primary constituents of both sexes are always passed on from both sides: the decision as to which of them are to become active depends on secondary factors, which have not yet been clearly recognised in any case. The male halves of the sexual double-determinants of the mother are just as capable of undergoing development as are the female halves, and *vice versâ*: — the 'law of sexual transmission,'* which was propounded by Haeckel some time ago, is not tenable. Expressed in a purely empirical manner, the facts have been more correctly formulated by Déjerine † according to Darwin's (?) views, in his valuable work on the heredity of nervous complaints: — 'the prepotency of *one* of the parents in transmission may be direct, and follow the sex, or may cross over, and become manifest in the opposite sex.'

For this reason the so-called 'transmission of sex' was entirely left aside in the section on the struggle of the parental characters during the formation of the child; transmission of the primary and secondary sexual characters occurs, but sex itself cannot be transmitted.

---

* Ernst Haeckel, 'Generelle Morphologie der Organismen,' Bd. II Berlin, 1866, p. 183.

† J. Déjerine, 'L'Hérédité dans les maladies du système nerveux,' Paris, 1886, p. 17.

2. Pathological Dimorphism: Hæmophilia.

In connection with the attempt to trace sexual dimorphism to its origin in the idioplasm, I will now make a few remarks with regard to a certain disease—or rather structural anomaly—which affects the human race, and which, I believe, will be better understood from this point of view; the analysis of it may, moreover, possibly throw a new light upon the causes of sexual dimorphism.

This anomaly is known as *hæmophilia*, and although of rare occurrence, it has been very accurately observed in a number of cases, and is known to be transmitted in a marked degree and in a very peculiar manner. It only occurs in the *male* members of a family, but is transmitted by the female members, and in this respect *resembles a secondary sexual character*.

The disease, however, is apparently not connected with the sexual organs, or with those parts which differ in the two sexes. It consists in an abnormal flaccidity of the walls of the blood-vessels, in consequence of which slight injuries cause serious hæmorhage, which cannot be easily stopped. As the blood-vessels are developed from certain cells of the so-called mesoblast or 'parablast,' hæmophilia is described in many text-books on pathological anatomy as an anomaly of the 'parablast' or 'Bindegewebekeim.' It must certainly be assumed that this disastrous variation in the blood-vessels—or rather of the cells from which the blood-vessels are formed—is due to some variation in certain mesoblast cells which cannot be more definitely defined at present. The determinants for the cells of the blood-vessels must in this case have varied in some way or other in the germ-plasm of the individuals affected, *and the anomaly must originally have occurred in a male individual.* There is apparently no reason why a similar variation of the determinants should not take place in a female, and cases of women exhibiting the hæmorrhagic diathesis may perhaps still be ascertained. The exclusive transmission of this anomalous condition to the male sex seems to me, however, to indicate that the determinants for the blood-vessels differ in men and women, in spite of the apparent similarity of the vessels themselves:—*these determinants must be double.*

On this assumption, we can easily account for the otherwise mysterious phenomena of heredity observed in these cases. As this disease only appears in men, the pathological variation must affect the 'male' half of the determinants for the cells of the blood-vessels of the person affected, and we may compare it with the variation which occurs in the cells constituting the larynx, the determinants of which must certainly be regarded as being double, and as undergoing variation in the 'male' half. The decision as to which halves of the double-determinants in the idioplasm are to be active during embryogeny and which passive, takes place simultaneously with that as to the sex of the embryo, as is proved by the case of hermaphrodite bees. It is therefore self-evident that this disease must remain latent—*i.e.*, no diseased formation of the tissues whatever can be produced—in the case of every female descendant of a 'bleeder,' for in them the 'female' untransformed halves of the determinants for the cells of the blood-vessels become active. If, however, the offspring develops into a male, the pathologically transformed 'male' halves of these determinants become active, and the disease can develop, provided that a stronger hereditary influence is not exerted in the formation of the blood-vessels of the healthy maternal side, so that the tendency to disease, which has been derived from the father, is overcome by the healthy tendency inherited from the mother. This was the case in four generations of a family of 'bleeders,' observed by Chelius, Mutzenbecher, and Lossen,—* the sons were not affected. In another case described by Thulasius-Grandidier, on the other hand, the disease was transmitted from the father to the male members of three generations. We can understand both cases from our point of view, for in no instance is an individual variation due to the variation of the corresponding determinants in *all* the ids of the germ-plasm, but only in the *majority* of them. But this majority may become reduced to a minority at every 'reducing division' and every time amphimixis occurs, the variation thus ceasing to manifest itself. As soon therefore as only a small majority of the ids contain 'hæmorrhagic determinants,' a considerable number and hereditary force of the healthy maternal determinants for the blood-vessels would preponderate over the morbid paternal ones, and consequently the male descendants would not inherit the disease. If, however, a considerable majority of 'hæmorrhagic determinants' were present

* Klebs, 'Lehrbuch der pathologichen Anatomie.'

in the germ-plasm of the father, a favourable reducing division is necessary if the son is to remain free from the disease. We can, moreover, even account for those cases in which several female members of a hæmorrhagic family in which the fathers were healthy, bear sons all of whom suffer from the disease. For the 'male' halves of the double-determinants in almost all the ids might have undergone a pathological change in the germ-plasm of the mothers, without producing any apparent result in them; in the sons, on the other hand, this would lead to the development of the disease, unless an unusually favourable reducing division had counteracted the marked preponderance of the morbid determinants. Hæmophilia, which remains latent in the mother, is just as liable to be transmitted by the mother to her male descendants as is any other masculine characteristic of the grandfather, such as the colour of his beard, or the quality of voice.

It seems to me that we cannot overlook the indication afforded by this agreement between the mode of transmission of ordinary sexual characters and of hæmophilia that *all, or nearly all, the determinants in the human germ are double*, half being 'male' and half 'female,' so that a determinant for any particular part may cause the development of the male or female type of the corresponding character.

The facts which Prosper Lucas * brought forward, and illustrated by numerous examples, concerning the occasional transmission of new characters to *one* sex only, — even when they have nothing to do with secondary sexual characters in the strict sense, — may be understood by this assumption of a wide distribution of double-determinants in the germ. The modification affects only the 'male' or the 'female' halves of these determinants of the germ-plasm in such instances. This is true as regards the disease we have just considered, inasmuch as it must have made its first appearance at some time or other, — as well as in numerous instances of colour-blindness, of the possession of supernumerary fingers, of the absence of certain fingers or of segments of fingers, and so on. Even the peculiar nature of the epidermis in the well-known case of Lambert, the 'porcupine-man,' was only transmitted to the male descendants.

* 'Traité philosophique et physiologique de l'hérédité naturelle,' Tom. II., Paris, 1850, p. 137.

In certain instances, polydactylism is known to be transmitted to the male members of a family only, while in others it passes from the mother to the daughters exclusively.

It appears, however, that such modifications of the one half in the double-determinants may in the course of time be transferred to the other half, even though, in the first instance, this only occurs to a slight extent; for cases are known in which an abnormality first arose (?) in the male sex, and afterwards passed over to certain individual female descendants. These cases have certainly not been followed out with sufficient accuracy to enable us definitely to deny that a modification of both halves of the double-determinants in question might possibly have taken place from the first, and that it only affected those in the one half (the homologous determinants) in a smaller number of ids.

Numerous instances in which an abnormality appears, sometimes in the male, and sometimes in the female members of a family, prove that both halves may become modified at the same time. Prosper Lucas mentions several instances of this kind, such as that of the family Ruhe: — in the first generation observed, the mother transmitted her polydactylism to the daughter, and in the second this peculiarity was passed on from the mother to the son, while in the third it was transmitted from the father to the son. Numerous facts in zoology indicate the correctness of the assumption that modifications in one half of a double-determinant exert an influence on the other half, so as to result in a similar transformation. It is well known that many secondary sexual characters of the male in birds and insects appear in a slighter degree in the female. Amongst the *Lycænidæ* — which are called 'blues,' on account of the preponderance of this colour in the members of the family — some species exist in which both sexes are brown, while in most the male is blue and the female brown, and in a small number of southern species, again, both sexes are blue. There can be no doubt that brown was the original colour of these species, and that the blue tint first appeared in the males, while the females remained brown; and that, finally, in certain species, the females also became blue, although not so markedly so as the corresponding males. The males therefore preceded the females as regards the change of colour; and if, with Darwin, we attribute the impulse to this change to sexual selection, it follows that the blue colour of the females must have been introduced mechanically, owing to its previous existence in the males, and that this secondary sexual character affected the other sex in the course of a great number of generations. According to our theory, this must have been due to the modification of the 'male' halves of the determinants having gradually caused a similar, if less marked, modification of the 'female' halves. We can thus also to some extent understand how it has been possible for certain females to precede the others as regards this modification, as the influence which produced it would not take effect to the same extent and at a similar rate in all individuals. In many species of *Lycænidæ* in which the females are brown, individuals of this sex occur more or less frequently which are clouded with blue, or even exhibit this colour in a marked degree.

3. POLYMORPHISM

Sexual trimorphism is of frequent occurrence in the animal kingdom. I will here refer to certain instances amongst butterflies, which were first discovered by Alfred Russel Wallace, and will begin with a case in which apparently the first step towards polymorphism has been taken.

The male of a common North American butterfly, *Papilio turnus*, resembles the ordinary 'swallow-tail,' having yellow wings ornamented with black transverse stripes; while the females sometimes resemble the males, and are sometimes quite black, and may thus differ markedly from one another. The yellow females occur in the eastern and northern parts of the United States, and the black ones in the west and south; we must therefore suppose that two local varieties of this butterfly exist, in the northern of which the two sexes have a similar coloration, while the southern form is dimorphic. This indicates, in terms of the idioplasm, that the determinants for the wing-scales are single in the northern, and double in the southern variety. Describing these determinants according to the colour which they produce, we may say that the last-named variety possesses double-determinants, the 'male' half of which is 'yellow' and the 'female' half 'black'; while the single-determinants of the northern form are 'yellow.' This species is properly speaking, therefore, not trimorphic, but includes two local varieties, one of which is dimorphic. If the two varieties interbreed,—as is actually the case at the junction of the districts in which the two forms are respectively distributed,—the double-determinants of the southern form will meet with single-determinants of the northern form in the germ-plasm of the offspring. The male descendants of such a cross would remain unmodified, but the females would be either black or yellow according to the power of transmission of the 'female' halves of the determinants, or—as was observed by Edwards *—a combination of both these colours. Such combinations might either arise from the cross between a yellow female and a yellow male of the dimorphic variety, or from that between a black female and a yellow male of the monomorphic form; for in the dimorphic variety the germ-plasm contains double-determinants in the males as well as in the females. If we suppose that these crosses occur frequently, the number of females of an intermediate form in the borderland of the two districts would gradually increase, and might ultimately result in the production of a constant intermediate female form. But if the males exhibited a preference for the females corresponding to them, the female forms would remain essentially distinct from one another. This seems to be the case in *Papilio turnus*, at least Edwards states that the intermediate forms are rare.

The dimorphic and polymorphic females in many butterflies may perhaps be looked upon as belonging to *sexually dimorphic local forms which have subsequently spread and occasionally crossed.* In places where the varieties merely exist side by side without interbreeding, each of them also contains either single- or double-determinants, according to whether it is sexually monomorphic or dimorphic; but when interbreeding occurs, the determinants of the two races come together, and then several homologous double-determinants may even meet in the same germ-plasm,—some in certain ids and some in others.

In *Papilio turnus* the case is not quite so simple as I have stated it: as a matter of fact, this species exhibits a *double dimorphism*, for the yellow females do not exactly resemble the males, but differ considerably from them as regards the shade of yellow and the pattern on the wings:—thus they possess an orange-coloured eye-spot on the posterior wings, which is absent in the male. We must therefore assume that double-deter-

* W. H. Edwards 'Butterflies of North America'

minants are present in the yellow variety also. If we imagine that the two sexes were identical as regards the wing-marking in the immediate ancestors of *Papilio turnus*, as they are in the closely allied European species *Papilio machaon*, and that this monomorphic ancestral form had persisted — in California, let us say, — we should have an instance of that kind of polymorphism which Wallace has described in the case of *Papilio memnon*, in which there are one male and three female forms. In this case we must suppose that the first and oldest form possessed *single*-determinants for the wings; while in the second and third forms the determinants were double, and their 'male' halves retained their original nature, the female halves becoming modified in two different directions.

It is therefore not necessary, as might have been supposed, to assume the existence of triple determinants from the fact of the trimorphism of a species alone, or of quadruple or quintuple ones in the case of polymorphism.

The *polymorphism of animal and plant stocks* rests on a different basis, as it concerns the physiologically dissimilar members of a higher stage of individuality — that of the stock. This kind of polymorphism has already been treated of as a phenomenon of development in connection with alternation of generations. We must, however, take the closely allied *polymorphism of animal communities* into consideration.

The differences between the male and female individuals in bees has already been referred to the existence of double-determinants. But a third form of individual, the *worker*, occurs in the honey-bees. These workers differ from the females in the slight development of the ovaries, the ovarioles not only being fewer in number than in the 'queen,' but even frequently containing no eggs at all, and at most only very few. The receptaculum seminis is also more or less reduced, and the abdomen is much shorter and thinner than that of the queen bee. If these were the only differences between the two forms, there would scarcely be any need to assume the existence of special determinants for these parts in the germ-plasm of the workers: we might imagine that the determinants of the ovarioles, for example, were so constituted as to become active in consequence of abundant nourishment, and to cause the development of ovarioles; while a smaller supply of nutriment would not always be sufficient for this purpose, and thus the complete formation of the sexual organs would be prevented. We know, indeed, that a fertilised egg may develop into a queen or a worker, according to whether the larva arising from it is fed on royal diet or with the less nutritious food supplied to the workers.

This explanation, however, even if correct as regards the degenerated parts of the workers, does not sufficiently account for the other differences between the two kinds of females. For the workers are not inferior to the queen bee in all respects: on the contrary, the worker's sting is straighter, longer, and stronger, and is provided with more teeth, than the queen's; the wings, moreover, are longer, the tarsal segment of the hind-limb is provided with the well-known brush, and the tibia has a depression known as the pocket, for carrying the masses of pollen which the insect collects. These two characteristic parts are wanting in the queen. Important differences must also exist as regards the minute structure of the brain, for the instincts of the queen are very different from those of the workers. After fertilisation has taken place, the queen lays eggs, but she neither gathers honey from flowers, excretes wax, nor makes the honeycomb. It is therefore incredible that the queen and workers should be formed by the agency of similar determinants. The germ-plasm must contain double-determinants for certain parts of the body of the queen and workers respectively. But as we have already assumed the existence of double-determinants for the formation of male and female bees, or at any rate for the development of those parts which differ in the two sexes, we can only make the further assumption that *the 'female' halves of the double-determinants may themselves consist of two halves*, corresponding to the queen and worker respectively, and that each of these halves must naturally be looked upon as a complete determinant as regards size and structure. It is of no consequence whether they are regarded as being closely connected together, or as independent structures in close proximity with one another: in either case they must have arisen by the doubling or tripling of a single ancestral determinant. The terms '*double-determinant*' and '*half-determinant*' are simply used for the sake of simplicity. Their relation to one another is similar to that existing between the homologous but heterodynamous determinants of different ids.

In the case of bees, the factor that determines which of the two halves of the 'female' determinants is to become active, seems to be the quality of the food supplied to the larva, so that the critical moment only arrives long after the termination of embryogeny, and before the chrysalis stage is reached. It is well known that when the queen is lost, another one is produced by feeding a larval worker with royal diet. Thus the sex depends on the occurrence or omission of fertilisation, but the modification into a queen or worker takes place much later, when the animal has reached the larval stage. The idea of the trimorphism of certain determinants thus becomes much less difficult to realise. We must look upon them as double-determinants contained in the ids of the germ-plasm, the female hemisphere of which is again composed of two dissimilar quarter-spheres. If the egg becomes fertilised the male half becomes inactive, and we have already represented this figuratively as taking place by the 'female' hemisphere extending over the 'male' hemisphere, and enveloping it like a mantle. This 'female' 'determinant mantle' consists of two halves, representing the queen and worker respectively, and we may suppose that the subsequent determination during the larval stage as to which half is to control the cell, takes place in such a manner that the 'worker' half extends over the other when the nourishment is poor, while with more abundant food the 'queen' half grows more rapidly, and prevents the 'worker' half from exerting any influence on the cell. I naturally do not in the least suppose that this figurative representation of the process represents the actual facts of the case, but it at any rate shows that the existence of trimorphic determinants — or more accurately, of double-determinants each possessing a dimorphic half — is conceivable.

We might, however, also assume the existence of three independent determinants side by side, so arranged that they become active under other definite influences; and this conception would better agree with the unavoidable assumption that the three determinants which act vicariously are of a similar size.

The differentiation of the determinants into several equivalent parts, each of which prevents the others from becoming active, may take place to a still further degree *by the 'male' half of the double-determinants becoming differentiated into two dissimilar halves*. The Termites, in addition to the workers or stunted females, possess 'soldiers' or males, in which the sexual organs are stunted, which possess very strong mandibles, and differ in other important structural details from the ordinary males. In this case, therefore, four determinants must be present, each capable of being substituted for another, and only one of which can be active at a time.

Apart from local dimorphism, a temporary dimorphism occasionally occurs, and is especially well known amongst butterflies as *seasonal dimorphism*. In this case the individuals of the same generation, hatched at the same time of year, are alike, but the summer and spring generations differ from one another.

In the European species *Vanessa levana-prorsa*, the individuals of the spring generation are characterised by a yellow and black pattern on the upper side of the wings; while the summer form (*prorsa*) has black wings, with a broad white transverse band, and delicate yellow lines running parallel to the margins. Were we to superpose these two patterns, it would be seen that the black parts in *prorsa* do not correspond to the yellow ones in *levana*, and that the white band in the former does not correspond to a yellow or black part in the latter. This band is, on the contrary, entirely wanting in *levana*, and is represented by both black and yellow regions.

These cases of dimorphism can also, it seems to me, only be accounted for in terms of the idioplasm by the assumption of double-determinants, which, however, are concerned in this case merely with the wing scales, and essentially with those on the upper side of the wings only; for the lower surface, though not precisely similar, differs far less in the two forms than does the upper side. We will speak of the halves of these double-determinants as 'winter' and 'summer' determinants, and may suppose that the influences of temperature which affect them at the beginning of the pupal stage determine which of the two halves is to predominate over the other. Nearly twenty years ago I showed that it is possible to compel the pupæ of the summer-generation to assume the winter form by exposing them to a low temperature, so that the butterfly emerged as a *levana* instead of a *prorsa*. The converse experiment was also occasionally successful, the pupæ of the winter-generation being forced to assume the summer form by the influence of a higher temperature during, or shortly after, pupation. We may perhaps therefore suppose that an increase of temperature prevents the 'winter' halves of the double-determinants in question from developing, while it is beneficial to the 'summer' halves; and that conversely, the development of the 'summer' halves remains stationary when the temperature is lowered, while the 'winter' halves continue to develop. This may be illustrated, as in the case of ordinary sexual dimorphism, by supposing a determinant of the germ-plasm to be spherical, and to consist of a 'summer' and a 'winter' half. This determinant would remain unchanged throughout embryogeny, and even during the entire caterpillar stage; and the increased or diminished temperature would only determine which half should outgrow the other and prevent it from controlling the cell, at the beginning of pupation, when the wings are formed.

In *Vanessa levana* the males and females resemble each other so closely in the pattern of the wings that they cannot be distinguished from one another with any degree of certainty; but in many other seasonally dimorphic butterflies, sexual dimorphism is exhibited in the pattern and coloration of the wings, and we must therefore assume that they possess double-determinants consisting of 'male' and 'female' halves, each of which is again subdivided into a 'summer' and a 'winter' half. We do not know what factors determine the sex of butterflies, but in many cases the determination is effected early, for ovaries and spermaries can be distinguished from one another in the full-grown caterpillar. Thus, as in the case of Termites, the decision concerning the subdivisions of the double-determinants takes place subsequently to that with regard to the primary halves.

4. Dichogeny in Plants

De Vries has made use of the term 'dichogeny' to describe that form of dimorphism which becomes manifest when a young vegetable tissue, under normal conditions, is capable of developing in different ways according to the external influences to which it is exposed. Shoots of ivy bear leaves on the side which is exposed to the light, and roots on the opposite side; but if the plant is rotated, the same shoot will grow leaves on the side which previously bore roots, and *vice versa*. The stimulus due to light therefore apparently causes a group of cells, which would have formed roots if they had been in the shade, to give rise to leaves.

The assumption, which is in accordance with de Vries's ideas, that all the hereditary tendencies of the species are contained in every cell of the ivy-shoot, and that those which concern the leaves only undergo development in response to the stimulus caused by the light, and those corresponding to roots only to that produced by the shade, does not materially help us in the solution of the problem. In point of fact, the *same* cells are not capable of forming roots and leaves: the leaves are much less numerous than the short and closely aggregated roots, and therefore a large number of cells, or groups of cells, which give rise to roots when shaded, do not develop leaves when exposed to the light; they consequently contain no 'leaf-determinants.' Hence the same idioplasm cannot produce roots when shaded from, and leaves when exposed to, the light; but the determinants for the roots or leaves respectively must be distributed very differently in the cells.

The predisposition to form either of these two structures is obviously determined in the growing point or apex of the shoot. The cells which are continually being produced by the apical cells are destined at a very early stage to form the rudiments of roots or leaves; only a certain number of cells on the illuminated side are provided with leaf-determinants, while much more numerous cells on the shaded side are furnished with root-determinants. The determination therefore takes place at a very early stage, when the shoot is actually in an embryonic condition; and the degree of illumination determines which side is to be provided with leaf- and which with root-determinants, — each kind being distributed in accordance with a different law, — as well as the side to which the groups of each kind of determinants are to pass during the nuclear divisions in the cells arising from the apical cells. This case is analogous to that of the inversion of the viscera in man, except that we do not know the cause of this change of position. Some influence, however, must in this case also be exerted during the early embryonic stages, and cause the liver to take up a position on the left side, and the spleen and heart on the right, long before these parts are actually formed. No subsequent influences could cause the liver to become shifted from the right to the left side, or could transform the liver into the spleen; just as in the ivy-shoot no influences can lead to the formation of leaves on the shaded side when it is once covered with roots.

It therefore seems to me that the *same* cells cannot give rise to two kinds of structures in this case, but that some groups form roots, and others leaves; and that, moreover, the idioplasms of these groups of cells differ from, and cannot be transformed into, one another, but that the sides of the shoot which produce roots and leaves respectively, are determined by the influence of the light as soon as the first embryonic rudiments of the shoot is formed.

The fact that rudiments of roots can actually be proved by the aid of the microscope to exist in the tissue of the stem in some plants, seems to me substantially to support this view. This is the case in the willow, in which the power of producing roots from cuttings exists in a high degree. But numerous other plants are also capable of multiplying by means of cuttings, and we may suppose that in them, in spite of the absence of visible root-germs, the determinants for the tissue of the roots are present, and are ready to develop into germs whenever external influences leading to the formation of roots come into play.

All cases of dichogeny, however, cannot be explained in this manner. In some instances the *same* cells may actually develop in different ways, and the idioplasm may therefore become transformed by external influences. A transformation as regards the dorso-ventral arrangement of the parts in a young shoot of *Thuja* takes place when it is turned round so that the relative positions of the upper and under surfaces are reversed.* The cells which, under ordinary conditions, take on the form of palisade cells, then assume the structure of the cells on the lower surface, and *vice versa*.

It appears to me that this fact is to be explained by supposing that the determinants of these two forms of cells are both present in each cell, but that only one or the other of them becomes active, according to the degree of illumination. I can, however, form no idea as to why such an arrangement is met with in this case.

---

* Compare the statements made on this subject by Detmer, Biolog. Centralblatt, Vol. VII., No. 23.

CHAPTER XII

DOUBTFUL PHENOMENA OF HEREDITY

1. 'Xenia' and Telegony

Although it is certainly unnecessary in a theory of heredity to discuss all the possible kinds of phenomena which are with doubtful justice included under this head, I should not like to pass over in silence the consideration of certain presumptive observations, as they have so often been discussed, and were considered worthy of notice by so eminent an authority as Darwin. These refer in the first place to the so-called '*xenia*,' and to the phenomenon generally known as '*infection of the germ*,'—which, in case it really exists, I should prefer to speak of as *telegony*.*

Focke has used the term 'xenia' to describe those cases in which 'hereditary characters are supposed to have been transmitted by the pollen to the tissues of the fruit as well as to the fertilised egg-cell and the embryo arising from it.'

Darwin mentioned many instances of this kind, and attempted to account for them by supposing that an emigration of 'gemmules' takes place from the sperm-cells (pollen-tubes) to the surrounding tissue of the fruit. Focke has collected all the known cases, and on reading them, one receives the impression that they may very likely be deceptive. Blue grains occasionally occur amongst the yellow ones in cobs of the yellow-grained maize (*Zea*) after fertilisation with the pollen of a blue-grained species. It is possible that previous crossings of the two species may have produced this result, which might wrongly be ascribed to the immediate influence of the pollen of another species on the fruit. J. Anderson Henry even thought he had observed that all the flowers in an inflorescence of a white *Calceolaria* were reddened by the influence of the pollen from a red kind on a *single* flower of this inflorescence!

---

* From τῆλε—at a distance, and γόνος—offspring.

As such eminent botanists as Focke,* and more recently de Vries,† have expressed much doubt with regard to these observations — or rather interpretations, — we must wait until these cases have been critically reinvestigated before attempting to account for them theoretically. The chief difficulty we should meet with in any such explanation would be due to the fact that we are here concerned with the influence of the *germ-plasm* of the sperm-cell on a tissue of another plant which only constitutes *a part* of this plant. It would thus be necessary to assume that all the determinants of this germ-plasm are not active, and that only those take effect which determine the nature of the fruit.

The uncertainty of the observations is still greater in the instances of so-called *infection of the germ*. If the case recorded by Darwin — but not observed by him personally — is reliable, and has been accurately described, all doubts must be set aside. A mare belonging to Lord Morton 'bore a hybrid to a quagga,' and subsequently 'produced two colts by a black Arabian horse; these colts were partially dun-coloured, and were striped on the legs more plainly than the real hybrid, or even than the quagga. One of the two colts had its neck and some other parts of the body plainly marked with stripes,' and the hair of the mane is said to have resembled the short, stiff and upright mane of the quagga, instead of that of the horse.‡ Similar cases of the influence of a previous fertilisation on the structure of subsequent offspring are related of several domestic animals, — viz., of cows, sheep, pigs, dogs, and pigeons, as well as of human beings when crosses occur between white races and negroes.

Up to the present time no experiments have been made with this special object, and it would be necessary to use every conceivable precaution in conducting such experiments, or they would be valueless; they could, therefore, be best made in zoological gardens, not only because of the undoubtedly pure material which might be used for the purpose, but also because

---

* Focke, 'Die Pflanzen-Mischlinge,' Berlin, 1881, p. 510, *et seq.*

† Hugo de Vries, 'Intracellulare Pangenesis,' Jena, 1889, p. 206.

‡ I have quoted this case from Darwin's 'Variation of Animals and Plants under Domestication,' 2nd ed., Vol. I., p. 435. Darwin does not seem to have known of the drawings of these colts which will be mentioned in a subsequent paragraph. I have not seen them, and only learnt of their existence from Settegast's book.

it would be possible to isolate the animals, and for the keepers to exert a strict control over them for considerable periods of time.

The philosopher Carneri mentions a case which came under his own notice. He kept a herd of cattle of the dark grey Mürzthal breed. On one occasion he put one of the cows to a 'light-coloured Pinzgau' bull instead of to one of the same breed. The cow threw a calf with the characteristic brown and white patches of the Pinzgau breed, as well as with distinct traces of the 'dark grey Mürzthal cross.' The cow was subsequently covered by a Mürzthal bull, and, contrary to expectation, the second calf was also a 'hybrid,' being for the most part grey, 'but possessing *large* brown spots like those of the Pinzgau breed.'

Both the above mentioned cases are not so conclusive as they appear to be at first sight. A drawing by Agasse of the foal possessing the characters of the quagga is to be seen at the Royal College of Surgeons in London, and shows indistinct dark stripes on the neck, withers, and legs. Similar stripes are, however, not very uncommon on purely bred foals, and ordinarily disappear as the animal grows older. No further resemblance to the quagga can, however, be detected in these pictures.*

I must not omit to mention that before having heard of the hypothesis of 'infection,' Carneri accounted for the case of the two breeds of cattle described above by supposing that 'a drop of Pinzgau blood' must have previously got into the Mürzthal herd without his being aware of it.

Thus even the best of these 'cases' are not reliable and actually convincing. We may, however, at any rate suppose that this so-called 'infection,' if not altogether deceptive, only occurs in rare instances, and by no means regularly, or at most only in some cases. Experienced breeders, like Settegast and Kühn of Halle, do not believe in it; for although they have frequently crossed various domestic animals, they have never observed an instance of it. Such cases could only be accounted for from our point of view by supposing that spermatozoa had reached the ovary after the first sexual union had occurred, and had penetrated into certain ova which were still immature. The

* According to Settegast ('Thierzucht,' Breslau, Bd. I., 1878, pp. 223–234).

immediate fertilisation of the latter is rendered inconceivable by the fact of their immaturity, and the sperm-cell must have remained in the body of the ovum until the maturation of the latter, with the nucleus of which it then united in the process of amphimixis. If this occurred sometime after the first of the offspring was born, it might easily have coincided approximately with the second *coitus*, from which the fertilisation would then apparently be due. If the 'infection' were proved beyond a doubt, a supplementary fertilisation of an egg-cell in this manner must be considered possible; we certainly might then reasonably ask why mares, cows, or sheep, should not occasionally become pregnant without being covered a second time. *But this has never yet been known to occur*, and I incline to Settegast's view that *there is no such thing* as an 'infection' of this kind, and that all the instances which have been recorded and discussed critically by him are based upon a misconception.

2. The Influence of Temporary Abnormal Conditions of the Parents on the Child

Although I do not consider that the cases which come under the above heading have anything to do with heredity, I should not like to leave them entirely on one side.

It has often been supposed that drunkenness of the parents at the time of conception may have harmful effect on the nature of the offspring. The child is said to be born in a weak bodily and mental condition, and inclined to idiocy, or even to madness, &c., although the parents may be quite normal both physically and mentally.

Cases certainly exist in which drunken parents have given rise to a completely normal child, although this is not a convincing proof against the above-named view; and in spite of the fact that most, or perhaps even all, the statements with regard to the injurious effects on the offspring will not bear a very close criticism, I am unwilling to entirely deny the *possibility* that a harmful influence may be exerted in such cases. These, however, have nothing to do with heredity, but are concerned with an *affection of the germ by means of an external influence*.

The experiments of the brothers Hertwig show that the development of the fertilised egg in lower animals may be considerably retarded by the action of various chemical substances, such as chloral, quinine, and morphia; and we also know that the ova of sea-urchins, if kept too long in the sea-water before being fertilised, tend to lose their vital energy, and consequently many spermatozoa, instead of a single one, are likely to enter each of them. A similar result may follow from the effects of the above-mentioned chemical reagents and in both cases an abnormal development of the egg, such as a duplication of parts, may be the consequence.

It does not appear to me impossible that an intermixture of alcohol with the blood of the parents may produce similar effects on the ovum and sperm-cell. According to the relative quantity of alcohol, either an exciting or a depressing influence might be exerted, either of which would lead to abnormal development. A depressing influence exerted on *both* germ-cells would certainly retard, or even quite prevent, the process of fertilisation; while if the egg-cell were *alone* affected, super-fertilisation (polyspermy) might result; and the same might occur by an excitation of the sperm-cells alone. The entrance of several spermatozoa into the small human ovum, which contains only a small amount of yolk, might produce an abnormal development just as much as in the case of the eggs of the star-fish or sea-urchin. A high degree of excitation in both germ-cells might, on the other hand, cause the complicated processes of the increase of the germ-plasm in the ovum and the subsequent conjugation of the two germ-plasms to take place in an inexact manner, owing to their being passed through too quickly, and would then produce an irregular development.

New predispositions can certainly never arise owing to such deviations from the normal course of development, and therefore a modification of the process of heredity itself is out of the question. It is, however, conceivable that more or less considerable abnormalities may affect the course of development, and either cause the death of the embryo, or else produce more or less marked deformities. The question as to whether such deformities really result in consequence of the drunken condition of the parents can only be decided by observation.

3. The Supposed Transmission of Diseases

There is no doubt that some diseases are passed on from one generation to another. All such cases are not, however, connected with heredity, and many of them are in all probability to be explained as the result of infection of the parental germ-cell with microscopic parasites, and ought consequently to be described as *infections of the germ*.

In man such a transference of disease has only definitely been proved to occur in the case of syphilis.* The father, as well as the mother, is capable of transmitting this disease to the embryo, and the only possible explanation of this fact is, therefore, that the specific bacteria of syphilis can be transmitted by the spermatozoon. Amongst the lower animals the 'pebrine' of the silkworm is an example, which has been well known for several decades, of the transference of a fatal disease from one generation to another through the egg: the germs of the fungus which produces the disease penetrate into the yolk. It is not known why these germs do not develop and multiply within the egg, and thus destroy it, but this is, however, the case. The fungi only begin to multiply in the young caterpillar † when it is half- or full-grown, or the disease may, again, only be fatal in the butterfly stage.

As we now know that many diseases of man and other mammals are due to such low forms of parasites, it is natural to suppose that the transmission of such diseases results from infection of the germ-cell with microbes, and not from inheritance in the true sense of the word—that is, from the transmission of an anomalous state of the germ-plasm itself.

I have elsewhere attempted to trace the 'heredity' of 'epilepsy,' produced artificially in guinea-pigs, by supposing that in this case a similiar process occurs. The slow development of this form of 'epilepsy,' resulting from an injury to the spinal cord or one of the larger nerves, seems to me, indeed, to support the conclusion that its symptoms, which resemble those of true epilepsy, are due to the migration of microbes, which advance from the injured part along the nerves in a centripetal direction until they reach the brain, where they set up the state of irritation characteristic of the disease. The great inconstancy of the symptoms, and the variety of forms of nervous diseases which the offspring exhibit, also indicate that a true heredity is

---

* *Cf*, *e.g.*, Dohrn, 'Zur Frage der hereditären Infection,' Deutsche med. Wochenschrift, Sept. 15, 1892.

† *Cf.*, F. Haberlandt, 'Der Seidenspinner des Maulbeerbaums u. seine Krankheiten Wien, 1871.

not concerned in the process, and that the transmission is in this case due to infection of the germ with the microbes by which the disease is induced.*

The 'transmission' of carcinoma might be accounted for in a similar way, — if, as has recently been supposed, this disease is really due to microbes.

It is, however, also conceivable that both causes — the transmission of abnormal predispositions, and infection of the germ — might combine to bring about the transference of a disease from one generation to another. Without desiring to encroach upon the domain of pathology, I am inclined to suppose that this is the case as regards 'hereditary' tuberculosis: there is no doubt about the occurrence of a 'tuberculous habit,' — that is, a certain complication of structural peculiarities which is commonly connected with the disease, such as a narrowness of the chest, for instance. These peculiarities must result from the structure of the germ-plasm, in which a definite variation of certain determinants and groups of determinants must have taken place, and they are therefore certainly transmissible. The disease itself, however, is not due to this 'habit,' but is caused by the presence of specific parasites, the tubercle-bacilli, which have a harmful effect upon the various living tissues. They may be introduced artificially into the blood, and then produce the disease even in perfectly normal individuals. They may, moreover, enter the body 'spontaneously,' *e.g.*, by some natural means, and will then also give rise to the disease. But in the latter case the probability of infection seems largely to depend upon the susceptibility or power of resistance of the individual, and at the present day pathologists are of opinion that persons exhibiting the 'tuberculous habit' already referred to have a much slighter power of resistance to the parasites which have passed into the body than strongly-built people. The inheritance of the disease would accordingly depend on the transmission of a constitution very liable to infection.

Without wishing to deny the existence of such a predisposition to infection, I do not believe that the transmission of tuberculosis is due merely to the inheritance of a greater degree

---

* A more detailed account and proof of this view concerning the infectious nature of traumatic epilepsy in guinea-pigs is contained in my essay on 'The Significance of Sexual Reproduction,' Appendix IV.

of susceptibility. A large number of facts seem to me, on the contrary, to support the view that *infection of the germ plays the chief part* in the process. It would be out of place to enter into particulars and attempt to prove this view here—the question belongs to the province of the pathologist: I merely wished to point out in this connection that a combination of hereditary transmission and infection of the germ is perfectly conceivable. The *phyletic origin* of such constitutional diseases is presumably to be explained as being due to the occurrence of certain individuals possessing constitutions which were abnormally susceptible to a certain kind of microbe. Such persons would be more readily attacked from without by this particular disease. If, however, it once attacked them, and were it of such a kind as to cause death only after some time, a further and much surer opportunity was offered to the microbes for transferring themselves to other hosts than was previously the case when they passed into the body from without:—they settled in the germ-cells of the individual affected, and were thus transferred to the descendants of this individual. Although the presence of parasites in the germ-cells has not yet actually been proved in the case of tuberculosis, in my opinion it by no means follows that such an infection does not nevertheless take place: we do not even know whether such microbes are of the ordinary form and size. In any case they must possess different vital qualities; for did they multiply in the egg- or sperm-cell in the same manner as in the tissues in which they are known to occur, the germ-cells would soon be destroyed. Numerous adaptations to the host may have occurred in this case as in that of other parasites; and, moreover, *latent periods of development* may have arisen during which the parasite does not undergo multiplication. It seems improbable that such arrangements should not be met with, and that the parasite should not make use of the favorable opportunity of becoming distributed with the greatest certainty. Latent periods very commonly occur in the germs of animals and plants whenever they are useful, and hence this arrangement must come about without any great difficulty.

Even although our most eminent pathologists, such as Ernst Ziegler, are now of opinion that tuberculosis is not transmitted by infection of the germ, because such a transmission has not been directly proved, and because, on the other hand, an infection from without cannot be conclusively disproved in any individual instance, I am inclined to believe that they have been too cautious in their conclusions, of which only a negative proof is furnished by either factor. For neither of these in the least proves that infection of the germ does *not* take place: from a more general, biological point of view, indeed, it seems to be far more probable that it *does*.

It will, I think, at any rate be conceded that a 'constitutional' disease cannot be taken as a proof that the processes of heredity are therein concerned until we can determine whether we are actually dealing with heredity, — *i.e.*, the transmission of a constitution, — and not only with a transference of microbes; and the main object of this section was to make this clear. But at the same time I have stated my reasons for using so few facts from the domain of pathology in support of my theory.

PART IV

THE TRANSFORMATION OF SPECIES: ITS ORIGIN IN THE IDIOPLASM

CHAPTER XIII

THE SUPPOSED TRANSMISSION OF ACQUIRED CHARACTERS

I. Difficulties in the Way of a Theoretical Basis for this Assumption

By *acquired* characters I mean those which are not performed in the germ, but which arise only through special influences affecting the body or individual parts of it. They are due to the reaction of these parts to any external influences apart from the necessary conditions for development. I have called them '*somatogenic*' characters, because they are produced by the reaction of the body or soma, and I contrast them with the '*blastogenic*' characters of an individual, or those which originate solely in the primary constituents of the germ ('Keimesanlagen'). It is an inevitable consequence of the theory of the germ-plasm, and of its present elaboration and extension so as to include the doctrine of determinants, that somatogenic variations are not transmissible, and that consequently every permanent variation proceeds from the germ, in which it must be represented by a modification of the primary constituents.

I will first attempt to show how this conclusion is arrived at theoretically, and will then proceed to test it by ascertaining how far it is in agreement with actual observation, and whether the theory can be justified by facts.

Somatogenic variations may be classified according to their origin into three categories, — viz., *injuries*, *functional variations*, and variations depending on the so-called '*influences of environment*,' — which include mainly climatic variations.

The hereditary transmission of any of these three kinds of somatogenic variations could be accounted for theoretically only by the assumption that that part of the soma which had been changed by external influences, could modify the germ-plasm contained in the germ-cells of the same individual, so that its offspring would, from the germ onwards, undergo similar variations to those which had been acquired by the action of external influences on the parental part in question.

As far as I can see, there are only two ways in which such a variation could conceivably occur in the germ-plasm in consequence of a corresponding somatic variation. We should either have to assume the presence in all parts of the body of definite tracks along which each somatogenic variation might be transferred to the germ-cells, in the germ-plasm of which it would produce a corresponding change; or else that gemmules, such as Darwin supposed to exist, are given off from every somatic cell and are conveyed to the germ-cells, — either through the vascular system, when one exists, — or by some other means, and that they must then penetrate into these cells, and become incorporated in their germ-plasm. Thus either the presence of hypothetical tracks along which a modifying, though totally inconceivable, influence might be transferred to the germ-cells, or else the discharge of material particles from the modified organ, must take part in the formation of the germ-plasm: there is no third way out of the difficulty.

Both these hypotheses have already been used to explain the supposed transmission of somatogenic variations, — the former, it is true, only in vague allusions hinting at 'nerve-influences,' which are supposed to start from the modified part, and to produce a corresponding alteration in the hereditary substance of the germ-cells. But no one has yet ventured to state more precisely how nerve-excitation can modify the germ-plasm *materially*, and *in accordance with* the somatic variation. It would probably be useless even to expect an answer to the question as to how a part, such as a muscle, enlarged by functional hypertrophy, is capable of producing a specific nervous current signifying 'enlargement.' If such an explanation were attempted, we should be compelled to imagine that every cell in the body was placed in communication with every germ-cell of the ovary or spermarium by means of a large number of nerve-tracks, and was capable of continually sending information to the germ-cells of what was occurring in its own substance, and of the manner in which it was influenced, and also of giving instructions how each of the millions of units in the germ-plasm should behave. I believe that it would be impossible to avoid absurdities in explanations of this kind, and consider the whole idea inadmissible.

The second possible explanation appears to me to be less acceptable at the present day than when it was put forward by Darwin in the form of a hypothesis of pangenesis. And, as already stated in earlier essays, I believe that the talented author of this hypothesis of heredity did not look upon it as a well-grounded assumption, but considered it merely as a working hypothesis, only intended to lead to a better insight. Meanwhile, many changes have been made, and we have become acquainted with facts which compel us to reject the idea of a 'circulation of gemmules,' and I am surprised that this has not hitherto been done. This hypothesis is rendered inadmissible, not merely because we must imagine that the gemmules are *given off*, and then *circulate* through the body, but principally on accouut of the implied *addition of gemmules —i.e., of primary constituents—to the germ-plasm of the germ-cells!*

According to Darwin's idea, there must be a constant addition of 'primary constituents' or gemmules to the germ-plasm already present in the germ-cells, unless, indeed, it is assumed that the entire nuclear substance in the germ-cells is formed by gemmules which migrate into them. Such an assumption is, however, contradicted by the fact that the *hereditary substance of the germ-cells, which we observe in the form of nuclear rods or idants, receives no addition to its organised bodies, the primary constituents.* I have come to this conclusion, not from the fact that we have never observed an addition of this kind, but from the way in which the hereditary substance has been shown to behave during its multiplication. We know that the cell contains a most wonderful mechanism which apparently has the sole function of distributing the idants quantitatively and qualitatively, according to the primary constituents which they contain; and this is done as equally as possible, or, at any rate, in a definitely prescribed manner. Why should the centrosomes and spindle-threads be present, and the longitudinal fission of the idants occur, if myriads of primary constituents of the germ-plasm circulate separately through the body, and are capable of entering the germ- and other cells from without, as well as of becoming properly arranged in them in the order in which they subsequently undergo development? Why should nature be so scrupulously careful to divide the idants as accurately as possible, if their composition were open to alteration at any moment by the entrance of new primary constituents or gemmules? *The process of the fission of the idioplasm in nuclear and cell-division seems to me directly and conclusively to refute the whole idea of the circulation of gemmules.* For the very reason that these nuclear rods or idants can never receive an addition of new primary constituents from without, the most extreme care is required, during their multiplication by division, to prevent the different qualities of the mother-cell from being distributed improperly amongst the daughter-cells, and causing an irreparable loss of certain primary constituents to one of the latter and its descendants.

It is impossible to assume the transmission of somatogenic variations in any theory which accepts the nuclear substance of the germ-cells as germ-plasm or 'hereditary substance'; for it is theoretically impossible to account for these variations, no matter how ingeniously the theory is constructed.

At the present day I can therefore state my conviction still more decidedly than formerly, that *all permanent — i.e., hereditary — variations of the body proceed from primary modifications of the primary constituents of the germ*; and that neither injuries, functional hypertrophy and atrophy, structural variations due to the effect of temperature or nutrition, nor any other influence of environment on the body, can be communicated to the germ-cells, and so become transmissible.

This statement naturally implies the rejection of Lamarck's principle of variation; for those factors which this talented philosopher and investigator believed to be all-important in the modification of species, — viz., the use and disuse of parts, — can have had no direct share in the process. I am by no means the only one to hold this view at the present day; and although the truth cannot be decided by a consensus of opinion, it is nevertheless a significant fact that the views of such naturalists as Ray Lankester,* Thiselton Dyer, Brooks, Meynert,† van Bemmelen,‡ and others, coincide with my own.

The fact, however, that we deny the transmission of the effects of use and disuse, does not imply that these factors are of no importance; and I have already attempted to show in former essays that both use and disuse may lead indirectly to variations,—the former wherever an increase as regards the character concerned is useful, and the latter in all cases in which an organ is no longer of any importance in the preservation of the species, and in which, so far as the disused organ is concerned, 'panmixia' occurs.

Want of space prevents me from discussing these questions in detail; their consideration belongs rather to a work on the theory of descent than to one on that of heredity, and I need only refer to my former essays,§ in which, I think, sufficient proof is given to show that the gradual degeneration of organs which are no longer of use does not require the assumption of the transmission of somatogenic variations, and that consequently the facts do not compel us to adopt a hypothesis which we seem unable to accept theoretically.

It therefore remains to be seen whether we are not acquainted with other facts which are explicable only on such a hypothesis: one side of this question will now be treated of in particular.

2. The Hypothesis Tested by Facts

A few words will suffice concerning the hypothesis of the *transmission of injuries and mutilations*, which has been accepted for so long a time, and is obstinately defended even at the present day; for since the appearance of my essay on 'The

---

* Ray Lankester, 'The History and Scope of Zoology,' 'Enc. Brit.,' Vol. xxiv.

† Meynert, 'Mechanik der Physionomik,' a Lecture held at the Meeting of German Naturalists at Wiesbaden, 1887.

‡ J. F. van Bemmelen, 'De Erfelijkheid van verwooven Eigenschappen,' s'Gravenhage, 1890.

§ *Cf.*, 'Uber die Verebung,' Jena, 1883, and 'Die Continuität des Keimplasma's als Grundlage einer Theorie der Verebung,' Jena, 1885. English translation, 'Essays upon Heredity,' pp. 71 and 165.

Supposed Transmission of Mutilations,' * no new observations on this point have appeared.† The old arguments, on the scientific worthlessness of which I then expressed my opinion, are constantly being brought forward.—in part altered, and in part with a new interpretation. It is now even less necessary than ever to return to the matter, as even among those observers who supported the view of the transmission of functional variations, a few agree with me in denying the transmission of mutilations. As an instance, I may mention Osborn, who, however, goes a little too far when he compares the contest of the old view of the transmission of mutilations with Don Quixote's celebrated fight with the windmills.‡ Only a few years ago, at a meeting of the German association of naturalists,§ two 'tailless' cats were exhibited, in which the absence of the tail was supposed to be due to their mother having accidentally lost hers; and biologists of such eminence as Ernst Haeckel ‖ have accounted for similar cases in the same way. Since such men still regard the inheritance of mutilations as possible, the exposition of the subject has not been a superfluous task.¶

There is, however, *a third kind of somatogenic variation*, produced by *the influence of environment*, *the mode of nutrition*, *the climate*, and so on; this often appears to be transmissible, and consequently capable of becoming increased in the course of generations. I myself called attention to this fact a number of years ago in the following words:—'I only know of *one* class

---

* Jena, 1889, 'Essays upon Heredity,' Oxford, 1889, p. 421.

† None, that is to say, which are in opposition to my views. My experiments with mice have been confirmed by Ritzema Bos and by Rosenthal. I have now continued these experiments to the nineteenth generation—always with the same negative results; cutting off the tails has no influence on the tails of the descendants. A similar result was obtained by both the above-mentioned observers from experiments on rats (*cf.* 'Biolog. Centralblatt,' Vol. xi., 1891, p. 734, &c.).

‡ Osborn, 'Are Acquired Variations Inherited?' Boston, 1890, p. 3.

§ At the meeting at Wiesbaden, 20th September 1887.

‖ Haeckel, 'Naturliche Schöpfungsgeschichte,' 3rd ed., 1889, p. 194. English edition, 'The History of Creation,' London, 1876, Vol. i., p. 214.

¶ See the Section on the Transmission of Mutilations in Eimer's book entitled 'Die Entstehung der Arten,' &c., Jena, 1888. (*Cf.* 'Organic Evolution as the Result of the Inheritance of Acquired Characters,' &c., translated by J. T. Cunningham, London, 1890.) A whole collection of 'proofs' are there given.

of changes in the organism which is with difficulty explained by the supposition of changes in the germ; these consist in modifications which appear as the *direct consequence of some alteration in the surroundings*. But our knowledge on this subject is still very defective, and we do not know the facts with sufficient precision to enable us to pronounce a final verdict as to the cause of such changes.' Mention was then made of a few of the large number of cases which have been repeatedly quoted, and I attempted to show that none of them stood criticism, that they could not be explained in the way some investigators supposed, and that somatogenic variations are only apparently hereditary; for in reality a change must first be brought about in the germ-plasm by the influence of the surroundings before such a variation can be produced. I then continued:—'It must be admitted that there are cases, such as the climatic varieties of certain butterflies, which raise some difficulties against this explanation. I myself, some years ago, experimentally investigated one such case, and even now I cannot explain the facts otherwise than by supposing,' as I did then, that somatic variations were transmissible. 'It must be remembered, however, that my experiments,' which have been repeated upon several American species by H. W. Edwards, 'were not undertaken with the object of investigating the question from this point of view alone. New experiments, under varying conditions, will be necessary to afford a true explanation of this aspect of the question.' *

Since 1883 I have waited in vain for some skilled entomologist or for one of the numerous advocates of the transmission of acquired characters, to carry out the proposed experiments. In the meantime, as far as the time and material at my disposal permitted, I have myself made a start on this line of research, and now possess the results of a series of new experiments, which, though not so numerous, complete, or exhaustive as I could have wished, are nevertheless sufficient to form a more trustworthy basis for a theory dealing with variations of this kind. Some of these are described in the following section, a more detailed account of them being left for another occasion.

---

* *Cf.* my essay, 'Uber Vererbung,' Jena, 1883. English edition, Oxford, 1889, pp. 98–99.

3. Climatic Variation in Butterflies

Polyommatus phlœas, a butterfly belonging to the family *Lycœnidœ*, is distributed over the whole of the temperate and colder parts of Europe and Asia. It also occurs on the shores of the Mediterranean, in Maderia, the Canaries, and in part of North America. Before the glacial epoch this species must have inhabited the more northern circumpolar regions, and have been driven southward during that epoch; subsequently it must have again migrated towards the north. In our latitudes the upper surface of the wings of this form is of a beautiful reddish-gold colour, and hence it has received the popular name 'Feuerfalter' (fire butterfly). Further south, the reddish-gold colour is more or less thickly dusted with black, and specimens from Sicily, Greece, or Japan often display only a few reddish-gold scales, the general appearance being almost black. In Germany this butterfly is double-brooded, and the two generations are similar; but in certain districts of Southern Europe, such as the Riviera di Levante, the first generation is reddish-gold, — the second, which flies in midsummer, and is known as the variety *eleus*, having the wings well dusted with black. As in Germany, during exceptionally hot summers, individuals with a blackish tint have repeatedly been caught together with the ordinary form, and as, moreover, in the extreme southern limit of their range — so far as my experience extends — both generations have a blackish colour, it would appear at first sight that the modifications are merely due to the effect of heat; — the butterfly becomes red when exposed to a moderate temperature, and black when the heat is greater.

The following experiments, however, prove that this conclusion cannot be a correct one. Caterpillars were raised from the eggs of the German form of *P. phlœas*, and the pupæ were then exposed to a much higher temperature till the emergence of the butterfly. The result was that many of the butterflies were slightly dusted with black, *but none of them resembled the darkest forms of the southern variety eleus*. I then made the reverse experiment, subjecting caterpillars which had just entered the pupal stage, and had been raised from the spring generation of the Neapolitan form, to a very low temperature.*

* I must take this opportunity of expressing my warmest thanks to Dr. Schiemenz, of the Zoological Station at Naples, for the kind and generous

Many butterflies were thus obtained which were not so black as those which had emerged from pupæ kept at a higher temperature, *but none were so light-coloured as the ordinary German form. The difference between the Neapolitan specimens which had become light-coloured from exposure to cold, and the normal German form, on the one hand; and that between the German specimens artificially darkened by warmth, and the normal Neapolitan form, on the other, is too great to be attributable to the incompleteness of the experiments.* The German and the Neapolitan forms are therefore *constitutionally distinct*, the former tending much more strongly towards a pure reddish-gold, and the latter towards a black coloration.

Both experiments, however, prove the correctness of the old assumption of Lepidopterists that the action of heat on a single generation is capable of giving the German form of a blackish tint; and since, moreover, it is clear that the development a single generation at a lower temperature can render the colour of the Neapolitan butterfly less black, it appears that the two varieties may have originated owing to a gradual cumulative influence of the climate, the slight effects of one summer or winter having been transmitted and added to from generation to generation. *This would then seem to be an instance of the transmission of acquired characters.*

I do not believe, however, that this is the correct interpretation of the facts. If it were, there could be no region in which the species is seasonally dimorphic, as I have myself ascertained it to be on the Ligurian coast. The germ-plasm would then contain either the primary constituents of the red variety, if the colony had been exposed for many generations to a low temperature; or those of the black one, if a high temperature had influenced it for the same length of time. It would then make no difference to what degree of temperature a single generation were exposed at the present day in artificial breeding, for the colour would have already been determined in the germ-plasm, which would contain, to use my own phraseology, either 'reddish-gold' or 'black' determinants for the wing-scales in question. Hence it would be quite impossible for the spring generation to develop reddish-gold, and the summer one

---

assistance he has given me in my efforts. Without his help I should have been unable to obtain the necessary living specimens.

black scales, for the germ-plasm would only contain either 'red' or 'black' determinants for a certain spot on the wing.

The theory of determinants will, I believe, supply a very simple explanation of this apparently complicated case, which I consider of great value, because it confirms this theory. Instead of supporting the doctrine of the transmission of somatogenic characters, this example shows how *such a process may apparently be brought about*, and on what it depends. A somatogenic character is not in this case inherited, but the modifying influence — the temperature — *affects the primary constituents of the wings in each individual*, — *i.e.*, a part of the soma, — *as well as the germ-plasm contained in the germ-cells of the animal.* It modifies the *same* determinants in the rudiments of the wings of the young chrysalis as in the germ-cells, — namely, those of the wing-scales. The variation cannot be transmitted from the wings to the germ-cells, but only affects the coloration of these organs of the individual in question; whereas it is transmitted from the germ-cells to successive generations, and consequently controls the coloration of their wings in so far as this is not again modified by *subsequent* influences of temperature; for the same determinants which are now present in the germ-cells of generation I are afterwards passed into the rudiments of the wings in the caterpillar and chrysalis of generation II, and the change which they underwent while lying in generation I may be increased or weakened by the influence of the temperature to which they are exposed after entering into generation II.

Since warmth affects the whole body, it is not surprising that the determinants which are modified by it should undergo these modifications, whether they are contained in the germ-plasm of a young egg or sperm-cell of the caterpillar, chrysalis, or butterfly, or in certain cells in the rudiments of the wings in the chrysalis or caterpillar. This, however, does not imply that they must undergo the *same amount* of variation in both places, for they have not by any means the same environment in the two situations. In the germ-plasm they are grouped amongst thousands of determinants of the species, all of which constitute the germ-plasm; while in the rudiments of the wings, they are associated with only a few other kinds of determinants, and a time must come when *each of them controls a cell by itself*, and transforms it into a red or a black wing-scale.

We know, however, of a fact which definitely proves that the susceptibility of the scale-determinants to the influence of temperature is greatest at a certain stage in the development of the butterfly — much greater than either before or afterwards. I have frequently noticed in seasonally dimorphic species like *Vanessa prorsa-levana*, that the modifying influence of heat or cold only acts at the *beginning of the pupal stage*. Although I have not yet been able to ascertain the time at which this occurs more precisely, it can be definitely stated that the winter pupæ of *Vanessa levana*, for instance, which have been exposed to a high temperature even only a month after entering the pupal stage, are never transformed into the *prorsa* form; they all emerge as *V. levana*.

This is not due to the fact that the colour of the wings is already deposited a month after the insect has entered the pupal stage, for at this time there is no trace of colour whatever. *There must, consequently, be a period in the disintegration of the determinants when they are most susceptible to the influences of temperature:* subsequently this is no longer the case. and although they are susceptible *before* this period, I nevertheless venture to suppose that they were so to a *far slighter extent*. This may be due to their connection with other determinants, or to other causes which we are not yet able to discover.

If, then, the determinants for the scales are only influenced very slightly by the temperature as long as they are situated in the germ-plasm, and are subsequently greatly affected by it at a certain period in the development of the wings, the above-mentioned phenomena admit of a simple explanation. The germ-plasm of the southern colony of *P. phlæas* must contain many determinants among those for the wings, which, in consequence of the exposure of thousands of generations to heat, have been adapted for the production of black scales, together with a large number of others which only require a small increase of temperature during pupation in order to give rise to a black colour. These latter kind cause such fluctuations in the coloration as occurred in my experiments; while the former produce the black coloration of the wings, which has become fixed in the constitution of the southern colony, and can no longer be removed by the action of cold on the young chrysalis.

In this case it is taken for granted that the ancestral form possessed pure reddish-gold wings, and that it inhabited high northern latitudes, — an assumption which alone enables us to understand the present distribution of the species, and which has been adopted by Hofmann * in his splendid researches on the origin of European butterflies. This, however, is of no great importance in the present question, but we must assume that either the reddish-golden tint, or the deep black dusting is the primary colour. The seasonal dimorphism and the occurrence of blackish specimens in Germany in hot summers are easily acounted for on the former assumption.

In consequence of the increase in temperature of the habitat of the species, many scale-determinants in the germ-plasm would gradually become so modified that the action of only a slight further increase on the rudiments of the pupal wings would lead to the production of black scales. In Germany the species has attained this point in its phyletic modification; and if the weather happens to be hot when the second annual brood enter upon the pupal stage, some butterflies of a blackish tint will be produced. This will be more likely to happen as the internal transformation of the determinants in question advances further, and the blackish tint will become more conspicuous as the scale-determinants which have reached this stage of modification in the germ-plasm become more numerous. These two conditions will obtain most often in districts where the summer is usually tolerably warm; and the fact is thus accounted for that dark specimens of *P. phlœas* are rarely caught in northern Germany, and in the far north not at all, although very dark forms occur comparatively often in the warm valleys of Valais.

In still warmer districts, like the Riviera, the summer brood of *P. phlœas* is almost always exposed to a high temperature, and hence the transformation of the determinants for the scales has become so great, that with the help of the usual summer heat at the time when the caterpillar enters the pupal stage, the variety *eleus* has been produced. This variety does not appear in the spring brood, because the additional heat required for the complete transformation of the determinants for the scales is absent during the pupal stage.

---

* Ernst Hofmann, 'Isoporien der europaischen Tagfalter,' Stuttgart, 1873.

If the area of distribution of the species extended uninterruptedly from the Polar regions to South Italy or North Africa, all the intermediate forms would occur, from the pure reddish-gold single-brooded form in Lapland to the black double-brooded variety *eleus:* there would thus be — first, two similar reddish-gold broods; then similar ones, — those butterflies which are exposed to a higher temperature during the pupal stage having a tendency to develop a black tint; and then seasonal dimorphic forms, the butterflies being black in the summer and reddish-gold in spring, as is actually the case in Genoa. A still longer action of a higher temperature would at first change a small and then an increasing number of determinants for the scales into the 'black' variety, so that finally two broods, both consisting of the black form *eleus*, would occur. The case of the Neapolitan colony is somewhat similar to the last-mentioned one, for many black specimens certainly occur amongst the spring brood, although there are also many light-coloured ones; none, however, are as light as the northern reddish-gold form of *P. phlæas*. I do not know whether a complete change of colour has been attained by *both* annual broods in any locality; but if it has. I should expect it to be seen in Southern Japan rather than elsewhere, for the butterflies which I possess from the neighbourhood of Tokio display an unusually dark colour.

This case has been discussed at length because it appears to me to be especially significant, not only in the explanation of the climatic varieties of butterflies, but also as regards *the theory of heredity, and the assumption of material determinants which exist in the germ-plasm and are passed on from one generation to another*. The facts are so evidently in favour of this assumption that no other explanation seems possible. It must, however, be remembered that the artificial modification of the colour on the wings does not take place if the change in temperature occurs only when the scales begin to become coloured. The colouring matter is consequently not produced by the *direct* influence of chemical transformations, but by an indirect influence, which we may suppose to be due to a mutual disarrangement and rearrangement of the 'biophor-material' of which the determinants consist, by the co-operation of which latter the chemical process forming the colour is derived.

The occurrence of seasonal dimorphism alone, shows with certainty that the determinants for the scales are influenced by temperature in the germ-plasm to a much slighter extent than in the rudiments of the wings. If the modifying influence had the same effect in both cases, the germ-plasm in the germ-cells of a butterfly of the summer generation would be modified as much as the wings of the same individual; and consequently the offspring, even if exposed to a low temperature, would necessarily display a greater tendency towards the summer coloration, because the latter was already potentially contained in the germ. This, then, would only be the case if the influence of the cold were stronger than that of the heat. In any case, however, a coloration intermediate between that produced by cold and by heat respectively would result, and would be transmitted to both generations, even when the two influences were equally strong. If we indicate the winter and summer colorations respectively by A and B, the coloration of each generation would then be 1 A + 1 B. It is only when the germ-plasm is modified to a much smaller extent than the determinants which have already entered the rudiments of the wings, that an alteration of coloration can become permanent.

In many other animals and plants influences of temperature and environment may very possibly produce permanent hereditary variations in a similar manner; but it is difficult—in fact almost impossible—to identify such cases with anything like certainty from the observations which have hitherto been made. Thus we find it stated that 'in Cashmere dogs soon become covered with a woolly hair;'* but we do not know who observed this, or who ascertained that such a change—if it really does occur—is transmitted. 'Merino sheep lose their fine wool when they are transported to a tropical climate;' but I have not been able to discover whether this loss occurs in the first, or in the course of several, generations. We are thus left in uncertainty as to the possibility of a direct climatic variation of a somatic part having taken place in these instances, which would again disappear in the next generation provided that the descendants were placed under the original climatic conditions. The same applies to the races of naked dogs from the tropics, such as the

* These statements are quoted from an essay by Giard, who takes them as a proof of the transmission of somatogenic modifications. *Cf.* 'L'Hérédité des modifications somatiques,' *Revue Scientifique*, December 6th, 1890.

Guinea dog, for 'they do not become covered with hair when transported to a temperate climate.' *

Many climatic varieties of plants may also be due wholly or in part to the simultaneous variation of corresponding determinants in some part of the soma and in the germ-plasm of the reproductive cells, and these variations must of necessity be hereditary. Temperature, and nutrition in its widest sense, affect the whole body of the plant,—the somatic-cells as well as the germ-cells. It cannot, however, at present be stated whether the determinants in the soma are in this case influenced more strongly than those which are still in the germ-plasm. It is conceivable, and, I am inclined to think, even most usual, that certain determinants are affected to the same extent whether the influence of the environment happens to act on them in the germ-plasm, or in any stage of somatic transformation. In this case the change may have been perhaps scarcely or not at all noticeable in the first generation, and may gradually have become apparent, and also transmissible, in the course of subsequent generations. On the other hand, there are probably many influences of environment which produce a considerable change in the body of the plant, without, however, modifying the corresponding determinants in the germ-plasm. The experiments made by Nägeli and many others on the genus *Hieracium* at any rate support this view, though they have hardly been carried on long enough to exclude the possibility of a very faint and gradual alteration occurring in the germ-plasm.

The question as to which influences are capable of simultaneously modifying the developing and growing soma and the corresponding determinants in the germ-plasm, even in a very different degree, can only be solved by future experiments. The cases of an apparent inheritance of somatogenic variations are due to this coincidence;—no others are, it seems to me, conceivable. All those influences, however, such as the use and disuse of a part, which can only affect this part itself in a specific manner, are incapable of producing a *corresponding* change in the respective determinants of the germ-cells, and consequently cannot lead to hereditary modifications. In such cases the

* Compare also the cases of degeneration in the descendants of the European dog in India, which have been carefully collected by Darwin in his 'Variation of Animals and Plants under Domestication,' Vol. I., p. 45.

external influence affects only the fully-formed organ, — such as a muscle which has become enlarged by exercise; for the influence consists in the increased activity of the organ, which takes place within it alone; the germ-plasm of the germ-cells, and even the determinant in the germ-plasm for the muscle in question, are not thereby affected. In all cases of functional hypertrophy or atrophy, the external influence affects none of the determinants, but only the fully-formed organs, — *i.e.*, groups of specific cells produced from determinants. In my opinion it is very probable that such twofold modifying influences of environment as we meet with in *P. phlœas*, can only occur when the determinants which have not yet been transformed into the organ, as well as the germ-cells, have been affected by the modifying influence. And this will be most likely to happen in those structures which, like the scales on the wings of butterflies, are formed at a later stage of the animal's existence, *and the determinants of which are consequently stored in an undeveloped condition in the idants of certain somatic cells during a great part of the ontogeny*. The wings of the butterfly arise as outgrowths from the hypodermis of the caterpillar. Before these outgrowths can be formed, the determinants for the wing-scales must be contained in the idioplasm of some of the cells of the hypodermis, but after their appearance they would be found in some of the cells in the rudiments of the wings. The wings at first are small, and contain by no means so large a number of cells as when full-grown, so that inactive determinants for several wing-scales must be contained in the idants of *one* nucleus. At a certain period in the course of further growth, however, the number of cells increases to such an extent that each determinant constitutes the idioplasm of a particular cell, and the modifying external influences seem then to have the greatest effect on these determinants. By means of experiments it may be possible to ascertain exactly when this occurs.

It might have been expected that in this section I should enter into the whole question of the possibility of the transmission of acquired variations, about which there has been o much dispute of late years, and that all the arguments and facts which have been put forward in favour of the theory should be discussed. But, as I have already remarked, this seems to be out of place in a theory of heredity, the object of which is to show whether this form of transmission is or is not possible from a theoretical point of view, and to ascertain further whether in the latter case such an *apparent* transmission might not possibly occur under certain circumstances, and to account for this theoretically. I have always emphasised the fact that it is easier to explain the transformation of species on Lamarck's principle; but this is no reason for the retention of a theory which cannot be accepted on theoretical grounds, unless no other explanation can be given for the facts. So far, my opponents have been unable to prove that this is the case.

The above explanation of the causes of the climatic variations of butterflies may perhaps convince some of those who have till now opposed my views that we are here not dealing blindly with mere principles, but with inductive methods. The view of the non-inheritance of acquired modifications has been especially opposed in America, principally by the palæontologists. It can certainly not be denied that certain facts in palæontology, such as the development of the feet and teeth in Ungulates, furnish us with an extremely fine and uninterrupted series of forms which may apparently be very easily explained on the assumption of the inheritance of acquired modifications. But is not this exactly what would be expected, in case phylogeny essentially depends on selection — that is, on an increasingly complete adaptation to certain external conditions of life of a purely general nature? It appears to me that neither the completeness of the developmental series, nor the close relation of the nature of the modifications to function, give any clue to the causes which have produced these series. They may quite as well have originated by continued selection alone, as by continued transmission of functional variations.

The eminent American naturalist, Lester Ward,* is, however, in error if he supposes that the proof that climatic influences are capable of modifying the germ-plasm, contains all that is required by the neo-Lamarckian school. Further details will be given in the next chapter as to the manner in which I now suppose variation has originated: but quite apart from this, the supposition that climatic influences can produce modifications of the germ-plasm, has certainly nothing to do with the view that

* *Cf.* the essay directed against my views by this author ('Neo-Darwinism and Neo-Lamarckism,' Washington, 1891), which is written from the thoroughly objective and truly scientific point of view.

functional modifications of any particular organ can cause a corresponding change in the germ-plasm. I believe I have here furnished a proof that the former supposition is a correct one: the onus of proof of the latter lies with the neo-Lamarckians.

CHAPTER XIV

VARIATION

1. Normal Individual Variation

Heredity is the transmission of the physical nature of the parent to the offspring. We have seen that this transmission affects the whole organism, and extends to the most trifling details; and we also know that it is never complete, and that the offspring and parent are never identical, but that the former always differs more or less from the latter. These differences give rise to the phenomenon of *variation, which thus forms an integral part of heredity, for the latter always includes the former.*

A theory of heredity must therefore include a theoretical substantiation of variation, such as I will now attempt. Why is the offspring never an exact duplicate of its parent, even when it possesses but one parent, as is the case in parthenogenesis and reproduction by budding? And what forms the basis of the constant '*individual variations*' which, after the precedent of Darwin and Wallace, we regard as the foundation of all processes of natural selection, and as the means which rendered possible such a rich development of organic forms of the most diverse kinds on the face of the earth?

Darwin himself considered the difference between parent and offspring as due to the diversity of external influences, and I was essentially of the same opinion formerly, and stated that 'all dissimilarities of organisms must depend upon the individuals having been affected by dissimilar external influences during the course of the development of organic nature.' * At that time I attributed to the organism the virtual 'power of

* 'Studien zur Descendenztheorie,' II., Leipzig, 1876, p. 304. English edition, 'Studies in the Theory of Descent,' translated and edited by R. Meldola, with a prefatory note by Charles Darwin, London, 1882, Vol. II., p. 677.

giving rise by multiplication only to exact copies of itself'; but this capacity, in my opinion, did not produce accurate results, because the organism is also capable of reacting to external influences, and may therefore deviate from the inherited tendency in one or another direction, according to the nature of these influences.

Variation would consequently not depend upon a special force existing in the organism, but would simply be the result of external influences, which, either directly or indirectly, are capable of preventing the organism from keeping strictly to the inherited course of development.

Although I still consider this view to be in general correct, the origin of individual variation, on which the transformation of species is based, cannot be deduced so easily from the action of dissimilar external influences as seemed possible when I wrote the passage quoted above. I have discussed this question in full on a previous occasion, and will here only briefly refer to it.*

At that time, no one supposed that any difference existed between the modifications which may be brought about in the soma and those which proceed from the germ-plasm. Since then, however, we have been compelled—at least in my opinion—to consider that only those variations which are '*blastogenic*.' and not those which are '*somatogenic*,' can be transmitted. We can no longer regard the direct influence of external impressions on the soma as a means of producing hereditary individual variations. It therefore remains to be seen what is the origin of these variations, upon the existence of which we imagine the entire development of organic nature to depend.

This development could be accounted for most easily on Nägeli's hypothesis, according to which the idioplasm is so constituted, that in the course of generations it could exert a definite and regular transforming influence upon itself, and by this means could convert one species into another. Many reasons may, however, be urged against this hypothesis. In the first place, development by internal forces only is contradictory to the close adaptations of organisms to their conditions of life; and secondly, we should not make use of unknown

---

* *Cf.* 'Die Bedeutung der sexuellen Fortpflanzung für die Selectionstheorie,' Jena, 1886. English Edition, Oxford, 1889, p. 255.

forces for the explanation of natural phenomena until we have proved that the known ones are insufficient.

Such a phyletic principle of development would naturally not produce ordinary individual variations, but would lead to modifications *which affect all the individuals of a species in the same manner*. All the individual modifications which actually exist would, in that case, have to be considered insignificant in phyletic development; and in plants Nägeli actually looked upon them to be transitory, and not hereditary habitat-modifications ('Standorts-Modificationen'). But this view is opposed by the transmission of an immense number of individual differences in man and animals.

If we recognise that the processes of selection are the chief factors in organic development, we must attach the greatest importance to these hereditary differences in individuals, and must try to discover their origin.

Brooks accepts this view, and has propounded a theory of heredity which is based on Darwin's hypothesis of pangenesis, and according to which variation is in the main dependent on sexual reproduction.* Variability, in his opinion, is caused by every 'gemmule' of the spermatozoon uniting with 'that particle of the ovum which is destined to give rise in the offspring to the cell which corresponds to the one which produced the germ or gemmule. . . . When this cell becomes developed in the body of the offspring *it will be a hybrid, and it will therefore tend to vary*.' Brooks, moreover, assigns different functions to the two kinds of germ-cells, and represents them as being charged or filled to different degrees with 'gemmules,' the egg-cell containing a much smaller number than the sperm-cell. In his opinion the egg-cell is the conservative principle which controls the transmission of purely racial or specific characters, whereas the sperm-cell is the progressive element which causes variation.

Brooks has ingeniously advanced every argument which could be brought forward in favour of his theory, but I doubt whether he still holds to it, for many new discoveries have since been made which contradict it. Although the view that 'acquired' characters are not hereditary is not universally admitted, it has

---

* W. A. Brooks, 'The Law of Heredity: A Study of the Cause of Variation and the Origin of Living Organisms,' Baltimore, 1883.

been accepted by Brooks; and the assumption of a diversity of the hereditary substances contained in the male and female germ-cells, on which his theory is based, is rendered untenable if it is recognised that the idants of both cells remain precisely the same as regards number and quality during the process of amphimixis. It will not, therefore, be rash to conclude that the few observations which seem to be in favour of the dissimilar effect of male and female germ-cells are not convincing, although we may not be able at present to explain them fully. To these belong certain rare cases, which have, perhaps, not been very accurately observed. and possibly form exceptions to the rule that hybrids of two species are *identical*, whether the father or mother belonged to the species A or B.

If we are forced to reject the assumption of the dissimilar action of the two germ-cells, together with that of an internal force of transformation, we can only refer hereditary individual variability to the inequality of external influences; and it then remains to be seen how such influences can produce *hereditary* differences if *somatogenic* modifications are not hereditary: for external influences act directly, and often exclusively, upon the *body*, and not on the *germ-cells*.

In a former essay I have already attempted to show that the constant occurrence of individual variability and the continual transformation in the intermixture required by selection is brought about by amphimixis, although it is not the primary cause of this variability; and that the accomplishment of sexual reproduction is even based in almost all the known organic forms on this necessity of preserving and continually remodelling the hereditary variability of individuals. I am convinced that *the two forms of amphimixis — namely, the conjugation of unicellular, and the sexual reproduction of multicellular organisms — are means of producing variation*. The process furnishes an inexhaustible supply of fresh *combinations* of individual variations which are indispensable to the process of selection.

Hatschek* has contested this view of the significance of sexual reproduction, and states that 'transformations of species occur far too rarely to admit of their explaining such an uninterruptedly active process as sexual reproduction.' It seems to me, how-

* B. Hatschek, 'Lehrbuch der Zoologie,' Jena, 1888, p. 10.

ever, that he has overlooked the fact that the *transformation* of a species, as well as the *preservation* of its constancy, are based upon natural selection, and that this is incessantly at work, never ceasing for a moment.

From what was said in the chapters on the struggle of the determinants of the two parents in ontogeny, and on the 'reducing-division' of the germ-plasm which is indispensable in amphimixis, it follows that by means of the latter process fresh combinations of the possible variations in a species must constantly be produced. On the one hand, the germ-plasm of a new individual produced by amphimixis never receives more than half the ids of each parent—and these are differently selected and arranged in each case; and on the other, the co-operation of the ids of both sides would not always strike an average in all parts of the new organism, but each part would resemble that either of the father or the mother in proportion to the number and controlling force of the individual homologous determinants; the resultant of the co-operating forces may be different in different parts.

Although the process of amphimixis is an essential condition for the further development of the species, and for its adaptation to new conditions of existence amongst the higher and more complicated organisms, *it is not the primary cause of hereditary variation*. By its means those specific variations which already exist in a species may continually be blended in a fresh manner, but it is incapable of giving rise to new variations, even though it often appears to do so.

When it first occurred to me that sexual reproduction was necessary to produce the variations required for the action of selection, I imagined that its influence upon the germ-plasm was still more powerful. Since all differences—even the qualitative ones—are ultimately of a *quantitative* nature, and as the union of the primary constituents of the parents may either strengthen or weaken a 'character,' I imagined that the combination of very strong primary constituents for the same part in both parents would not only cause the part to reappear especially markedly in the child, but would also double the strength of the primary constituents of the part in the germ-cells of the child; and thus the continued intercrossing of offspring in which this part is strongly developed, might cause it to be increased more and more, so that it exhibited far more than an ordinary individual difference. If we suppose this process to go on in the various parts of the body, the transformation of a species would thus be accounted for.*

The production of races by artificial selection appears, in fact, to be in part due to such an accumulation of parental 'characters'; but I shall show later on that it is not accompanied by an actual *variation* of the determinants, which alone could gradually lead to a transformation of the species. We know tha the paternal and maternal idants do not fuse in the process o amphimixis, and the immense number of cases of perfect transmission proves that the determinants of both sides undergo no alteration by being brought together. The modification of the determinants is a process which is not directly connected with sexual intermingling, but follows its own course, and must be due to special causes.

This is made still clearer if we reflect that the lower organisms —*e.g.*, sponges and polypes—must possess a very small number of determinants compared to those in higher forms, such as birds and mammals. *The number of determinants in an id of germ-plasm has therefore increased considerably, and even enormously, in the course of phyletic development.* A single peacock's feather may possibly be controlled by as many determinants as an entire polype. Amphimixis alone, however, could never produce a multiplication of the determinants.

The cause of hereditary variation must lie deeper than this; it must be due to the direct effect of external influences on the biophors and determinants, which I imagine to take place in the following way.

The entire substance of the earliest organisms must have consisted of equivalent biophors, the nucleus and cell-body not having yet become differentiated. In these lowest forms, whether they exist or not at the present day, the perfect constancy of the composition of the body may occasionally have been disturbed by external influences of different kinds, and these modifications must have been preserved, as they persisted in the two parts resulting from reproduction by binary fission.

When the morphoplasm and idioplasm subsequently became differentiated, and the latter was enclosed in the nucleus as the

* Compare my essay 'Die Bedeutung der sexuellen Fortpflanzung,' Jena, 1866, p. 40. English edition, Oxford, 1889, p. 279.

hereditary substance and controlled the body of the cell, variations which had been produced solely by the direct action of external influences on the *body of the cell*, could no longer be transmitted to the offspring as an inheritance, for they were dependent on the part of the morphoplasm in which they originated, and did not pass into the idioplasm, which comprises all the primary constituents of the species. Every hereditary variation must therefore have originated in the nucleus, even in unicellular organisms — which in this respect therefore resemble the Metazoa and Metaphyta, the sole difference being that in the unicellular form we are concerned with the characters of one cell and not of many. In many unicellular organisms — *e.g.*, the higher Infusoria — the cell-body is very highly differentiated: it may possess a complex arrangement of cilia, undulating membranes, trichocysts, and flagella, each having a definite position in the body, — as well as a protective case of a definite form, with a lid capable of being closed; the animal, moreover, is capable of reconstructing all these parts should any of them become injured. We are therefore forced to admit that this minute organism must have a centre in which the latent germs for all these structures slumber, and from which reproduction may take place. This centre is the nucleus, and modifications of the nuclear matter can alone give rise to transformations of a *hereditary* nature in the cell-body.

The fact that species have remained sharply defined as long as we have known them, proves that these transformations occur neither easily nor rapidly.

The germ-plasm of multicellular forms is obviously also very constant, and the biophors constituting it are capable of nourishing themselves and of growing, so as to furnish exact duplicates of themselves when they divide. We could not otherwise understand how it would be possible, in spite of the enormous growth of the germ-plasm from one generation to another, for the specific characters, and even the most minute individual characteristics, to be preserved through so many generations.

The difficulty with which the germ-plasm becomes changed is not so clearly proved by the instances of constancy displayed by some species of ancient Egyptian animals — the ibis and crocodile — through thousands of years, which were formerly advanced by Nägeli and myself for this purpose. It may be objected that these species were always subject to the control of natural selection, which eliminated every case of deviation from the perfectly adapted form. If insignificant individual characteristics which are of no use to the species can, however, be preserved in the human race for several generations, this must be owing to the fact that the corresponding determinants have very little tendency during their growth and multiplication to suffer any marked variation, and that, on the contrary, they reproduce exact duplicates of themselves. I was therefore quite justified in supposing that the germ-plasm possessed a great power of remaining constant.*

We can none the less avoid assuming that *the elements of the germ-plasm — i.e.*, the biophors and determinants — *are subject to continual changes of composition* during their almost uninterrupted growth, and that *these very minute fluctuations, which are imperceptible to us, are the primary cause of the greater deviations in the determinants, which we finally observe in the form of individual variations.*

The assumption that such very minute fluctuations occur, naturally follows from the impossibility of a complete uniformity as regards nutrition existing during growth; and in fact, though underestimating its importance,† I formerly made this assumption, in correctly supposing that the influences producing these fluctuations 'are mostly changeable, and occur sometimes in one and sometimes in another direction.' I had not then taken into consideration the fact that the fluctuations accumulate in consequence of the process of amphimixis. If a single determinant increases to a hundred thousand during the multiplication of the germ-cells of an individual, it is not likely that the nutrition of all these determinants during the process will be absolutely the same in strength and kind. If this is not the case, minute differences could not fail to appear in the subsequent determinants. These minute fluctuations may undoubtedly again disappear, as I formerly assumed, provided that the modified determinant is exposed to counteracting influences, and *alone* they are quite incapable of producing an individual variation of any perceptible character; but they may become cumulative. For the germ-plasm always consists of a large number of ids, each of which contains one of the homologous determinants in

*Cf. my essay 'Die Bedeutung der sexuellen Fortpflanzung,' Jena, 1886, p. 28. English edition, Oxford, 1889, p. 271.

† *Loc. cit.*, p. 272.

question, and the co-operation of all of these determines the character. *A hereditary individual variation will therefore arise if many of the homologous determinants vary in the same way.*

We can thus understand the process of *doubling* of the determinants, which must have occurred repeatedly, as more complicated structures arose in phylogeny. More abundant nourishment would cause a determinant to grow and multiply more rapidly, and if the first multiplication occurs before the 'reserve germ-plasm' for the next generation has become separated off, this double determinant will be permanently retained by the species. The corresponding part of the body, however, will only display a perceptible variation when the majority of the homologous determinants have become doubled.

Minute fluctuations must thus continually occur in the composition of the biophors and determinants. Their variability depends on the same principle as the systematic disintegration of the determinants in the germ-plasm, and is due to the *dissimilar composition of the elements of the growing substance.* If the determinants consisted of masses which were all exactly alike, inequality of nutrition could never transform determinant A into A_1:—it could only alter its rate of growth. They are, however, composed of biophors of different kinds, which react dissimilarly under different conditions of growth. This renders possible a disarrangement of the proportional numbers of the different biophors in a determinant, and consequently also the variation of the latter. It is therefore quite conceivable that all the qualities of a cell are not affected by these influences, but that only certain of them vary, and that only a few amongst a large number of similar determinants need become modified.

The facts already mentioned with regard to the *climatic variation of butterflies*, prove that such an alteration can actually occur. In these insects the determinants of certain coloured wing-scales are slowly changed in the course of generations, in consequence of the rise in temperature of the climate, and thus the colour of the scales becomes considerably modified. Such cases of conspicuous variation are not often met with; all species of butterflies, at any rate, are not affected in this way by changes of temperature, and those that are thus influenced do not display the variation in all the different kinds of scales. This indicates that the determinants have a strong tendency to remain constant, and that the deviations to which they are subject on account of inequalities of nutrition are as a rule so infinitesimal that their effect is unnoticeable.

These deviations are nevertheless of great significance, *for they form the material from which the visible individual variations are produced by means of amphimixis together with selection;* and new species arise by the increase and combination of these variations.

The hypothesis that the germ-plasm consists of ids is quite indispensable in this case Every determinant is represented in the germ-plasm as many times as there are ids in the latter, for every id contains all the kinds of determinants. The determinant N, for example, is represented a hundred times if the germ-plasm consists of a hundred ids. Most of these ids must differ slightly from one another, for in the course of generations they are continually brought together in new combinations by the process of amphimixis. On the occurrence of this process, however, the diversity of the ids persists, even if we go back to the origin of the multicellular forms, or of the unicellular ones from the primordial organism. A condition in which all the ids could be considered similar is never found; in fact, as already stated, *the dissimilarity of individuals must be traced to the primordial organism*, at a period when neither the process of amphimixis nor idioplasm had come into existence, and in which every individual organism derived its individuality directly from the dissimilarity of external influences. From these organisms the dissimilarity was transmitted to the unicellular forms, which cannot all have originated from one primordial organism, but each species must have arisen polyphyletically from a large number of similarly modified organisms. This point has often been misunderstood, and I have been asked to explain, for example, how the adaptations of flowers, fruits, and seeds in Phanerogams, could have been derived from a combination of characters acquired by the shapeless primordial ancestors. *The characters were not inherited from the primordial beings, but variability, or the dissimilarity of individuals.*

We might, however, be inclined to believe that external influences must affect *all* homologous determinants of a germ-plasm in the same way, and must cause them to vary; but this would be erroneous.

Since reproduction is connected with amphimixis in the

multicellular organisms, and the latter process is never entirely wanting in any species, the germ-plasm of these forms consists of many ids from different sources, — one half of them being derived from the father, and one half from the mother, each half, again, containing ids from the grandparents in varying proportions; the ids of the grandparents, moreover, are derived from one, two, or even three great-grandparents, and so on. As I have shown above, the proportions in which the individual ancestors may be represented by ids may vary very much; and consequently the germ-plasm of different individuals, even when they are closely related, must always differ.

Each id contains all the determining elements of the species, but in a manner peculiar to the individual. If a determinant N differs slightly in every id, it will also vary a little during growth if exposed to modifying influences; so that, for example, the determinant N may remain unaltered in id A, while N^1 varies in B. On the other hand, the modifying influence of nutrition may very likely be slightly different in A and in B, and may produce a variation in N, while N^1 remains unaltered. Sufficient factors would thus be present to cause a variation of one or more homologous determinants in certain, but not in all ids. In the only carefully observed cases of blastogenic variation, due to the direct influence of external conditions, — viz., those of the climatic variations of *Polyommatus phlæas*, — it is evident that the influence of temperature was not quite uniform. Some of the captive Neapolitan specimens of this species are darker and others lighter, although they were raised in a room under conditions as nearly the same as possible; and in those caught in the neighbourhood of Naples, the amount of black-dusting — the character modified by the climate — varied considerably. The same influences, even when they act during many successive generations, do not necessarily cause the individuals of a species to vary to the same extent; and I account for this by the fact that the ids of every animal contain different variants of a particular determinant N, some of which are more susceptible to heat than others. The germ-plasm, as a whole, will vary to a greater or less extent in proportion to the number of the determinants which are more or less variable.

Many enigmatical points, however, still remain. Thus the transformation by heat of many of the original reddish-gold wing-scales of *P. phlæas* into black ones does not occur evenly, so as gradually to change the entire reddish-gold upper surface of the wing into black; but certain parts first become darker, and then other adjacent ones, the whole surface being blackened only in the very darkest specimens. The margin and base of the wing first turn black, and the change then spreads slowly towards the middle, which, however, remains unaltered in most individuals. Since we must suppose that scales of the same colour arise from similar determinants, why are they affected to such an unequal extent by the modifying influence of heat?

The explanation previously used can, however, be applied in this case also.

It was shown in the chapter on reversion that although new specific characters are produced by the modification of certain determinants or groups of determinants, this modification never affects the homologous determinants in *all* the ids of the germ-plasm simultaneously. It must, on the contrary, be assumed that variation affects only a small majority of modified determinants at first, but increases by the selection and preference of the most modified individuals, until finally a predominant majority of all the ids contain modified determinants.

This evidently implies that *new and old specific characters are respectively represented by a small and a large majority of modified determinants*. If this statement be applied to the case of *P. phlæas*, we obtain a principle by the aid of which the dissimilar effect of heat on the determinants for the middle and the margin of the wing respectively may be understood. It is very improbable that the surface of the wing of the brown ancestral form of *P. phlæas* has changed uniformly into reddish-gold, and it is much more likely that a lighter spot in the middle became modified first and took on a reddish-gold colour, in consequence of sexual selection, and that this then gradually extended towards the margin. If this were so, the reddish-gold scales of the centre must be represented in the germ-plasm by a greater majority of homodynamous determinants than are those at the margin: and in this way we can understand why the black-dusting of the wings affects the sides first, and the middle last of all. This must be so if the old 'brown' determinants under the influence of heat give rise to black scales more easily than do the 'reddish-gold' ones. Whether this explanation be correct in this particular case or not, it is nevertheless true that *the diversity of the extent to which the determinants for the same character are represented in different regions of the body, affords us a principle by which we can understand the unequal effect produced by similar modifying influences upon various regions of the body.*

Even though it can no longer be doubted that climatic and other external influences are capable of producing permanent variations in a species, owing to the fact that, after acting uniformly for a long period, they cause the first slight modifications of certain determinants to increase, and gradually affect the less changeable variants of the determinants also, the countless majority of modifications is not due to this cause, but to the processes of selection. The question then arises as to the origin of variations which are sufficiently considerable for natural selection to act upon them. External influences at first produce only very slight fluctuations in the determinants — presumably not only in some, but in all; *a continual supply of the most minute variations of the different determinants will thus always be present.*

I do not, however, believe that variations, as we perceive them, are the direct result of these minute fluctuations on the part of individual determinants: they can only *be produced by the accumulation of a large number of fluctuations of this kind.* This is an immediate consequence of the theory. Since the germ-plasm consists of many ids, each of which contains the same number of homologous determinants, and as, moreover, any character is the result of the interaction of all its homologous determinants, the variation of a single determinant would be imperceptible; a character can only be modified to an appreciable extent when a majority of the determinants are equivalent, or at any rate are similarly modified.

In my opinion, a variation of this kind is produced by solitary homodynamous determinants in different ids and individuals being brought together in *one* germ-plasm by means of the processes of 'reducing division' and amphimixis, so that they can thus form a majority.

I will illustrate this by a simple example. A small brown European butterfly, *Lycæna agestis*, Hb., has a small black spot formed by a few scales in the centre of the wing. Let us suppose that this spot is controlled by a *single* determinant F, and that the germ-plasm in this species contains a hundred ids, and consequently a hundred determinants F; and that, in consequence of a change in nutrition, some of these determinants F in different individuals constantly vary in such a manner that if they were in the majority in the germ-plasm they would give rise to a white spot instead of a black one. It will sometimes happen that they are in the majority; for, in consequence of the process of amphimixis, the modified determinants F may become accumulated in the course of generations, so as to exceed fifty in one or more individuals. The black spot will then turn white, and among thousands of individuals belonging to this species some will exhibit this variation.

The subsequent course of the phyletic development of this white spot depends upon its *physiological value* to the species. Even if it is only of slight importance, it will gradually come to be possessed by an increasing number of individuals, and will ultimately be transmitted to *all* of them: *i.e.*, it will become a specific character. This extension could hardly occur without amphimixis; for by its means any minority of determinants F^1, wherever produced, may accumulate so as to constitute a majority; and otherwise they could not have been effective, since a minority could never have produced a white spot.

This example was selected because it is based on facts. A variety of *Lycæna agestis* actually exists which possesses a milky-white spot on each of the four wings in the place of the black one; it is known as the variety *artaxerxes*, and occurs in the north of England. It is immaterial whether sexual selection or a protective resemblance has caused the dominance of this modification.

Many variations of one species or another are merely due to the modification of a few or a large number of determinants; changes in the colour of individual parts or of the whole body may occur without an accompanying increase in the total number of determinants of the germ-plasm, but, on the other hand, as was shown above, many modifications do result from an *increase in their total number*. We have already seen that the doubling of a determinant of the germ-plasm may be referred to the influences of nutrition, and no difficulties therefore arise in the application of the principle just propounded to the multiplication of the determinants. The important modifications in species, including all enlargement of parts and higher differentiation of organs, must be connected with this increase: and the accumulation of double determinants of single ids, as well as their purely qualitative modifications, may be increased by means of the 'reducing division' and amphimixis, till the variation becomes perceptible, and natural selection comes into play.

The '*increase*' of a character derived from two parents which merely possessed it to a slight extent, is an entirely different matter. A *fusion* of the primary constituents of a character common to both parents, and a consequent increase of these primary constituents, such as was hitherto supposed to take place, evidently do not occur: such an assumption is contradicted by ordinary experience; for if the primary constituents of the parents could accumulate in this way, all parts of the offspring must be twice as large, or at any rate larger than, those of the parent, and this is not the case. If, however, it were maintained that it is merely a question of the differences in the primary constituents, and that the offspring receives half the sum of the characters of the two parents, it could be replied that this might be approximately true in some cases, but no explanation would be given as to how an increase of a characteristic can occur, and may even be produced artificially by pairing animals which exhibit a certain tendency to vary in the desired direction. If two animals are paired which possess the character a in the degree $2a$, the offspring would contain $\frac{2a + 2a}{2}$ —*i.e.*, $2a$, and no increase would be produced. Moreover, the real meaning of the term 'halving' would then remain perfectly vague. Hensen * thought that 'the effect of transmission was to *halve* the characters of the parents,' since 'a similar whole can only result from two equal half-transmissions.' I believe that this is to a certain extent true, but not in the sense which would entail the halving of an *indivisible* primary constituent.

The solution of the problem is to be found in *the multiplicity of the ids and determinants*, and in distinguishing carefully between the vague idea of the '*character*' and the definite one of the *determinate* or hereditary part. Each determinate is controlled by as many determinants as there are ids in the germ-plasm; but half the ids of each parent, together with half the determinants of *every kind*, are removed from the germ-plasm by the reducing division of the germ-cell. Half the primary constit-

* V. Hensen, 'Physiologie der Zeugung.'

uents are, however, not removed in this process; on the contrary, the occurrence of 'pseudo-monogonic heredity' proves that every parent-organism transmits the whole of them to the offspring; *every primary constituent is, however, represented by only half the number of determinants.*

From this theoretical basis, the results arrived at by breeders are easy to understand. '*Like begets like*,' is their chief maxim, and this is true whenever two individuals are coupled which possess a certain character as an inheritance transmitted through a long series of generations. For such a character must be represented by homodynamous determinants in the great majority of ids in the germ-plasm; and as it consequently will not be entirely removed from either half of the latter by the reducing division, it will in most cases be represented in a majority of ids in the offspring.

The idea that the '*increase*' of a 'character' can be produced simply by crossing, is due to an inaccuracy of expression. 'Like begets like,' but not something else; and in this respect theory and practice agree. It is theoretically inconceivable that precisely the same part — that is to say, the same determinate or group of determinates — may be increased merely by the pairing of parents possessing it. If, for instance, two individuals of *Lycæna agestis* pair, and each of them possesses a white spot instead of a black one, in the centre of the wing, none of the offspring could exhibit a spot twice as large as, or even any larger than, that of the parents. For the spot is controlled by one or more homologous determinants, and if those of the white variety are in the majority, the spot will be of this colour; but the adjacent determinants cannot thereby become modified. The spot can at any rate only become pure white in case it was previously merely grey, owing to the relatively larger number of 'black' determinants which took part in the control of the spot in the parents having now been completely excluded from the control of the cells by a preponderating majority of 'white' determinants.

The term 'quality' or 'character' ('Eigenschaft') is the real cause of confusion in this instance. In the Introduction to this book I pointed out that this term may have several different meanings with regard to heredity. *Whiteness* in plumage, for example, is a 'quality' which a breeder tries to obtain by always selecting the whitest bird for breeding purposes. From a breed of blue pigeons, for instance, a bird with a white head is paired with another possessing a white tail, and in this way some young may be obtained with white heads and tails. The 'qualtities' of the white head and white tail are here, therefore, combined. The feathers of both parts of the body, however, have their own determinants; and in this cross the process which took place in the idioplasm was not the accumulation of homologous determinants, but the acquisition of a majority by the 'white' determinants for the feathers of the head and tail over the 'blue' ones. This is, therefore, not a case of the summation of like with like, but a victory of similar determinants in different parts of the body.

Surprise has often been expressed at the case, mentioned by Darwin, of two crested canaries which produced young none of which possessed larger crests, while many, at any rate, were bald. As Darwin himself pointed out, the crest in birds is due to a sparser covering of feathers on the head, and this peculiarity may be increased so as to result in baldness in the offspring. This is also an instance of the increase of a 'quality,' but only that of baldness, and not of the crest which is valued by the breeder. In terms of the idioplasm, this may be explained in the same way as was done with regard to the increase of the character of whiteness in pigeons. It is due to the arrangement of 'bald' determinants,—if I may use such an expression,—some of which are derived from the father and control the region *a*, the others from the mother controlling the region *b*. In the father the region *b*, and in the mother the region *a*, still possessed a 'feather' determinant; in the offspring the 'bald' determinants for both regions were accidently brought together owing to the reducing division and the subsequent amphimixis. It is shown by our theory that bald-headed offspring need not invariably result in such a case; and, as a matter of fact, all the young birds did not exhibit this peculiarity.

The increase in the number of feathers in any particular part, such as has occurred, for instance, in a fantail pigeon, is another case of this kind. The tail of this bird consists of about forty quill-feathers, instead of twelve, as in the original form; and the breed has undoubtedly been produced by artificial selection, those pigeons always being chosen for breeding purposes which possessed an extra feather in the tail. The young frequently, or at any rate in some cases, would thus exhibit more tail-feathers than either of the parents. But this would not by any means be an 'increase in the force' of a 'character,' but simply *an increase in the number of new feathers in the individual.* Let us suppose that the male bird possessed two extra feathers which were situated at a^1 and d^1 between the normal feathers *a* and *b* and *c* and *d*, and that the mother had two supernumerary feathers in the positions f^1 and h^1: a majority of the determinants for the extra feathers of the two parents might then be united in the germ-plasm of one of the offspring, so that the latter would possess all the four new feathers. This 'increase' in the character of additional feathers in the tail, therefore, depends upon the constitution of the germ-plasm, and can consequently be transmitted to the next generation.

This example clearly shows that *all really new structures are not merely the result of transmission, but are due to the variation and frequent multiplication of the determinants.* The mere *extension* of a 'character' over larger regions or the whole of the body, even if we choose to speak of it as an 'increase,' may be produced by pairing individuals which possess the desired 'character' in different parts. But an increase which is connected with the formation of new structures, and consequently with the *multiplication of the determinants in the germ-plasm*, can never be produced by such means alone. When this results, the cause of the modification must be the *variation of the determinants themselves.*

Thus in the case of the fantail, a new feather can never be produced by transmission alone: the offspring can merely possess new combinations of such feathers as were present in the parents. All really new structures can only originate in a previous modification of the germ-plasm.

Let us now take an example from among *sexually dimorphic* forms, in which we are quite certain of the phyletic modification, apart from the complication arising from the existence of sexual double-determinants. *The long tail-feathers of male humming birds* have arisen by a gradual lengthening of the ordinary tail-feathers, such as are possessed by the females at the present day. As already remarked, this lengthening is the result of a considerable multiplication of the determinants which give rise to the feather: the process of lengthening implies that variations in the ids which possessed a larger number of determinants than those of the original feather were exposed to selection. It was stated above that fluctuations in the structure of the determinants, caused by inequalities of nutrition, may also be produced in consequence of their more rapid growth and earlier division. If, then, certain determinants for the feather of the original form underwent division at an early stage in individual males, so that their number became doubled, a lengthening of the feather must have resulted as soon as the doubling occurred in the majority of the ids. The majority need not have appeared in one individual from the first, but, like every other simple variation of a determinant, it may have originated sporadically, in separate ids of different individuals, and have become increased by the repetition of the process of amphimixis in every generation. *The modification would become apparent* and the processes of selection could take place, as soon as these ids had attained a majority in any individual.

2. *Pathological Variation*

The above theoretical explanation of the 'increase' of a character possesses the important advantage of *accounting for the sudden appearance of more extensive variations.* If more abundant nourishment can cause the doubling of a determinant in the germ-plasm, it is possible, and even probable, that many or all the contiguous determinants for the same feather will become doubled. The feather must consequently *at once* increase to twice the size. Doubts have often been justly raised as to whether the process of sexual selection at first produces very slight variations, which would scarcely be noticed and preferred by the selecting sex. We learn from the doctrine of determinants that it is unnecessary to take such minute variations into account, and that more extensive ones may suddenly arise directly from the germ-plasm.

The doubling by division, not only of biophors and determinants, ids and idants, but even of *individual groups of congruent determinants*, such as must be contained in the germ-plasm in the form of primary constituents of any organ — *e.g.*, a feather — is theoretically possible. But this is one of the more special questions, the details of which may be reserved for subsequent investigations. I think it is highly probable that many *congenital deformities*, such as the occasional doubling of the tarsus in the hind-limbs of beetles and other insects, are due to the doubling of a group of determinants, and perhaps the much-discussed and debated problem concerning *supernumerary fingers and toes in human beings* may be explained in a similar way. There is nothing impossible in the assumption that the latter phenomenon is due to reversion 'to an extremely remote, lowly-organised, and many-fingered ancestor,' for we know of other cases of reversion to very distant ancestral forms. But in none of the reliable instances does reversion to ancestral characters extend through such an enormous lapse of time or immense number of generations as must be assumed in this case. The striping on mules points back to an early equine ancestor, and we are led to the conclusion that at the present day the germ-plasm of horses and asses still contains solitary 'zebra' determinants. Reversion may occur to yet more remote ancestors of the modern horse,—even to those possessing three toes; but cases of reversion to still earlier ancestors can hardly be proved with any degree of certainty; nor is it probable, from a theoretical point of view, that any groups of determinants of such extremely remote primitive mammals should have been preserved in the germ-plasm of human beings. Moreover, it is not at all certain that the primitive mammals possessed more than five fingers and toes; and it would be necessary to go back to much more remote ancestors before obtaining any support for the explanation of human polydactylism given by Darwin, and formerly accepted by Bardeleben, Wiedersheim, and others.

Not only is there no firm foundation for this latter assumption, but there appear to me to be very weighty reasons against it. We must not, in the first place, overlook the fact that these primitive ancestors did not possess 'human' fingers; supernumerary fingers are, nevertheless, real fingers, and though they are not always perfect, they are furnished with the form of nail typical of the human finger, and not with claws. In my opinion, we are not justified in assuming that such a supernumerary finger is represented in the germ-plasm by a group of determinants derived from the primitive ancestor, and that this group has in the meantime become transformed into the type of the human finger.

Apart from polydactylism, cases of the doubling of the limbs are known, which, from their nature, cannot be looked upon as atavistic: insects, for example, have never possessed a double tarsus. There must, therefore, be another way in which this doubling might originate.

The sudden appearance of polydactylism and its great tendency to transmissibility can, moreover, be easily understood if we suppose that excessive local nutrition has caused the group of determinants in question to become doubled.* For when this doubling has once occurred in several ids in the germ-plasm, it must be capable of being transmitted on account of the continuity of the germ-plasm; and the degree of certainty with which this will take place will increase as the number of ids in which it has occurred becomes greater. I am entirely of Ernst Ziegler's opinion that polydactylsm is due to a germ variation: this *must* be so whenever it is hereditary, for it would not otherwise be transmissible.

In this way we can also understand why polydactylism, after it has once arisen and has been transmitted through several generations, may finally disappear; for at every fresh 'reducing division' the number of abnormal ids is increased or decreased, and in the latter case their effect may be entirely obliterated, in consequence of their meeting with perfectly normal ids during the process of amphimixis; and in the next generation they may be for ever eliminated. Like all individual variations, they may be absent from one generation and appear again in the next; but in case of continual crossing in normal human beings, theoretically they may be expected once more to disappear completely. This agrees with fact, for supernumerary fingers have never been observed in more than five consecutive generations.

We know that variation consists not only in the addition of parts, but also in their disappearance. *The process of degeneration of parts* must be attributed to the disappearance of the respective determinants from the germ-plasm. In the chapter on reversion, I have already attempted to show that regressive transformations need not occur in *all* the ids of the germ-plasm *at the same time*, and that reversion to long-lost ancestral characters may be ascribed to the preservation of a minority

---

* Dr. R. Zander has recently declared himself in favour of the view that supernumerary fingers, &c., are formed by the mechanical constriction of the rudiments of the embryonic fingers by amniotic threads. But if this is true, they would not be transmissible, and another explanation must be made for the doubling of the tarsus in beetles. (*Cf.* 'Ist die Polydactylie als theromorphe Varietat oder als Missbildung anzusehen?' *Virch. Arch.*, Bd. 125, 1891, p. 453.)

of the determinants in question, which gradually decreases in course of time. The cause of the regression of a determinant is to be looked for in insufficient nutrition, — which condition may occur in a determinant quite as likely as that of more abundant nutrition. If this occurs in the majority of the ids either directly, or in consequence of the accumulation produced by amphimixis, the character controlled by these determinants becomes regressive in that particular individual. If, however, it no longer has a physiological value, it becomes slowly but surely suppressed by panmixia in an ever-increasing number of individuals until it disappears. Specific characters which have long been unrecognisable externally, instead of disappearing entirely, may still be retained in individual ids in the form of incompletely-degenerated determinants; and, as already mentioned, these may cause the reappearance of a character under particularly favourable circumstances.

3. *Summary of Sections* 1 *and* 2, *and Conclusions*

The above remarks may be briefly summarised as follows: —

The origin of a variation is equally independent of selection and of amphimixis, and is due to the constant recurrence of slight inequalities of nutrition in the germ-plasm which affect every determinant in one way or another, and differ even in the same germ-plasm, — not only in different individuals but also in different regions. These variations are at first infinitesimal, but may accumulate; and, in fact, they must do so when the modified conditions of nutrition which gave rise to them have lasted for several generations. In this way deviations may occur in the structure of single determinants or of groups of them, — never, perhaps, in all ids at once, but at any rate in several or even many of them. A doubling of certain determinants of the germ-plasm may originate in the same way. The process of amphimixis has an important share in the accumulation of these modified determinants, for it may raise the minority previously existing in the two parents to a majority by combining their halved germ-plasms. Then, and then only, does selection begin to take place.

The extreme importance of sexual reproduction in processes of transformation only becomes evident, however, when we realise that adaptations are usually concerned with several variations at the same time, and rarely or never arise in connection with a single one. *The process of amphimixis alone rendered it possible for such manifold combinations of characters to be offered to selection, so that the proper choice could be made.* If the view which I have long held is correct, a *single* character is never alone acted upon by natural selection, but the whole aggregate of specific characters is incessantly exposed to this process. The constancy as well as the transformation of the existing specific characters, the removal of superfluous ones, and the development of new characters, is due to the incessant and uninterrupted control of selection. This is only rendered possible by the continual intermingling of all the existing varieties of characters, which can only be effected by amphimixis. Hence, although the latter process is not the primary cause of individual variation, it is nevertheless an indispensable factor in selection, for by its means alone can the material from which variations arise be so arranged that selection can operate.

The theory of variation here propounded also affords a more satisfactory explanation of a further difficulty than can be obtained by any other. In considering the unlimited number of adaptations of organisms to the conditions of existence, we must be surprised at the wonderful plasticity of the species. It gives us the inpression that every variation, however unexpected, might be produced by a species as soon as the species has use for it. On reflecting how certain animals and plants, or parts of plants, are imitated in colour, form, and marking by other animals, we might be inclined to suppose that every part of an animal may assume any required form, colour, or marking, according to requirement.

This must not, however, be taken literally; an organism cannot assume every form, though it may become adapted in so many ways that we cannot possibly attribute its immense number of adaptations to *rare, fortuitous variations, occurring only once.* The *necessary variations* from which transformations arise by means of selection, *must in all cases be exhibited over and over again by many individuals.*

The presence of such an ever active material for primary variations is a direct consequence of the theory here propounded, according to which every part or 'determinate' of a species must present every possible variant in different individuals in the course of generations, and will sometimes be represented in a larger, and sometimes in a smaller majority of modified ids. Since an absolute equality of nutrition in homologous determinants, either in different individuals or in the different ids of the same germ-plasm, is inconceivable, and as every minute variation of a determinant does not disappear of itself with the individual in which it is present, but is transmitted directly to the germ-plasm of the next generation, there can be no scarcity of variations of every determinant, and the presence of the material required for all the *possible* variations of all parts seems to be supported theoretically.

Before considering the modifications which the germ-plasm must undergo *as a whole* during the transformation of species, I should like to meet an objection which might be raised. If all determinants are incessantly exposed to slight inequalities of nutrition and consequently suffer slight variations, what is the cause of the extraordinary pertinacity by means of which the species is preserved without the type undergoing variation,—*what is the cause of the constancy of species?* We might expect that all organic forms must be in a constant state of transition, and that no form, and no organ, could be retained for any length of time.

Several points are disregarded in such a question. In the first place, every species is under the uninterrupted control of natural selection, as is clearly shown by the degeneration of parts which have become useless. And since the old hypothesis of the transmission of somatic variations must, it appears to me, be definitely rejected, this process of degeneration can only be explained as the result of panmixia, *i.e.*, the cessation of the control of natural selection over that part which is no longer of use. We may, however, conclude from the fact that such degeneration is universal, that the determinants are in a constant state of fluctuation; and as degeneration takes place very slowly in all cases, I further infer that, in spite of the frequency with which these fluctuations occur, *they only increase very gradually so as to give rise to perceptible variations.*

As at first stated, the individual fluctuations of the determinants must be regarded as excessively small. Natural selection could produce no *direct* result from an individual variation, for it could not produce a cumulative effect; an accumulation can only be produced by amphimixis, and I am inclined to assume that much of the importance of the latter process is due to this fact. It can cause minorities of modified determinants to accumulate into majorities by mingling the halved germ-plasms of two individuals. By the aid of the 'reducing division' it can also level and equalise matters by fortuitously dispersing the homologously modified determinants of an individual.

It must not be forgotten that slight primary variations of a determinant need not always continue in the same direction: influences of nutrition in the reverse direction will frequently cause them once more to disappear. Only after a determinant has been modified to a considerable degree by the action of a uniform influence during a long period, and the determinants of many ids have become similarly and simultaneously modified, can a variation become visible — after being first accumulated by amphimixis. And even then it by no means forms a permanent specific character, for the question as to whether it will or will not give rise to one is decided by natural selection.

Thus several powerful influences prevent the constant variation of the specific type.

An answer to the question as to *what variations the idioplasm undergoes in the transformation of species* will be found in the chapter on reversion, and it will here only be necessary to summarise what I have already said.

The transformation of species is due to the variation of some, and frequently even of most, of the determinants. Many species do not possess a single character which resembles that of an allied species, and in this case *all* the determinants must be different. But this only implies that all the determinants $a - x$ have been modified *in the majority of ids*; a minority of the latter will contain unmodified ancestral determinants. As the transformation of a species proceeds, the number of modified determinants increases together with the number of ids in which they occur. Nevertheless the dominating principle of selection only permits the transformation of *all* the ids to occur very gradually, so that the germ-plasm of a young species may often contain completely unmodified ids of the ancestral species; and even older species may contain solitary groups of unmodified determinants in many of their ids. This, and this alone, renders reversion possible.

It has been recently maintained that, as a consequence of my theory, I must adopt one of two alternatives, and assume either that the germ-plasm of the higher animals consists of ids of the primitive protozoan ancestors,* or that every id is constructed in accordance with the existing character of the species: my real view, however, is intermediate between these two. I believe that the germ-plasm of a species always consists to a great extent of specific ids, amongst which, however, some more or less unmodified ancestral ones are present, the number being largest when the species is young. The germ-plasm varies in different species, and must differ very much in the higher and lower forms; but its transformation does not go on at the same rate in all the ids, for some are modified only gradually, or are transmitted unaltered through long lapses of time till they are at last casually removed by a 'reducing division.'

This might be regarded as a defect in the process of the transformation of species, for the possibility of reversion, as well as the retention of inactive ancestral ids in the germ-plasm, can scarcely be considered useful to the species. But in nature no contrivance is absolutely perfect, — not even the marvellously developed human eye: all structures are only as perfect as *possible*, — that is to say, as perfect as they need be in order to perform their required functions. This statement also applies to the mechanism for the transformation of species: — it approaches perfection as nearly as is necessary for the performance of its function.

4. Variations on a Larger Scale

a. The Origin of these Variations

We have till now chiefly confined our attention to the question of general individual variation, as exemplified by those minor hereditary differences which distinguish individuals from one another. But there can be no question that variations sometimes occur on a larger scale, and these usually appear suddenly, are met with only in single individuals, and are as a rule hereditary. Darwin has given a large number of instances of this kind. Although the special peculiarity of the black-shouldered peacock, for example, may be due to reversion to an unknown ancestral form, there are many well attested cases in

* Compare Marcus Hartog, 'Nature,' Vol. 44, December 1891. The deductions made by this author from my former views are logically correct, but are no longer justifiable, since in the meantime I myself have gained further insight into the problems concerned.

which a given structure in a species has suddenly become considerably modified. How far such modifications have to do with the formation of new species need not concern us at present: we have only to consider the causes and method of origin of these modifications.

A large number of cases of this kind have been observed, especially amongst plants. Not only *fruits*, but *leaves*, *blossoms*, and *entire shoots* have been found to vary suddenly, and in a striking manner. Several varieties of fruits must be included in this category — such as the variety of peach known as the nectarine, as well as the moss-rose, copper-beech, copper-hazel, the varieties of the beech, hornbeam, and oak, the fern-leaved variety of the maple, and numerous other plants cultivated in our gardens.

These varieties in some cases first appeared as seedlings, *i.e.*, as entire plants. and in others as simple branches or shoots, in which latter case they are usually known as *bud variations*.

Let us first take into consideration the *varieties which have originated from seeds*. These occur most frequently in cultivated plants, *i.e.*, in species which have existed for some length of time under conditions which differ more or less from the natural ones. We are therefore undoubtedly justified in attributing the cause of the variation to the influence of changed external surroundings. But a wild plant, which is transplanted into a garden soil, does not always begin to vary at once: it has, in fact, been shown by Hoffmann's experiments, of which we have already given an account, that many generations often elapse before conspicuous variations occur. And even then they do not appear in all the seedlings, occurring perhaps only in one among several hundreds or thousands.

As in the case of ordinary individual variations of a minor kind, the modification begins a long time before it becomes apparent. A few determinants are first changed, and then a gradually increasing number, until at last, by means of the 'reducing division' and amphimixis, they occur in such numbers in certain germ-cells that they form the majority. The fact that these variations occur on a larger scale than the ordinary ones is due to the *permanent* action of *uniform* changes in nutrition, which give a constant direction to the modification of susceptible determinants, so that an increase is effected. This is wanting in the case of the ordinary and constantly varying nutrition.

'*The accumulative action of changed conditions of life*,' suggested by Darwin,* is consequently theoretically supported to a certain extent by the theory of the continuity of the germ-plasm: those determinants which varied in the first generation continue to do so in a similar direction in the second and third.

Professor Hoffmann has for many years been making very interesting experiments in the Botanical Garden at Giessen which bear on this point, some of which I will now describe.

Various plants, bearing flowers of the normal structure, were exposed during a number of generations to greatly modified conditions of life; they were, for example, grown crowded together in small pots, so that each plant restricted the amount of food obtainable by the others, and were thus scantily nourished. Under this treatment some species—such as *Nigella damascena*, *Papaver alpinum*, and *Tagetes patula*—bore a number of non-typical *double* flowers more or less frequently. The fact that these deviations from the ordinary type in no case appeared in the first generation, proves that they were due to an influence exerted upon the germ-plasm, and not to the *direct* influence of the abnormal conditions of nutrition upon the soma of the plant. Seeds of normal wild flowering plants of different species, when grown in cultivated soil or even when thickly sown in pots, *never produced plants possessing even one double flower*. Only in the course of several, and often many generations, did any of these wild plants exhibit a greater or less number of double flowers, or occasional modifications in the leaves or in the colour of the flowers. It seems to me that only one explanation can be given of this fact, viz., that the altered conditions at first only produced *imperceptible* variations in the germ-plasm of an individual plant,—such, for instance, as alterations in the determinants for the leaves or flowers in *individual* ids, but not in *all* of them at once. These modified determinants were transmitted to the next generation in consequence of the continuity of the germ-plasm; but since the causes of variation continued to operate, the homologous determinants in several other ids also became modified, and thus the number of modified determinants for the leaves and flowers continued slowly to increase, until finally they exceeded the normal determinants in number, and the

* Darwin, 'Animals and Plants under Domestication,' Vol. II., p. 249.

abnormality became apparent as a variation in the flower or in the leaf.

The reason why I lay special stress on these cases is, that natural selection plays no part in them, for we are here dealing with artificial, and not with the natural conditions of life. The imperceptible variations of the germ-plasm naturally did not increase continuously, for the processes of 'reducing division' and amphimixis exerted their influence on every fresh generation, and helped to bring about either a marked decrease of the modified determinants, or their sudden increase — a doubling, or a still greater augmentation. In this way it is easy to understand all the particulars connected with the occurrence of the abnormal flowers and leaves. As a rule, the number of these abnormalities increased in the course of generations with a fairly constant regularity. The following results, for example, were obtained from four generations of *Nigella damascena*, when sown close together: —

1883. No double flowers.

1884. " "

1885. Six double and twenty-three typical flowers, or 24 per cent.

1886. One double and ten typical flowers, or 10 per cent.

The number of double flowers was, however, not always constant, and in some cases they again disappeared entirely. *Papaver alpinum*, for instance, which Hoffmann had cultivated uninterruptedly since 1862, displayed 'a trace of variability in the form of the leaves, and a more decided variation in the colour of the flowers in 1882.' These experiments were continued from 1882 to 1886, and yielded the following proportion of double and normal flowers: —

In 1881	the proportion	was	40·0	per cent.
In 1882	"	"	4·0	"
In 1883	"	"	5·3	"
In 1884	"	"	13·0	"
In 1885	"	"	0	"
In 1886	"	"	0	"

The complete disappearance of abnormal flowers in the two last of these years seems inexplicable at first sight; but it is easily accounted for if we bear in mind the fact that natural, and not artificial fertilisation occurred, — *i e.*, the plants were crossed at random, — and that every reducing division and subsequent amphimixis were capable of reducing the modified determinants to a minority, and even of removing them completely from the germ-plasm. The modification of the determinants manifestly occurs very gradually, as the small number of modified flowers in most of the years shows; and the chances must have been greatly in favour of the union in amphimixis of germ-cells which contained few or no modified determinants.

Had fertilisation been produced artificially, and abnormal flowers always employed for the purpose, it would have been easy in the course of a moderate number of generations to modify the species entirely, and these cases would then have served as an illustration of the process of natural selection. It would have been still easier to produce the contrary effect by selection, — that is to say, to keep the species constant, and suppress any subsequent variation in this direction. For in the present instance, variation evidently occurred slowly, and most of the determinants were not easily affected by it; and these experiments furnish additional proof of the truth of the statement propounded above that the elements of the germ-plasm only change slowly, and with difficulty: they merely fluctuate to a very slight degree, and only undergo an important change *of any duration* when uniform influences continue to act on them in one direction for a longer time.

It by no means follows from what has just been said that influences of environment and nutrition exist, which, when they have acted for a long time, are able to modify the majority of the determinants for certain parts of the body, and thus to produce purely climatic variations, in the origin of which natural selection has no share. Many — perhaps even most — of the 'climatic' varieties are rightly so called.

Such sudden variants, however, appear not only in *entire* plants which have been produced sexually, but also in the individual shoots of a plant. These *bud-variations* are rarely met with; but in cases in which they occur, they can be propagated by cuttings or grafting, and often even by seeds.

When I ventured some years ago to suggest that sexual reproduction has come into force in organic nature in order to preserve the variability which had existed since the time of the primordial beings, facts concerning bud-variation were put forward by several persons to prove that variability may occur in the absence of sexual reproduction. At that time I certainly did not attach sufficient importance to the variation of the germ-plasm in consequence of influences acting directly; but the existence of bud-variations does not prove that variation occurs without amphimixis. For all those plants in which bud-variation has been observed are reproduced sexually, and their idioplasm therefore contains ids and determinants which differ individually: the different intermingling and behaviour of these in the process of growth would alone form a basis for variations.

In my opinion, indeed, this heterogeneous composition of the germ-plasm produced by amphimixis is an essential factor in bud-variation, notwithstanding the fact that it may not in this case give the first stimulus to variation, any more than it does in ordinary individual variation. Plants which have for a long time been propagated by means of buds and shoots, like the potato and sugar-cane, must possess a germ-plasm consisting of different kinds of ids; for they were formerly reproduced sexually, and the complex intermingling of their determinants thus produced, cannot have undergone an appreciable modification during the period in which they have multiplied asexually. Their germ-plasm must therefore present far more favourable conditions for variation than would one composed of identical ids or of one kind of id only, — did such a germ-plasm exist.

The primary cause of bud-variation must be the same as that of variation from seeds, and must be due to *inequality of nutrition in the germ-plasm*; — the term 'nutrition' being used in its widest sense, so as to include differences in temperature, &c. This view not only receives support from theoretical considerations, — for theoretically it could only be replaced by the assumption of an internal phyletic developmental force, — but it is also supported by observation. For all recorded observations go to prove that bud-variations are most likely to occur when the plant is placed under abnormal conditions, and especially when it is cultivated. As the direct modification of the soma caused by these conditions is not hereditary (*Nägeli*), and *cannot* be so, — for somatic variations are only hereditary when they proceed from the germ, — we are obliged to assume that the modification of several or many determinants in the germ-plasm is due to inequality of nutrition.

That bud-variations are produced by the same causes as those which occur in reproduction by seeds, is borne out by the fact that the former occur most frequently in those species which have already varied greatly in multiplying by seeds.* Expressing this in terms of the idioplasm, we may say that it occurs oftenest *in those species in which the homologous determinants already exhibit considerable differences.*

We attribute these variations to influences of nutrition, which at first bring about slight, and then more marked deviations in certain determinants of the germ-plasm during the course of their growth and multiplication, if these influences continue; but this alone does not fully account for the process. The question then arises as to how the modifying influences can cause a *particular* bud to undergo variation while all the rest remain unchanged, although they are exposed to the same influence. Some other influence is therefore required before a modification of this kind can appear.

If we remember that bud-variations sometimes occur in wild plants, or in those which, like the forest-trees of our parks, exist under practically the same conditions as many wild ones, it will appear still more probable that the inequalities of nutrition, while constituting the primary cause of bud-variation, cannot *alone* bring it about.

My own conception of the process is as follows. Just as in the case of ordinary individual variation, bud-variation is primarily due to those slight, fluctuating, structural changes which all determinants undergo in consequence of minute and inevitable fluctuations in nutrition. As in the former case, the homologous determinants of the various ids are not all affected to the same extent; some become greatly modified, and others little or not at all. A difference is, however, seen in the fact that in this case the *same* influence of change—*e.g.*, generally improved nutrition—occurs for a considerable time, and throughout several generations. As in the case of variations in plants raised from seeds, a modification of greater extent can thus be produced in these determinants.

Up to this point the process is quite similar to that of the spasmodic variation of seedlings. A difference, however, results owing to the non-occurrence of amphimixis in the case of gemmation, for the variable bud does not arise from the germ-plasm of a seed, but from '*blastogenic*' *germ-plasm* ('Knospen-Keim-

* Darwin, *loc. cit.*

plasma'). The latter is derived directly from the former, from which the plant in question — or one of its ancestors, if the plant itself was raised from a cutting — arose. If, during the growth of a tree, one of the determinants, N, varied in the same way in certain ids, and tended, for example, to produce red leaves instead of green ones, red leaves would nevertheless not be produced till all, or at any rate a majority, of the determinants N had become transformed into the red variety. If a 'reducing division' intervened between the germ-plasm of the bud and the growth of the latter into a shoot, a minority of 'red' determinants might give rise to a majority in one of the two daughter-cells; but this reducing division is exactly what does not take place in ordinary cell-multiplication.

Since bud-variations are of very rare occurrence, and only *one* of many thousand buds on the *same* plant varies so as to produce red leaves, for instance, while the altered conditions simultaneously affect all the buds, I conclude that the modified determinants which the tree contains *may sometimes attain a majority in consequence of an abnormal differential nuclear division*. Should this take place in the apical cell or cells of a bud, the resulting shoot might, to return to our former example, bear both green and red leaves; for, according to our pre-supposition, the 'green' and 'red' determinants of the apical cells would have been separated during the cell-divisions in such a manner that in some leaves the green determinants, and in others the red ones, might be in the majority. If the separation occurs at an earlier stage, before the apical cells are formed, — that is to say, in the cambium, — a shoot bearing red leaves alone might arise from the cell which receives the 'red' group at the differential division of the nucleus.

There is nothing impossible about this assumption; for during the process of mitotic nuclear division, irregularities might occur in the complex apparatus by which this process is effected, and individual cases of such irregularity have actually been observed: — even the possibility of a *direct* nuclear division cannot be entirely overlooked. I am, however, far from considering this hypothesis as established, and merely offer it as a suggestion.

Nägeli was of the opinion that all variations are slowly prepared in the idioplasm in the course of generations before they become apparent, and in stating this view he also specially referred to bud-variations. I fully agree with him in this respect, and in the course of this book have repeatedly shown how these gradual modifications ('Umstimmungen') of the germ-plasm, or of individual parts of it, are to a certain extent the natural result of its assumed structure. In the case of bud-variations these invisible modifications may occur in a much earlier generation than that in which they appear; and hence it is easy to understand why this form of variation mostly occurs in those plants which, like the rose and Azalea, have already varied in reproduction by seed. For modified determinants are more readily accumulated by amphimixis; and a germ-plasm which has inherited such determinants from its ancestors may, after these have been still further modified, give rise to a bud in which, by a fortuitous differential nuclear division, the modified determinants are in the majority, and can thus become effective.

Unfortunately it has not been observed whether complicated modifications, like that of the moss-rose, owe their origin to bud-variation. This is quite possible theoretically, for the invisible preliminary to variation — the modification of certain determinants — is just as likely to affect a single determinant as a whole group; even the formation of two or more new determinants, by the multiplication of a primary one, is quite as possible here as in the case of the ordinary transformation of species. The sudden appearance of such modified groups of determinants is due therefore to fortuitous differential nuclear division. Such cases prove that the preparation for the modification is a slow process, for it appears impossible to conceive of any cause producing a sudden variation of an entire group of determinants by any method.

The *extent* of any spasmodic variation will depend on the extent to which the various groups of determinants have been permanently exposed to abnormal nutrition. It would doubtless be as yet premature to bring their *qualitative* characteristics into any causal relation with definite influences producing variation. We can only state the necessity of assuming, *a priori*, that the extraordinarily complex germ-plasm is provided with special means for the transmission of nutrient fluid, the increase or diminution of which must produce purely local differences of nutrition; and that, on the other hand, the vital units must, owing to slight changes in their structure, modify the characteristics of the part of the body which they represent, in some inexplicable way. It is not yet possible, for instance, to give an exact account of the changes in the group of determinants for the legs which led to the sudden production of a bow-legged 'otter'-sheep, or of the modifications of certain determinants in the rudiments of the leaves which caused the formation of the notches characteristic of a certain variety of birch.

b. The Transmission of these Variations

The transmission of 'sports' in plants has hitherto been a very obscure problem. Seed-variations are often propagated by means of seeds, but cannot always be thus reproduced, or at least only in individual cases. Bud-variations can usually only be propagated by cuttings or grafting, but some instances are known in which they have also been reproduced by seeds, though only in a certain percentage of cases. The origin of these irregularities was unknown, and no previous theory of heredity could offer a reason for them; all that could be said was, that in these instances transmission was very capricious. The theory of the germ-plasm, however, offers a very simple explanation of them. We will first consider certain facts recorded by Darwin, which are very valuable in respect of this theory.

Darwin states that 'when a new peculiarity first appears, we can never predict whether it will be inherited.'* If both parents exhibit the variation, 'the probability is strong that it will be transmitted to at least some of their offspring.' Bud-variations are propagated to a much smaller extent than seed-variations, but the power of transmission often appears to be very capricious, inasmuch as one and the same modification on a single plant is transmitted by seed in one case and not in another. Thus vain attempts were for a long time made to propagate the weeping ash by seeds. Over twenty thousand seeds developed into trees of the ordinary form, but offspring with hanging boughs were finally raised from the seeds of another specimen of this variety. The same ash-tree, however, did not transmit this character to all its offspring, but only to a certain

* 'The Variation of Animals and Plants under Domestication,' Vol. I., London, 1888, p. 460.

percentage of them; and Darwin mentions a famous weeping oak at Moccas Court which transmitted its special character to all its seedlings, though in *varying degrees*.

According to our theory, the transmission of a variation by seeds depends on the presence of a corresponding modification in the majority of the determinants in the germ-plasm of the seeds. If the germ-plasm contains a hundred ids, the controlling forces of which are equal, more than fifty determinants N must be transformed into N^1 before the modification would be perceptible in the seedling. Since therefore, as was remarked above, new variations probably never appear simultaneously in *all* the determinants, but only in a varying percentage of the ids, the chances are greatly against all the seedlings produced by the transformed plant exhibiting this modification. For every germ-cell has undergone a 'reducing division,' and hence many of them will always only possess a minority of the modified determinants N^1, and this may even be a very small one if the majority in the germ-plasm of the parent-plant was small. When two such cells unite in the process of amphimixis, the resulting germ-plasm contains only a small minority of modified determinants N^1, and the modification is inappreciable. This accounts for the fact that the seedlings of a variety hardly ever reproduce the variety in *all* cases; and that in rarer instances, such as that of the weeping oak already mentioned, all the seedlings may exhibit the corresponding variation, although in varying degrees. For the composition of the germ-plasm must differ in each of them in consequence of the processes of 'reducing division' and amphimixis, even when the parent-plant contains only a small but ever varying minority of ancestral determinants.

We can, moreover, easily account for the fact that seedlings of a variety, such as the balsamine, may resemble the parent-plant, without transmitting the character of the variety to their offspring. In this case all the daughter-plants of the variety in question must retain a majority of the modified determinants, but in very varying degrees. In those which contain a very large majority in their germ-plasm, it necessarily follows that the larger number of the germ-cells produced must contain a majority of modified determinants; but in the case of those in which the proportions are more equal, the chances are in favour of the seedlings containing only minorities of these determinants.

Darwin mentions that only thirty per cent. of the seedlings of the wild variegated variety of *Ballota nigra* possessed the variegated leaves of the parent-plant; but sixty per cent. in the second generation were variegated. This is also quite in accordance with our theory; for the determinants for the leaves in the mother-plant cannot all have been modified, but only the majority; these would then become differently grouped by means of the reducing divisions of the germ-cells. A seed reproduces either the variety or original form according to whether modified determinants, which were brought together in fertilisation, constituted the majority or minority. As in the instance first quoted, variegated plants were alone retained for cultivation in the second generation, and consequently a greater number of the determinants of the variety were brought together at fertilisation: *the percentage of the variety was therefore bound to rise in the third generation.* This proportion would have increased still further if a fourth and fifth generation had been raised in the same way; for in those seeds which yield variegated plants the germ-plasm must of necessity contain more modified determinants than in the case of those producing the original form. Hence with every generation the chances of an increased majority of modified determinants become greater; and I have no doubt that in this instance, by the constant selection of the most variegated plants for further cultivation, a 'pure' variegated race might eventually have been obtained, which would have transmitted its character to the great majority of its offspring — or as we usually express it — 'to *all* its offspring.'

The fact that many weeping ashes transmit their special character to many, but not all, the seedlings is also attributable to the diverse effects of the 'reducing division' and amphimixis, the former of which causes the introduction into the germ-plasm of a large majority of modified determinants, and the latter of only a slight one. The *manner in which the variety arises* must here, however, be taken into account. *Bud-variations are much more rarely reproduced by seeds than are variations arising from seeds*; but, on the other hand, they can almost always be propagated by grafting, budding, or by cuttings. Hence the capacity of a certain individual for transmitting the variation to its seeds may be owing to its having arisen from a seed, while in another case the same variation arose in a bud, and it could therefore rarely or never be transmitted by seeds.

The most difficult point to explain theoretically, is why bud-variation is *not* generally, but only occasionally, transmitted by seeds. Even this may to a certain extent be accounted for by the present theory.

A bud is a growing point, enveloped by scales. It arises from the apical cells, which produce the other cells of the shoot by means of continual division, and these form the interfoliar parts, leaves, and flower-stocks. In accordance with the principle of the continuity of the germ-plasm, a part of the 'blastogenic' germ-plasm of the apical cells must be transmitted in an 'unalterable' ('gebundenen') condition to certain cells of the shoot as accessory idioplasm ('Neben-Idioplasma') or '*reserve germ-plasm*.' From these cells it is passed on to the sexual organs, where it is used for the formation of germ-cells. This reserve germ-plasm remains undisintegrated, and is perfectly distinct from, and independent of, the 'blastogenic' germ-plasm, which is gradually distributed during the ontogeny of the shoot into groups of determinants.

The fact that bud-variations are so rarely transmitted by seeds seems to me to be owing to the cause just mentioned; for the majority of modified determinants required to make a modification apparent may evidently be present in the 'blastogenic' germ-plasm, though absent in the reserve germ-plasm. If we remember that these variations have been prepared long beforehand in the germ-plasm, and that at first a few, and then gradually a larger number of determinants become modified in a similar way, and that finally a fortuitous differential nuclear division must, on our assumption, intervene before the 'blastogenic' germ-plasm of a certain apical cell can contain a majority of modified determinants, it becomes comprehensible why the reserve germ-plasm contained in this cell may behave differently, and contain only a few or none of the determinants in question in a modified condition. We must not forget that the influences which produce the variation are not those of the nutrition of the bud in which the modification appears, but those which affected the determinants during their long course from the ancestral plant of a past generation to the bud in which they now appear.

This explanation seems to me to sufficiently account for the fact that the seeds which give rise to 'sports' indirectly, need not necessarily transmit the modification.

The question as to why this transmission is of *such rare occurrence* must, however, be considered more closely. I am inclined to seek the causes of this fact in the processes of 'reducing division' and amphimixis, to which the reserve germ-plasm, but not the 'blastogenic' germ-plasm, is subjected. Let us suppose that the two were precisely similar at first with respect to their contained modified determinants N^1, each possessing a small majority of them: the shoot must then exhibit the variation, but only certain germ-cells would contain a majority of N^1, the rest containing a minority of these determinants in consequence of the diversity resulting from the reducing division. It is true that two germ-cells containing majorities of N^1 might unite in fertilisation; but this extremely favourable case would only occur very rarely,—when, as assumed, the majority in the reserve germ-plasm was only a *slight* one; while the other cases, in which amphimixis leads to N^1 being in the minority, would take place much more frequently.

It is, however, by no means certain that reserve germ-plasm and 'blastogenic' germ-plasm must contain a similar percentage of N^1. The two may very well differ in this respect, and it is extremely probable that a larger percentage of N^1 in the reserve germ-plasm leads to the formation of seeds which give rise to seedlings exhibiting variation. We may therefore suppose that in the one case, when the 'blastogenic' germ-plasm contains even a small majority of N^1, the modification of the shoot will actually occur; and that in the alternative case, in which the reserve germ-plasm alone contains a majority of N^1, variation will take place in a larger or smaller number of seedlings in the following generations. When both contain a majority of N^1, the shoot must vary, and some of the seeds arising from it must transmit the variation.

This explanation is not so hypothetical as it might appear. Certain facts point with certainty to the conclusion that these 'sports' in many cases really contain only a very slight majority of modified determinants. In most bud-variations *reversions to the original form* are of frequent occurrence, both in the buds themselves and in the generation derived from them either by seeds or buds. This has already been mentioned in the chapter on reversion, and is so well known that a brief statement of a few cases will suffice. In the botanical gardens at Bonn, Professor Strasburger showed me an immense hornbeam with deeply notched leaves, like those of the oak, a large branch of which had reverted completely, and bore leaves of the ordinary form. I have in my garden a 'fern-leaved' beech, some of the leaves of which have the usual form; Darwin states, in fact, that some shoots of this variety may produce fern-like leaves, normal leaves, and various intermediate forms. Slight inequalities in the nuclear divisions may in this case displace the controlling resultant of the determinants N and N^1, supposing that the majority of N^1 is only a small one.

SUMMARY AND CONCLUSION

All the phenomena of heredity depend on minute vital units which we have called 'biophors,' and of which living matter is composed: these are capable of assimilation, growth, and multiplication by division. We are unacquainted with the lowest conceivable organisms, and do not even know if they still exist. But they must at any rate have done so at some time or other, in the form of single biophors, in which multiplication and transmission occurred together, no special mechanism for the purposes of heredity being present. A higher order of beings would then have been constituted by those organisms which were composed of a large number of similar biophors. Of these also we have no actual knowledge based on observation, but must suppose that they too required no special apparatus for the processes of transmission; for a reproduction by binary fission must result in two perfectly corresponding halves, each containing similar biophors, and each of which, simply by the multiplication of these units, is able to give rise to a complete organism exactly like the parent.

This simple form of transmission must have become modified when the biophors underwent differentiation in connection with a division of labour, and became combined in various ways to form the body of the organism. These two kinds of hypothetical beings might be respectively distinguished as homo-biophorids and hetero-biophorids. Not only might a firmer cortex and softer internal substance be present in the latter, but a differentiation into anterior, posterior, dorsal, and ventral regions might occur; several layers of the body substance, differing structurally and functionally from one another, might also be developed, together with motile and non-motile processes — such as flagella, cilia, spines, and hooks, — like those present amongst the Infusoria; and there might, moreover, be a permanent aperture in the firmer outer layer through which solid food passed into the interior, and so on.

When the body thus became constructed, in a more or less complex manner, of various kinds of biophors arranged in a definite manner, simple binary fission no longer sufficed for the transmission of the characters of the parent to the offspring. If the parts situated in the anterior, posterior, right, left, dorsal, and ventral regions differed from one another, all the elements — *i.e.*, all the kinds and groups of biophors — could not, by any method of halving, be transmitted to both the offspring resulting by division so that they could develop by mere growth into an organism resembling the parent. Special means must then have been adopted to render such a completion and consequent perfect transmission possible; *and this was attained by the formation of a nucleus*.

We may, with de Vries, regard the cell-nucleus as having originally served merely for the storage of reserve biophors, which were destined to become doubled on the division of the organism, each half rendering the completion to an entire individual possible when those kinds of biophors which were wanting were transferred to it. Subsequently — that is, in the multicellular organs possessing highly differentiated cells — the nucleus took on other functions, which regulated the specific activity of the cell, though it still retained biophors capable of supplying the characters of the cells which were still wanting, and therefore still served as the bearer of the biophors controlling the character of the cell.

If, therefore, a special apparatus for transmission became necessary in the hetero-biophorids or unicellular organisms, and appeared in the 'cell' in the form of a '*nucleus*,' it must have become still more complex on the introduction of the remarkable process of amphimixis, which, in its simplest and original form, consists in the complete fusion of two organisms in such a manner that nucleus unites with nucleus and cell-body with cell-body. In the higher unicellular organisms this process is in most cases restricted to the fusion of the nuclei, half the nucleus of one animal uniting with half that of another. The process of division shows that the nucleus has a structure precisely analogous to that of the nucleus in multicellular organisms; we may therefore assume that the hereditary substance here likewise consists of several equivalent groups of biophors, constituting 'nuclear rods' or 'idants,' each of which contains all the kinds of biophors of the organism, though they deviate slightly from one another in their composition, as they correspond to individual variations. Half the idants of two individuals become united in the process of amphimixis, and thus a fresh intermixture of individual characters results.

The apparatus for transmission in those multicellular organisms in which the cells have undergone a division of labour. is essentially similar to that seen in unicellular beings; although, in correspondence with the greater complexity of their structure, it is more complicated. As the process of amphimixis occurs in them also, and the fusion of highly-differentiated multicellular individuals seems only to be possible by a temporary return to the unicellular condition, we find that the so-called 'sexual reproduction,' which is of general occurrence amongst them, consists in all the primary constituents ('Anlagen') of the entire organism being collected together in the nuclear matter of a single reproductive cell. Two kinds of such cells, which are differently equipped, and mutually attract one another, then unite in the process of amphimixis, and constitute what we are accustomed to call the 'fertilised egg-cell,' which contains the combined hereditary substances of two individuals.

According to our view, this hereditary substance of the multicellular organisms consists of three orders of vital units, the lowest of which is constituted by the biophors. In the unicellular forms a more or less polymorphic mass of biophors, having a definite arrangement, constitutes the individual nuclear rods or idants, several of these making up the hereditary substance of the nucleus which controls the cells; and similarly in these higher forms, groups of biophors, arranged in a certain order, constitute the primary constituents of the individual cells of the body, and together form the second order of vital units, — the cell-determinants, — or simply, the '*determinants*.'

The histological character of every cell in a multicellular organism, including its rate and mode of division, is controlled by such a determinant. The germ-plasm does not, however, contain a special determinant for every cell; but cells of a similar kind, when, like the blood-cells, they are not localised, may be represented by a single determinant in the germ-plasm. On the other hand, every cell, or group of cells, which *is to remain independently variable*, must be represented in the germ-plasm by a special determinant. Were this not the case, the cell in question could only vary in common with other cells which are controlled by the same determinant.

The germ-cell of a species must contain as many determinants as the organism has cells or groups of cells which are independently variable from the germ onwards, and these determinants must have a definite mutual arrangement in the germ-plasm, and must therefore constitute a definitely limited aggregate, or higher vital unit, the '*id*.'

From the facts of sexual reproduction and heredity we must conclude that the germ-plasm contains many ids, and not a single one only. The formation of hybrids proves that the two parents together transmit all their specific characters, so that in the process of fertilisation each contributes a hereditary substance which contains the primary constituents of all parts of the organism, — that is, all the determinants required for building up a new individual. The hereditary substance becomes halved at the final stage of development of the germ-cells, and consequently all the determinants must previously have been grouped into at least two ids. But it is very probable that many more ids are usually present, and that in many cases their number far exceeds a hundred.

It cannot be stated with certainty which portions of the elements of the germ-plasm observable in the nucleus of the ovum correspond to ids, though it is probable that only parts of, and not the entire 'chromosomes,' are to be regarded as such. Until this point can be definitely decided, our further detailed deductions will be based on the view that the nuclear rods (chromosomes) are aggregates of ids, which we speak of as '*idants*.' In a certain sense, the latter are also vital units, for they grow and multiply by division; and the combination of ids contained in them, although not a permanent one, persists for some time.

The '*germ-plasm*,' or hereditary substance of the Metazoa and Metaphyta, therefore, consists of a larger or smaller number of idants, which in turn are composed of ids; each id has a definite and special architecture, as it is composed of determinants, each of which plays a perfectly definite part in development.

The development of the primary constituents, contained

in the germ-plasm of the reproductive cell, takes place in the course of the cell-divisions to which the ontogeny of a multicellular organism is due, in which process all the ids behave in an exactly similar manner. In the first cell-division every id divides into two halves, each of which contains only half the entire number of determinants; and this process of disintegration is repeated at every subsequent cell-division, so that the ids of the following ontogenetic stages gradually become poorer as regards the diversity of their determinants, until they finally contain only a single kind.

Each cell in every stage is in all cases controlled by only *one* kind of determinant, but several of the same kind may be contained in the id; and the 'control' of the cell is effected by the disintegration of the determinants into biophors, which penetrate through the nuclear membrane into the cell-body; and there, according to definite forces and laws of which we are ignorant, bring about the histological differentiation of the cell, by multiplying more rapidly at the expense of those biophors already forming the cell-body. Each determinant must become 'ripe,' and undergo disintegration into its biophors, at a definite time or at a certain stage of ontogeny. The rest of the determinants in the id of a cell, which are destined for subsequent stages, remain intact, and have therefore no effect on the control of the cell; but the mode of their arrangement in the id, and the special rate of multiplication of each kind, determine the nature of the next nuclear division — that is, as to which determinants are to be distributed to one daughter-cell, and which to the other. The histological nature of these two cells, as well as the control of their successors, is determined by this division; and thus the distribution of the primary constituents contained in the germ-plasm is effected by the architecture of the id, which is at first of a definite kind, but afterwards undergoes continual and systematic changes in consequence of the uneven rate of multiplication and gradual disintegration of the ids.

The apparatus for cell-division is only of secondary importance in the process; its chief part, the 'centrosome,' like the hereditary substance, is derived from the parental germ-cell or cells, but only constitutes the mechanism for the division of the nucleus and cell, and contains no 'primary constituents.' The rate of the cell-divisions cannot, moreover, be determined by the centrosome, although it produces the required stimulus: the apparatus for division is set in motion by the cell, which is controlled by the idioplasm. Were this not the case, the nuclear matter could not be the hereditary substance, for most of the hereditary characters of a species are due in a less degree to the differentiation of individual cells than to the number and grouping of the cells of which a certain organ or entire part of the body consists; these, however, again depend on the mode and rate of cell-division.

The processes occurring in the idioplasm which direct the development of the organism from the ovum — or to speak in more general terms, from one cell, the germ-cell, — do not in themselves furnish an explanation of a series of phenomena which are in part directly connected with the ontogeny, or else result from it sooner or later: the phenomena of *regeneration*, *gemmation*, and *fission*, and the *formation of new germ-cells*, all require special supplementary hypotheses.

The simplest cases of *regeneration* are due to the fully formed tissue, consisting of similar cells, always containing a reserve of young cells, which are capable of replacing a normal or abnormal loss. This, however, is insufficient in the more complex cases, in which entire parts of the body, such as the tail or the limbs, are regenerated when they have been forcibly removed. We must here assume that the cells of the parts which are capable of regeneration contain 'supplementary determinants' in addition to those which control them, and that these are the primary constituents of the parts which are to be formed anew in the process of regeneration. They are supplied to certain parts of the body at an earlier ontogenetic stage in the form of 'inactive accessory idioplasm,' and only become active when the opposition to growth has been removed in consequence of the loss of the part in question. The equipment of a cell of any part with supplementary determinants presupposes a greater complexity in their distribution, in correspondence with the greater complexity in structure of the part; and thus the capacity for regeneration is limited, for a part can no longer be provided with an apparatus for regeneration when its structure is too complicated. The ordinary assumption that the regenerative 'force' decreases as the complexity in structure increases, is therefore to a certain extent true, but not if it implies the existence of a special force which provides for regeneration, and which always diminishes in correspondence with the degree of organisation. Even if we imagine this 'force' to be a mechanico-physiological one, it could not be considered as a *primary* quality of the organism, and to some extent the inevitable result of life itself, but must be looked upon as an adaptation.

Reproduction by fission is closely connected with regeneration; it presupposes the existence of a similar apparatus in the idioplasm, which, however, has in most cases reached a higher stage of development; fission must have arisen phyletically from regeneration.

The origin of multiplication by *gemmation*, and the phenomena exhibited by this form of reproduction, are different from those concerned in fission. In plants and Cœlenterates, gemmation originates in one cell, which must consequently contain a combination of all the determinants of the species closely resembling that existing in the fertilised ovum. In the Polyzoa, however, this process does not originate in one cell, but in at least two, and probably more, belonging to two different layers of cells (germinal layers) of the body; and in Tunicata, again, the material for the bud is produced from all three germinal layers.

The first of these forms of budding must be primarily due to the admixture of 'unalterable' ('gebundenem') germ-plasm to certain series of cells in ontogeny in the form of *inactive 'accessory idioplasm,'* or *'blastogenic' idioplasm* ('Knospungs-Idioplasma'). In plants this is contained in the apical cells; and in hydroid polypes, in the cells of the ectoderm.

In the second group of animals mentioned above, we must assume that the 'blastogenic' germ-plasm becomes disintegrated into two groups of determinants at an early ontogenetic stage, and that each of these is passed on in an 'unalterable' condition, through various generations of cells, until the time and place of its activity are reached.

In the third group, the inactive 'blastogenic' idioplasm divides into three groups of determinants, one of which passes into the ectoderm, the second into certain cell-series of the mesoderm, and the third into others in the endoderm, until they reach the part in which they have to become active.

Gemmation must have originated phyletically by a doubling of the germ-plasm taking place in the fertilised egg, so that one half remained inactive, and was then either passed on as inactive 'blastogenic' germ-plasm, or else became divided up in the course of ontogeny into groups, which were passed separately to the same region, viz., that of the bud.

In all cases in which the power of budding was permanently retained by the species, the occurrence of this process of doubling of the germ-plasm seems to have persisted through the ontogenetic stages from an early period; for we find that the individuals arising by gemmation very frequently vary independently of one another, and often even to a great extent. But independent variation from the germ onwards implies the existence of special determinants in the 'blastogenic' germ-plasm. Medusæ could never have been produced from polypes by budding if independently variable determinants of the buds had not been present in the germ of the fertilised ovum. We therefore assume that *two kinds of germ-plasm exist in those species in which alternation of generations occurs*, both of which are present in the egg-cell as well as in the bud, though only one of them is active at a time and controls ontogeny, while the other remains inactive. The alternating activity of these two germ-plasms causes the alternation of generations.

The *formation of germ-cells* is brought about by the occurrence of similar processes in the idioplasm to those which cause gemmation. One part of the germ-plasm contained in the fertilised egg-cell remains inactive and 'unalterable,'— that is, it does not immediately become disintegrated into groups, but is passed on in the form of accessory idioplasm to certain series of cells in ontogeny, and thus reaches the parts in which germ-cells are to be formed. Thus the whole of the parental germ-plasm, with all its determinants, forms the foundation of the germ-cells which will give rise to the next generation, and the extremely accurate and detailed transmission of parental characters to the offspring is thereby rendered comprehensible.

In multicellular plants and animals, the germ-plasm becomes more complex in consequence of *sexual reproduction*, in which process the ids of two different individuals, the parents, are accumulated in the fertilised egg-cell every time amphimixis occurs. This has caused the occurrence of the 'reducing division,' which accompanies the formation of male and female germ-cells, and results in the number of ids and idants being reduced to the half. The reduction is important in elucidating the phenomena of heredity in forms which are reproduced sexually, for the ids of a germ-plasm are not by any means all alike, but differ to the same relative extent as do the corresponding individuals. As the reduction does not always occur in the same way, and the resulting halves contain different idants on different occasions, and these fall to the share of individual germ-cells, it is possible for the germ-cells of one individual to contain very different combinations of idants. This results in the dissimilarity between the offspring of the same parents, — or, to express it in more general terms, in the extreme diversity as regards the intermixture of individual differences.

During the development of a new individual from the fertilised egg-cell, the ontogeny is directed by the ids of the two parents which constitute the germ-plasm. Structures intermediate between those of the parents thus frequently arise — but only when perfectly homologous ids are opposed to one another, and have a similar 'controlling force.' This force depends not only on the similar rate of multiplication of the biophors transmitted by the controlling determinants into the cell-body, and on the suppression of those already present, but also on the number of precisely similar determinants derived from each parent. The larger the number of 'homodynamous' determinants, the greater is their controlling effect on the cell; and if a larger number of homodynamous determinants are opposed to fewer heterodynamous determinants of the other parent, the former gain the victory. The preponderance of one of the parents in transmission is thus rendered comprehensible, whether it concerns individual parts or the entire organism.

The type of the child is determined by the paternal and maternal ids contained in the corresponding germ-cells meeting together in the process of fertilisation, and the blending of parental and ancestral characters is thus predetermined, and cannot become essentially modified by subsequent influences. The facts relating to identical twins and to plant-hybrids prove that this is so. In the latter, the individuals produced by crossing two constant species display as constant an intermixture of characters as would be the case if they constituted a natural species. The ids of each species must be looked upon as perfectly homodynamous as regards the specific characters; two distinct groups of homodynamous ids are opposed to one another, and the preponderance of one or other parental group in any particular part of the plant depends on the presence of a larger number of homodynamous determinants representing the part in question, and on their possession of a greater controlling force.

Reversion to grandparents and great-grandparents, or to uncles and aunts, may be accounted for by the fact that, in the first place, the idants and ids are not formed anew in the germ-plasm of the parents, but are derived from the grandparents; and, secondly, that the combination of ids contained in the individual germ-cells of the parent becomes very diversified in consequence of the 'reducing division.' The usually accepted assumption of breeders that one-fourth of the 'blood' of the grandchild is derived from each of the four grandparents, and one-eighth from each of the eight great-grandparents, is therefore inaccurate. The number of ids of any particular ancestor which are contained in the germ-plasm of a ripe germ-cell depends entirely on the manner in which the reducing division occurs; and, under certain circumstances, a germ-cell might presumably contain half the entire number of ids of one grandparent, and none of those of the other three. The larger the number of ids derived from an ancestor, the greater is the probability that some of the characters of this ancestor will appear in the descendant; but this depends on the force of the ids of the other parent, which comes into play when amphimixis takes place, and also on whether the ids derived from this ancestor are the dominant ones which determined his 'type' ('Bild').

Reversion to an ancestor must consequently always occur when, in consequence of the 'reducing division,' the ids determining the type of this ancestor reach a particular germ-cell of the individual in question, — if they are not opposed by a stronger group of ids derived from the other parent in the process of amphimixis. This holds good for each individual part of the offspring, as well as for the entire organism, for the number of homodynamous determinants may be, and generally is, different in the various parts, — at any rate in the case of the *individual* differences between human beings.

From this theory, it could be predicted that hybrid-plants fertilised with their own pollen must produce very variable offspring, and that individuals of these hybrids must, moreover, revert to one or other of the ancestral species: both these statements are borne out by fact.

Although *reversion to more remote ancestors* is also brought about by the same factors,—viz., the 'reducing division' and amphimixis,—it requires further elucidation. The theory of selection requires that only a majority, and not *all*, the determinants of a part which is to be modified shall undergo a corresponding change. Old unmodified determinants of various parts are therefore retained in the germ-plasm of a species, and can only be removed from it very gradually by fortuitous 'reducing divisions.' This renders reversion to the characters of very remote ancestors *possible;* its *occurrence*, however, depends upon the reducing division and amphimixis taking place in a favourable manner. If the reduction causes similar groups of ancestral determinants to be brought together in *several* ids, and this germ-plasm, in the process of amphimixis, unites with that of another germ-cell, which also contains similar ancestral determinants in *several* ids, these may gain the victory over the modern determinants in the struggle of the ids during ontogeny. This, however, will chiefly depend on the kind and strength of the modern determinants which are opposed to them; and thus reversion to ancestral characters occurs very frequently in crosses between races (pigeons) and species (mules), in which the modern determinants are heterodynamous;—they do not co-operate, and their forces counteract one another, while the ancestral determinants are similar and their forces cumulative.

Numerous phenomena of reversion in plants and animals may be explained in a very simple manner on these principles, and from this point of view it is also possible to understand that form of reversion which occurs in gemmation and parthenogenesis. The more remote the ancestors to the characters of which reversion occurs, the more rarely will it take place. Reversion to the three-toed ancestors of the horse, for instance, is of extremely rare occurrence, for it is due to a retention of the ancestral determinants in question—which have certainly disappeared from all the ids in the germ-plasm of most existing horses—in single individuals of certain series of generations, and to the chance of the coming together of two germ-cells containing such ancestral determinants.

The remarkable phenomenon of *dimorphism*, which has been introduced so extensively—more especially into the animal kingdom—by means of sexual reproduction, must be due to the presence in the idioplasm of *double determinants* for all those cells. groups of cells, and entire organisms, which are capable of taking on a male and female form. But only one half of such a double determinant remains inactive, while the other becomes active. The sexual differentiation of the germ-cells must thus be due to the presence of spermatogenetic and oogenetic double determinants; and even all the secondary sexual characters must be traced to a similar origin in the idioplasm. The corresponding double determinants are contained not only in the germ-plasm, but are passed on through the cell-stages of ontogeny to that part of the body in which the two characters become separated from one another. One of the determinants then becomes active, its twin half remaining in an inactive condition in the nucleus of a somatic cell, and under certain circumstances becoming active subsequently. This, however, only occurs exceptionally, in such cases as that in which a female animal (*e.g.*, a hen or duck) develops male characters in consequence of castration. Hermaphrodite bees, in which the whole body consists of the most wonderful intermixture of male and female parts, furnishes an instructive proof of the presence of both kinds of characters in all parts of the body, and consequently of the truth of the assumption of double determinants.

Double determinants not only occur individually, but entire *groups* of male and female determinants are opposed to one another, and these are just as dependent on one another as are the two halves of the individual double determinant, one of which always remains inactive when the other becomes active. These groups may be very dissimilar; in many cases (*e.g.*, the olfactory organs of male crustaceans and the ornamental feathers of male birds) the male group contains many more individual determinants than the female. One half of the double group may also become degenerated, so that the corresponding organ (*e.g.*, the wing in many female butterflies) disappears in one sex.

The number of double determinants reaches its highest limit when the two sexes differ completely from one another in all their parts, as is the case in *Bonellia viridis;* even then, however, a number of single determinants may still be present, if, as in this case, the larval stage is similar in both sexes.

The assumption of double determinants is also able to throw some light upon certain enigmatical phenomena of heredity exhibited by human beings. It has long been known that *hæmophilia* occurs in men only, but is transmitted by women. If we assume that the *visible* sexual differences, as well as those existing in the system which we are unable actually to recognise, are due to the presence of double determinants, the peculiar limitation of this uncommon disease to one sex is explained. The disease, like a secondary sexual character, is only transmitted to the sex in which it first appeared, *for this half of the double determinants of the 'mesoblast germ' has alone been modified by the disease.*

The *sexual polymorphism* exhibited by certain butterflies may also be explained by assuming the presence of double determinants of several local varieties of the same species which interbreed with one another. The polymorphism of bees and other animals which form communities, requires, however, the assumption of triple or quadruple determinants. In these animals the female half of the double determinant again becomes doubled, and this may also be the case as regards the male half (Termites).

Lastly, the assumption of double determinants in the idioplasm accounts for *temporary* dimorphism, such as seasonal dimorphism.

The occurrence of dimorphism is in all cases attributable to the presence of two kinds of determinants; but the causes which determine which of the two is to become active, are extremely varied, and cannot in many cases be accurately indicated. The determining influences, however, are always external ones—such as fertilisation, nutrition, and the effect of light in cases of *dichogeny* in plants.

It is self-evident from the theory of heredity here propounded, that only those characters are transmissible which have been controlled—*i.e.*, produced—by determinants of the germ, and that consequently only those variations are hereditary which result from the modification of several or many determinants in the germ-plasm, and not those which have arisen subsequently in consequence of some influence exerted upon the cells of the body. In other words, it follows from this theory that *somatogenic or acquired characters cannot be transmitted.*

This, however, does not imply that external influences are incapable of producing hereditary variations; on the contrary, they always give rise to such variations when they are capable of modifying the determinants of the germ-plasm Climatic influences, for example, may very well produce permanent variations, by slowly causing gradually increasing alterations to occur in certain determinants in the course of generations. An apparent transmission of somatogenic modifications may even take place under certain circumstances, by the climatic influence affecting certain determinants of the germ-plasm at the same time, and when they are about to pass to that part of the body which they have to control. This is indicated by the climatic variations of the butterfly *Polyommatus phlœas*.

The primary causes of *variation* is always the effect of external influences. Were it possible for growth to take place under absolutely constant external influences, variation would not occur; but as this is impossible, all growth is connected with smaller or greater deviations from the inherited developmental tendency.

When these deviations only affect the soma, they give rise to temporary non-hereditary variations; but when they occur in the germ-plasm, they are transmitted to the next generation and cause corresponding *hereditary variations in the body*.

Since the germ-plasm undergoes a very considerable growth from the fertilised egg-cell to the germ-cells of the offspring, minute fluctuations continually take place in the composition of its vital units, the biophors and determinants. If permanent and constant influences, such as those of climate, act upon them, these minute fluctuations will become accumulated in the course of time and generations, and may thus give rise to appreciable individual variations, and then gradually to racial, and even perhaps to specific characters. If an influence acts in a certain direction for a short time only, it alone may or may not give rise to an individual variation in the soma, according to the number of ids of the germ-plasm affected by it. Whenever a *majority* of ids become modified, a corresponding variation must appear in the soma. As, however, an intermingling of the ids takes place twice, owing to the successive processes of 'reducing division' and amphimixis, minorities of modified ids may be increased to majorities; and sexual reproduction may then cause the fluctuating material for *invisible* variations in the determinants to give rise to *perceptible* somatic variations, and these are made use of by natural selection, aided by constantly recurring amphimixis. The latter process gives natural selection a choice of innumerable combinations of the most diversified variations, resulting from the constant minute fluctuations of all the units in the germ-plasm.

Strictly speaking, *the process of amphimixis alone* cannot bring about an *increase or decrease in the development of a character;* though it may, indeed, establish it more firmly in the germ-plasm by causing an increase in the number of ids, the determinants of which produce the character. An increased development, in the ordinary sense of the term, may, it is true, take place by the extension of a variation over *larger areas* of the body; the multiplicity of the ids, and the possibility of the constant production of new idic combinations by the process of amphimixis, accounts for the statement of breeders that the constancy of a character, as well as its increased development, may be affected by selection. The so-called 'individual potency' must, moreover, be due to the presence of a large number of homodynamous determinants for all the more important characters, and it probably results, not only from breeding a race true for a long time, — although this is of course necessary, — but also from favourable combinations of ids being produced by the processes of reducing division and amphimixis.

Variations do not, however, depend merely on modifications in the composition of a determinant or group of determinants, but frequently result from a doubling or further multiplication of the latter; and this must also depend primarily on modified external influences, such as those produced by changes in the nutrition of a part of the germ-plasm. The apparently sudden appearance of parts — such as feathers and other epidermic structures, as well as of certain pathological structures, such as the supernumerary fingers and toes of human beings — may be explained in this manner. All such variations do not, indeed, *actually* arise suddenly, but take place gradually in some of the ids, and only suddenly become apparent when they have accumulated to form a majority.

The *suddenness* with which variations appear is, in all probability, only apparent in most cases, as is well shown by Hoffmann's experiments on wild plants, in which variations were produced by abnormal conditions of life. The *degeneration* of parts which are no longer required, or have simply become useless, is due to the reduction and final disappearance of the corresponding determinants from the germ-plasm. But as this again depends on the fluctuating variations of these determinants in the different ids. it will not occur to the same extent in all the ids at the same time; and thus the remains of these reduced determinants are often preserved in individual ids through countless generations, and may occasionally cause reversion to take place when they have accumulated in consequence of the 'reducing division' and amphimixis.

Sudden variations of buds have only been observed in plants which also are, or have been, propagated sexually, and in which the structure of the germ-plasm is therefore just as complex as in species in which sexual reproduction *alone* occurs. These variations are also due to the effects of dissimilar modifying external influences on the determinants contained in the 'blastogenic' germ-plasm, which is passed on from cell to cell during the process of growth. It would, however, be impossible to understand why only a single bud out of millions should undergo transformation unless some other cause were also at work. This may be due to occasional irregular nuclear divisions, which would give rise to a similar result to that produced by amphimixis in reproduction by seeds, the modified determinants of individual ids occasionally becoming accumulated and then taking effect.

The power of transmission of '*sudden*' *variations in plants*, which is apparently very capricious, may be easily understood in principle. As the modification never occurs in *all* the ids of the germ-plasm, but only in *many* of them, and as this majority may be a slight or a considerable one, the transmission of the variation will depend on whether the majority is *often* obtained and even increased, or whether it becomes diminished, or even entirely lost, during the reducing divisions and amphimixis, when the plant is reproduced by seeds. In the case of the first alternative, the 'sport' will be transmitted; while in that of the second, transmission will only occur rarely or not at all. Even details of the apparently enigmatical phenomena of heredity of known 'sport' varieties — such as those of the balsamines, weeping ashes, the variegated variety of *Ballota nigra*, and others — thus receive a very simple explanation.

The capricious transmission of bud-variations by seeds, which only occurs in the smaller proportion of cases, cannot be explained so easily. It is due to the 'blastogenic' germ-plasm and the 'reserve germ-plasm'—destined to form germ-cells—following different courses, and consequently they will not always contain the same number of modified ids. The infrequency of the transmission of bud-variations by seeds may depend on the production of a new combination of ids of the 'reserve germ-plasm' from which the germ-cells are formed, whenever this formation occurs, owing to the 'reducing division' and amphimixis. No such occurrence takes place in the 'blastogenic' germ-plasm as long as merely asexual reproduction continues.

According to our view, the power of transmission—which is possessed by all organisms, and on which the development of the higher organic forms is based—therefore depends on simple growth merely in the case of the very lowest conceivable organisms with which we are not acquainted; while in all forms which have already undergone differentiation, it results from the possession of a *special apparatus for transmission*.

This apparatus first occurs in the unicellular organisms, in which it consists of a substance composed of the different kinds of vital units or biophors, which occur in the substance of the organism, and presumably in a similar proportion; there are, at any rate, numerous individual biophors of every kind, all of which are arranged together on a definite plan. This substance is surrounded by a membrane,—the nuclear membrane,—provided with pores, through which the biophors of the nucleus can pass into the cell-body, there to multiply at the expense of the nutritive materials—to which the vital particles of the cell-body themselves may become reduced under certain circumstances—and to become arranged in virtue of the forces dwelling within them.

To these processes is due the power possessed by the organism of giving rise by division to two complete individuals of a similar nature.

Even at this stage of differentiation, the hereditary substance is rendered more complicated by the process of amphimixis, or mingling of individual differences, in which this substance periodically becomes halved, and is then again completed by the hereditary substance of another individual. The result is, that every part of the organism is represented in the hereditary substance by different varieties of the same kind of biophor, and that, consequently, the individuals subsequently arising by division will be approximately intermediate in structure between the two parents.

In the multicellular forms exhibiting cell-differentiation, the apparatus for transmission becomes more complicated the more numerous and diversified are the kinds of cells composing the organism. Multiplication can only take place if each individual originates in, and again returns to, the unicellular condition. The division of the entire organism would only result in the production of two unequal halves, which could not of themselves undergo completion, and would require a special apparatus for the purpose. In order that this may arise, the previous *production of an apparatus for transmission*, adapted for reproduction by unicellular germs, is indispensable.

The production of this apparatus results from the formation of a *germ-plasm*, *i.e.*, a nuclear substance which contains reserve biophors for the construction of the corresponding cell-body, as well as for the formation of all the cell-bodies of the entire organism; and all of which are connected together into a definitely arranged structure, in such a manner that the constituent parts share regularly and successively, and not simultaneously, in the control of the cell-body. In order that this result may be produced, the smaller vital units or biophors are combined to form those of the next higher order — the determinants, — each of which controls *one* kind of cell, and consequently includes all the biophors required for the determination of this particular kind of cell. *The germ-cell contains at least as many determinants as there are different cells or groups of cells in the fully-formed organism which are capable of being individually determined from the germ onwards.*

Since the process of amphimixis or '*intermingling of individual differences*' is also retained in multicellular forms, the individual germ-cells must for this reason alone contain a mass of germ-plasm, each unit of which contains all the kinds of determinants of the species in close combination. The hereditary substance of the germ-cell thus came to be composed of ids and idants.

The multiplication of multicellular organisms by fission and gemmation results from a considerable increase in the complexity of the apparatus for transmission, in which process not only the determinants required for the control of their own nature, but also ids of germ-plasm in an *unalterable condition* —in which they are at the time incapable of undergoing disintegration—were distributed to certain cells of the body. This addition of latent germ-plasm to certain series of somatic cells also results in the formation of germ-cells in most multicellular forms; while the power of regeneration depends on a systematic addition of certain inactive determinants, or groups of determinants, to certain cells of the body.

Further complications of the germ-plasm produce the phenomena of alternation of generations and the polymorphism often connected with it, as well as the sexual dimorphism always occurring in a greater or a less degree in connection with 'sexual reproduction.' All determinants and groups of determinants which exist in two or more forms, must be present in a double or multiple condition in the germ-plasm, and be so arranged that each constituent part only becomes active in turn. This assumption is, however, insufficient in the case of alternation of generations, in which several kinds of germ-plasm must be present and become active in turn.

Thus an ever increasing complexity of the substance which renders the repetition of the organism possible is gradually produced in the phylogeny of living beings, and eventually reaches so high a degree that it is difficult to believe that such an infinite complexity of structure can actually exist in particles so minute. The more deeply, however, we penetrate into the phenomena of heredity, the more firmly are we convinced that something of the kind does exist, for it is impossible to explain the observed phenomena by means of much simpler assumptions. We are thus reminded afresh that we have to deal not only with the infinitely great, but also with the infinitely small; the idea of size is a purely relative one, and on either hand extends infinity.

INDEX

ACER negundo, reversion in, 338
Actiniæ, regeneration in, 127
Adler, on galls, 219
Adventitious buds, Sachs's view of, 211; formation by means of accessory idioplasm, 212, 218
Alternation of generations, in its relation to the idioplasm, 173; hereditary variability of individual generations, 174; in Daphnidæ, 175; in Aphis, 178; in Medusæ, 179; caused by two kinds of germ-plasm, 182, 457
Amphimixis, or the intermingling of individuals, 20, 24; its relation to reproduction, 232; resulting in complications in the composition of the germ-plasm, 235; the cause of all reversion, 337; the means of producing variation, 413; its relation to bud-variations, 439; the cause of an increase or decrease in the development of a character, 464
Ancestral germ-plasms — not analogous to Spencer's 'physiological units,' 11
Antirrhinum majus, Darwin's experiments with the peloric form of, 331
Aphis, alternation of generations, 178
Ascaris megalocephala, the chromosomes of, 86; diagram, fig. 18, page 233
Ascidians, regeneration, 139; eggs contrasted with those of frogs, 141; gemmation, 160; comparison of gemmation with embryogeny, 162
Auerbach, on nuclear division, 23
BALFOUR, phylogeny of gemmation, 169
Balsamine, transmission of variations of, 445
Baumann, chemical constitution of protoplasm, 40
Bees, polymorphism in, 376
Begonia, adventitious buds, 211
Beneden, Ed. van, observations on fertilisation of egg in *Ascaris*, 23; function of the centrosomes, 25; view concerning idants, 244
Biophors, or bearers of vitality, the smallest vital units, 40; a group of molecules on which the phenomena of life depend, 40; resemblance to Spencer's 'physiological units,' 40; difference between them and Nägeli's 'micellæ,' 41; de Vries's 'pangenes,' and Wiesner's 'plasomes,' 42; bearers of cell-qualities, 42; number of kinds unlimited, 44; all protoplasm constituted by, 45; their size and number, 85; combined anew in the formation of galls, 221, 448
Beyerinck on galls, 219; gall of *Nematus*, 222
Blending, of parental characters not a fundamental phenomenon of heredity, 8; of paternal and maternal idioplasm in ontogeny, 253; of parental characters in the offspring, 261, 266, 297
Blood-corpuscles controlled by determinants, 57
Blumenbach's 'nisus formativus,' 103, 104
Bonellia viridis, its dimorphism, 364
Bonnet, on the regeneration of eye of *Triton*, 125; on worms, 131

Born, development of eggs in a fixed position, 135

Boveri, observations on fertilisation of egg in *Ascaris*, 23; function of the centrosomes, 25; observations on the egg of the sea-urchin artificially deprived of its nucleus, 28; chromosomes of *Ascaris megalocephala*, 86; differentiation of somatic cells of *Ascaris megalocephala*, 191; number of nuclear rods in two species of *Ascaris*, 241; view concerning idants, 242

Brandza, examination of hybrids with regard to parental characters, 270

Brooks, W. K., on the laws of heredity, 9; on variation, 412; on the non-transmission of acquired characters, 396

Brücke, Ernst, on 'Elementar organismen,' 2; on ultimate vital particles, 20; on albumen, 38; on the organic nature of protoplasm, 39

Bryophyllum, budding in, 211

Bud-variation, 436; theoretical explanation of, 441

Bütschli, investigations on the processes of nuclear division, 23; the polar bodies correspond to aborted ova, 251

Butterflies, the markings on the wings of, as confirming the theory of determinants, 54, 174, 264, 265; *Kallima parallecta* as an example of relative perfection, 272; the origin of climatic variations in the idioplasm, 379, 399, 407, 418

CARDAMINE pratensis, budding in, 211

Carneri, observations on telegony, 385

Caspary, experiments in crossing *Cytisus laburnum* and *C. purpureus*, 343

Castration, as a cause of sexual reversion, 358

Cecidomyia poæ, formation of galls, 221

Centrosome, with its sphere of attraction, is an apparatus for the division of the cell and nucleus, 23; continuity of the centrosome, 48

Chabry, experiments on eggs of Ascidians, 136

Chromatin, the hereditary substance, 24, 26; must be different in each kind of cell, 32

Chromosomes correspond to the idioplasm, 10; their form, 24; spoken of as idants, 67

Continuity of the germ-plasm, historical account of, 198

Control of the cell, 45

Cuttings, development and origin in the idioplasm of, 212

Cyclops, observations on segmenting ova, 191

Cymathoidæ, temporary sexual dimorphism, 111

Cypris reptans, markings on the shell in their relation to the theory of determinants, 88; reversion in, 344; theoretical explanation of reversion in, 347

Cytisus adami, 339; theoretical explanation of the variation in, 344

DAPHNIA pulex, alternation of generations in, 175

Daphnidæ, embryogeny of, 184; secondary nature of the differentiation of germ-cells of, 186; germ-track of, 192

Darwin, Charles, 'pangenesis,' 2; 'gemmules,' 3; comparison of the 'gemmules' with Spencer's 'physiological units,' 6; on correlated variations, 84; on the white coloration in animals and plants, 277; on breeding, 291; on the force of heredity in cattle, 293; on reversion in plant-hybrids, 299; on reversion to remote ancestors, 316; on reversion in pigeons, 323; in mules, 328; in *Antirrhinum majus*, 331; experiments on crossing *Cytisus laburnum*

and *C. purpurea*, 340; on latent primary constituents, 352; on transmission of acquired characters, 384, 394; on variation, 406, 410; on baldness in birds, 426; on the transmission of bud-variations, 441; on the transmission of a variation in *Ballota nigra*, 446
Datura ferox and D. lævis, reversion of hybrids to an unknown ancestral form, 317
Davenport, on budding in Polyzoa, 159
Degeneration in its relation to the theory of determinants, 83
Déjerine, on the transmission of nervous complaints, 369
Determinants or determining parts, 57; control groups of cells, 57; are a group of biophors, 59; possess special qualities, 60; are definitely localised in the idioplasm, 61; constitute the ids, 62; their forces of attraction, 66; their behaviour in the course of ontogeny, 69; their disintegration into biophors, 69; their growth and multiplication, 71; their number, 89; supplementary determinants, 103, 112, 127, 131, 149; the separation of a group of, 153; their disintegration in gemmation, 160; their separation into groups in the development of somatic cells, 209; their modification in the formation of galls, 222; control of the cell by the combined influence of paternal and maternal determinants, 261; homologous and heterologous determinants, 264; their controlling forces, 269; number of homodynamous determinants varies in ontogeny, 271; homodynamous and heterodynamous determinants, 264, 278; proof of their disintegration into biophors, 348; in cases of dimorphism, 355; variations due to changes in their composition, 406, 418, 448
Determinants, the theory of, 53; summary, 225; reversion explained by, 335; applied to sexual dimorphism, 366
Determinates or hereditary parts, 57
Dianthus chinensis and D. barbatus, crosses between, 302
Dichogeny in plants, 114, 380, 462
Dicyemidæ, distribution of hereditary parts, 59
Digitalis lutea and D. purpurea, crosses between, 255
Dimorphism, normal, 352; its basis in the idioplasm, 354; pathological, 370; double, in *Papilio turnus*, 375; seasonal, 379, 462
Diptera, course of germ-track, 192; renewal of the alimentary epithelium, 57
Disarticulation of the limb of Triton, in connection with the hypothesis of supplementary determinants, 119
Diseases, transmission of, 387; hæmophilia, 370
Doubling of limbs in insects, 429; of whole groups of determinants, 428
Driesch, experiments with eggs of sea-urchins, 137; on the blastomeres of the frog, 141

ELSBERG, Louis, 41
Embryonic cells, 33
Eudendrium, budding of, 155, 156, fig. 6

FERNS, regeneration and gemmation, 215
Fertilisation, essential part of the process, 232
Fission in Naidæ, 146; in Microstomidæ, 149, 456
Flemming, on nuclear division, 23; on the splitting of the chromosomes in nuclear division, 25, 68
Focke, on plant-hybrids, 261; on hybrids of *Nicotiana*, 267; on reversion in hybrids, 299; on re-crossing, 302; on 'xenia,' 383

Fol, on the transference of the centrosome into the ovum during fertilisation, 29
Fraisse, on regeneration in Amphibia, 96; in *Salamandra*, 99; in lizards, 110, 116
Fucoidæ, polar bodies of, 251

GALLS, 218; definite new formations composed of modified cells, 221
Galton, Francis, acceptation of the 'gemmule' hypothesis, but rejection of that of their free circulation, 7; theory of the blending of characters of parents in the children, 7, 257; germ-substance composed of homologous gemmules, 72; supposed priority as regards the assumption of the continuity of the germ-plasm, 198
Gärtner, on plant-hybrids, 299; re-crossing of hybrids, 305
Gemmation in Cœlerentata, 154; 'blastogenic' idioplasm, 157; in Polyzoa, 158; in Tunicata, 160; in plants, 163; comparison of the process in plants and animals, 166; its phylogeny, 168; resulting from adaptation in polypes, 217; brief summary, 225, 456
Germ-cells, their formation, 183; shifting of place of origin in the course of phylogeny, 186; only certain series of cells contain the primary constituents of germ-cells, 189; distinction between germ- and somatic-cells, 208; the combination of ids in germ-cells, 245, 247; only half the number of parental ids contained in them, 257, 457
Germ-plasm constituting the immortal reproductive substance, 9; is the first ontogenetic stage of the idioplasm, 35; its fundamental units, 37; its composition out of biophors, 40, 48; its fixed architecture, 61; it forms a complete unit by itself, 62; summary relating to its structure, 75; magnitude of its constituents, 85; 'blastogenic' idioplasm, 166; its regular division, 171; accessory germ-plasm, 174; its continuity, 183; its composition, 37, 77, 186; its modification caused by amphimixis, 235; its composition out of paternal and maternal idants, 254; its partial variations, 249, 271; reversion due to its ancestral determinants, 336; transformation and the gradual production of reversion, 336; 'blastogenic' germ-plasm, 441; reserve germ-plasm, 447, 453
Germ-tracks, course taken by the germ-plasm from the ovum to the reproductive cell, 184; their course in the Metazoa, 192; their cells alone capable of giving rise to primary germ-cells, 194; each cell in them contains perfectly definite somatic elements, 210; brief summary, 228
Giard, view that the polar bodies correspond to aborted ova, 251
Godron, reversion in *Melandryum album* and *M. rubrum*, 306
Götte, regeneration in *Salamandra*, 99
Gruber, August, artificial division of Infusoria, 52
Guignard, on the reduction of idants in the germ-cells of plants, 250; on the union of nuclei, 30

HABERLANDT, on the relations between the functions and position of the nucleus, 46
Häckel, Ernst, on the 'perigenesis of the plastidule,' 41; on the law of sexual transmission, 369
Häcker, Valentin, on the phenomena of reduction in the germ-cells of Arthropods, 250
Hæmophilia, 370
Hallez, on the embryology of Nematodes, 137
Hatschek, 53; on variation and sexual reproduction, 413
Henking, on the phenomena of reduction in the germ-cells of Arthropods, 250

Hensen, Victor, on amphigonic heredity, 253; on the increase of a character, 424
Herbert, experiments with *Cytisus adami*, 343
Hereditary substance, not contained in the body of the cell, but in its nucleus, 10; composed of primary constituents, 15; forms but a small part of the substance of the egg, 22; composed of different qualities, 26; the extreme complexity of its structure, 29; its growth, 31; of unicellular organisms, 451; of multicellular organisms, 452
Heredity, its fundamental phenomena, 20; in unicellular organisms, 52; in reference to the structure of the germ-plasm, 69; homotopic and homochronic forms, 75; in monogonic reproduction, 92; in sexual reproduction, 230, 253; explanation of apparently monogonic heredity in plant-hybrids, 259; in Man, 259; as regards the colour of the eyes, 278; example of apparently monogonic heredity, 267, 280; force of, 290; as regards acquired characters, 392; mutilations, 396; 'sports,' 444; the apparatus for transmission, 466
Hertwig, Oscar, view of fertilisation as a conjugation of nuclei, 23; on the germinal layers of the Metazoa, 113; on the nuclear rods in two varieties of *Ascaris megalocephala*, 241
Heterobiophorids, hypothetical primitive organisms, 450
Heterokinesis, nuclear division resulting in parts which are dissimilar as regards their hereditary tendencies, 34
Hildebrandt, hybrids of two species of *Oxalis*, 255
His, theory of special regions in the germ giving rise to special organs, 134
Hoffmann, experiments with *Papaver alpinum*, 437
Homobiophorids, hypothetical primitive organisms, 450
Homœokinesis, nuclear division resulting in parts which are similar as regards their hereditary tendencies, 34
Hoppe-Seyler, chemical constitution of protoplasm, 40
Humming-birds, sexual dimorphism of, 427
Hybrids of various species of plants, 255; the theory of idants, 259; plant-hybrids, 260; apparent monogonic transmission in, 302
Hydra, regeneration, 127; gemmation, 155
Hydractinia echinata, formation of germ-cells, 189
Hydroids, formation of the bud from a single cell, 166; embryogeny, 184; shifting of place of origin of germ-cells, 186; germ-tracks of, 207

Id, the, in ontogeny, 60; the number contained in the individual idant, 241
Idants, constituting the hereditary substance in sexual reproduction, 234; composed of dissimilar ids, 238; constituting groups of ids, 241; their number, 245; become doubled by division, 246; combination of, 247; of ancestors contained in the germ-plasm, 257; their combination in plant-hybrids, 297, 302; intermingling of, in transmission, 312
Idioplasm, Nägeli's conception of the, 10; my view of the, 33; contrasted with morphoplasm, 38; composed of ids, 63; phyletic variation of, in the differentiation of species, 79, 434
Ids or ancestral germ-plasms composed of determinants, 62; their individual difference, 236; homologous and heterologous, 264; gradual transformation of, 271; number variable according to age of characters, 273; struggle of parental ids, 285

Increase of a character by crossing, 424
Individual prepotency, 291
Ivy, 111; dichogeny in, 380

JÄGER, Gustav, supposed priority as regards the hypothesis of the continuity of the germ-plasm, 200

KENNEL, regeneration of beak of stork, 125; division, 150
Kiwi (*Apteryx*), disappearance of wings, 82
Kleinenberg, development of *Lumbricus trapezoides*, 170
Kölreutter, crossing of *Nicotiana rustica* and *N. paniculata*, 261, 269; on recrossing, 305
Kowalewsky, on supernumerary toes of horses, 334
Kraepelin, investigations on hermaphrodite bees, 361

LAMARCK, on the transmission of acquired characters, 395
Lang, A., the regenerative faculty as an arrangement for protection, 93; budding in hydroids, 155
Leptodora hyalina, development of, 176; sexual characters of, and the theory of determinants, 363
Leuckart, Rudolph, discovery of parthenogenesis in bees, 356; on the combination of secondary sexual characters in hermaphrodite bees, 360
Leydig, regeneration in lizards, 110, 116
Liebscher, crossing of *Hordeum steudelii* and *H. trifurcatum*, 301
Loeb, gemmation in *Tubularia mesembryanthemum*, 216
Lumbriculus, regeneration in, 126, 151
Lucas, Prosper, on transmission of polydactylism, 373
Lycæna, as illustration of the theory of determinants, 58, 62; *L. adonis* and *L. agestis*, 87; colouring and dimorphism of, 359, 373; variation in, 422

MARRUBIUM vaillantii, 270
Marsh, on supernumerary toes in horses, 334
Medusæ, formation of germ-cells, 187; degeneration into a mere gonophore, 187
Mesohippus, 334
Meyer, O. E., on molecules, 86
Micellar theory, 41
Miescher-Rusch, on nuclein, 50
Morphoplasm, 38
Mules, transverse stripes of, indicating reversion, 316; theoretical explanation of stripes, 329; pathological variation, 429
Müller, Josef, 'gamomachia' and 'gamophagia,' 296

NÄGELI, Carl von, physiological theory of descent, 9; the hereditary substance constitutes only an infinitesimal part of the egg-cell, 22; the 'micella,' 41; on variation, 411; the ibis and crocodile in connection with variation, 416; bud-variation, 442
Nais, regeneration of, 126
Naudin, experiments with *Datura*, 321
Nicotiana rustica and N. paniculata, crosses between, 261; difference in the flowers of, 267; *N. alata* and *N. langsdorffii*, 300
Nipples, supernumerary, in human beings as examples of reversion, 333
Nitsche, O., budding in Polyzoa, 158
Nussbaum, artificial division in Infusoria, 52; idea of the continuity of germ-cells, 201

OKA, budding in Polyzoa, 159
Ontogeny, or development of the individual, 32; depends upon gradual changes in the germ-plasm, 32; part played by determinants in the course of, 80; can only be explained by 'evolution,' and not by epigenesis, 138; resulting from the union of the germ-plasm of two parents, 253

Otter-sheep, 291
Owen, Sir Richard, on cells, 198
Oxalis, crosses between different species of, 255

PANGENES, the, of de Vries as bearers of constituent qualities, 15; their miscibility, 15; compared with biophors, 42
Papilio turnus, 375
Pflüger, the hereditary substance only an infinitesimal part of the egg-cell, 22; influence of gravity on the development of frogs' eggs, 135
Phanerogams, budding in, 216
Philippeaux on regeneration in Tritons, 118
Phylogeny, 77; parallelism between phylogeny and ontogeny, 80; of the process of fission in the Metazoa, 151; of regeneration, 114; of gemmation, 168; of the apparatus for transmission, 451
Pigeons, reversion to the rock-pigeon, 323, 352; increase of a character by selection, 426, 427
Plasomes, the, of Wiesner, 20
Plastidules, the, of Haeckel, 41
Plumularia, budding in, 155
Podocoryne cornea, 158
Polydactylism, the transmission of, 373, 429
Polymorphism, 374; of animal and plant stocks, 376, 462
Polyommatus phlæas, 399, 420
Polyzoa, budding in, 158; formation of buds from different cells, 165; 'blastogenic' germ-plasm, 218
Primula acaulis, reversion to long-stalked variety in, 16
Protohippus, 334

RATH, O. vom, on threads of 'linin' between the idants, 244; on the phenomena of reduction in the germ-cells of Arthropods, 250
Rauber, supposed priority as regards the view of the continuity of the germ-plasm, 200
Ray Lankester on the non-transmission of acquired characters, 396
Reducing division of the germ-plasm, 11, 22, 235; consists in the extrusion of ids, 240; influence on the composition of the germ-plasm, 242; cause of difference between children of the same parents, 257; in its relation to reversion, 306
Regeneration, its cause and origin in the idioplasm, 93; constitutes an arrangement for protection, 93; physiological, 94; of the epidermis, 95; palingenetic, 105; cœnogenetic, 108; of the caudal region of the vertebral column, 109; phylogeny of, 114; physiological and pathological, 119; depends on adaptation, 119; in fishes, birds, and mammals, 120; difference of the capacity for in lower and higher animals, 124; acquired by selection, 125; facultative or polygenetic, 126; in plants, 132; formation of callus, 133; in animal embryos, 134; of segmentation-cells, 139; summary, 225, 455
Reisseck, on crosses between *Cytisus laburnum* and *C. purpureus*, 343
Reproduction, sexual, 230, 457; its effect, 240; in Man, and transmission of parental characters, 257; by fission 456
Reversion in plant-hybrids, 299; to the pure ancestral form, 305; in Man, and the crossing of different races, 308; to a grandparent, 309; to characters of remote ancestors, 316; in Datura, 317, 321; to ancestral forms of flowers in plants, 305; to rudimentary characters, 333; in gemmation, 338; in parthenogenesis, 344; sexual, 367; in bud-variations, 447, 458
Rhabditis nigrovenosa, segmentation of ovum, 195, fig. 15; germ-track of, 196
Roux, W., on the purpose of the apparatus f r divisi n of the nucleus, 26; on the struggle of

parts, 107: on gravity producing no effect in the differentiation of the egg into the embryo, 135; experiments on frogs' eggs, 136; on post-generation, 142
Rückert, J., behaviour of chromosomes during maturation in the ovum of the dog-fish, 50, 71, 247

SALHS, on the formation of callus, 133; on the growth of Chara, 164; on adventitious buds, 211
Sagitta, embryogeny of, 185; course of the germ-track of, 192
Salamandra, regeneration of limbs, 99
Salpæ, budding in, 162; alternation of generations in, 180
Schreiber, on the regenerative power in *Triton marmoratus*, 115
Seeliger, O., on budding—in Polyzoa, 158; in Clavelina, 160; in Salpæ, 162; in Pedicellina, 218
Self-differentiation of cells, 136
Semper, Carl, on the process of fission in *Nais*, 147
Settegast, on infection of the germ, 385
Siebold, von, on the determination of sex in bees, 356; on hermaphrodite bees, 361
Siren lacertina, regenerative power in, 115
Spallanzani, on the capacity for regeneration in different organs, 117; in Triton, 120; in the jaw of Triton, 125
Spencer, Herbert, 'physiological units,' 51; their relation to Darwin's 'gemmules,' 6; on heredity, 7; on regeneration, 104; regeneration compared with crystallisation, 128
Stimuli considered as causing an incitement to growth, 129
Strasburger, on fertilisation in Phanerogams, 23; on the essential similarity of male and female nuclei, 23; on the dynamical effect of nuclear matter, 45
Struggle of individual characters, 274; of the ids in ontogeny, 260
Supplementary determinants, 103; their change in the course of phylogeny, 112; in processes of regeneration in *Hydra*, 127; antimeral, 131; the origin of, 149
Syphilis, the transmission of, 388

TARDIGRADA, extent of the hereditary parts, 59
Telegony or infection of the germ, 383
Termites, polymorphism in, 379
Thuja, dichogeny in, 382
Transmission (*see* Heredity).
Triton, 100; diagram of regeneration of fore-limb in, 102, 115; experiment on regeneration of lung in, 117
Tuberculosis, and infection of the germ, 389
Tubularia mesembryanthemum, budding in, 216
Tunicata, budding in, 160
Twins, 140; identical and dissimilar, 254
Type in a technical sense, 313

VANESSA, levana and prorsa, 379
Variation, intercalary remarks on, 271; theoretical demonstration of, 410; normal individual variation, 410; cause of hereditary variation, 415; pathological variation, 428; summary of the present view with regard to variation, 431; variations on a larger scale in plants, 435
Verworn, M., view that the hereditary substance is also contained in the cell-body, 28
Vilmorin, on individual potency in plants, 291
Vines, view that the assumption of a special reproductive substance is unnecessary, 202; on embryonic substance, 204
Viola tricolor, reversion to the ancestral form, 320
Vöchting, H., on 'transplantation in the plant-body,' 129

Volvox, proof of that a difference exists between the somatic- and the germ-cells, 213; as an example of sexual dimorphism, 355
Vries, Hugo de, objection to the theory of ancestral germ-plasms, 12; 'intracellular pangenesis' compared with Darwin's 'pangenesis,' 12; objection to the view that the hereditary substance is contained in the nucleus only, 26; on the structure of protoplasm, 38; on Haeckel's 'plastidules,' 41; objection to Strasburger's and Haberlandt's views, 46; the 'pangenes' as bearers of cell-qualities, 42; on the continuity of cellular organisms, 56; on primary and accessory germ-tracks, 206, 211; on the continuity of the germ-plasm, 207; on galls as contradicting the latter, 218; galls of *Cecidomyia poæ*, 221; on the crosses between two species of beans, 297; on plant-hybrids, 300; on dichogeny, 380, 451

WAGNER, F. von, on gemmation and fission, 145; on fission in *Microstoma lineare*, 149; on gemmation, 154
Wallace, Alfred Russel, on *Papilio memnon*, 376; on variation, 410
Ward, Lester, Neo-Darwinism and Neo-Lamarckism, 408
Weeping Ash, 444, 446
Wiesner, Julius, on the elementary structure and growth of living substance, 19; on the constitution of protoplasm, 39; on the continuity of the constituent parts of the cell, 48
Willow, rudiments of roots in the shoots of the, 382

ZANDER, R., on polydactylism, 430
Ziegler, Ernst, on the law of the specific character of the tissues, 112; on tuberculosis, 390; on polydactylism regarded as a germ-variation, 430

Norwood Press:
J. S Cush ng & Co — Berw ck & Smith.
Boston, Mass, U S A.

Printed in Dunstable, United Kingdom

91220123R00281